Women Physicians and the Cultures of Medicine

Women Physicians and the Cultures of Medicine

EDITED BY

Ellen S. More, Elizabeth Fee, and Manon Parry

The Johns Hopkins University Press
Baltimore

Printed in the United States of America on acid-free paper
9 8 7 6 5 4 3 2 1

The Johns Hopkins University Press
2715 North Charles Street
Baltimore, Maryland 21218-4363
www.press.jhu.edu

Library of Congress Cataloging-in-Publication Data

Women physicians and the cultures of medicine / edited by Ellen S. More, Elizabeth Fee, Manon Parry.

p. ; cm.

Original work presented at a symposium, "Women Physicians, Women's Politics, Women's Health: Emerging Narratives," hosted by the National Library of Medicine in 2005. This symposium was organized in conjunction with the NLM exhibition, Changing the Face of Medicine: Celebrating America's Women Physicians.

Includes bibliographical references and index.

ISBN-13: 978-0-8018-9037-6 (hardcover : alk. paper)

ISBN-10: 0-8018-9037-3 (hardcover : alk. paper)

ISBN-13: 978-0-8018-9038-3 (pbk. : alk. paper)

ISBN-10: 0-8018-9038-1 (pbk. : alk. paper)

1. Women physicians—United States. I. More, Ellen Singer, 1946– II. Fee, Elizabeth. III. Parry, Manon.

[DNLM: 1. Physicians, Women—history—United States. 2. Physicians, Women—psychology—United States. 3. Culture—United States. 4. Gender Identity—United States. 5. History, 19th Century—United States. 6. History, 20th Century—United States. 7. History, 21st Century—United States. 8. Power (Psychology)—United States. WZ 80.5.W5 W8724 2008]
R692.W677 2008

610.82—dc22 2008010932

A catalog record for this book is available from the British Library.

Special discounts are available for bulk purchases of this book. For more information, please contact Special Sales at 410-516-6936 or specialsales@press.jhu.edu.

The Johns Hopkins University Press uses environmentally friendly book materials, including recycled text paper that is composed of at least 30 percent post-consumer waste, whenever possible. All of our book papers are acid-free, and our jackets and covers are printed on paper with recycled content.

Contents

Preface

This collection represents the newest work in the history of women physicians in American medicine from the nineteenth century to the present. In their original form, the essays composing this volume were presented at a symposium called "Women Physicians, Women's Politics, Women's Health: Emerging Narratives," hosted by the National Library of Medicine (NLM) in 2005. The symposium was organized in conjunction with the NLM exhibition, *Changing the Face of Medicine: Celebrating America's Women Physicians*, curated by Ellen S. More and Manon Parry.[1] The exhibition was intended to promote a better understanding of the accomplishments and struggles of women physicians since 1849, when Elizabeth Blackwell attained the first collegiate medical degree granted to a woman in the English-speaking world. In this book, as in the exhibition, we present a diverse range of three-dimensional figures who have tried to resolve the tension between the personal and the political forces in their lives. These chapters reveal the complex individual choices they faced as women and as physicians. Their challenges of identity formation, self-presentation, and performance are faced by many professional women in Western industrial societies, however inflected they may be by cultural specificities. Indeed, one woman lawyer of the nineteenth century called the challenge one of "double consciousness."[2]

Some of the authors of these chapters are seasoned scholars who have already written significant works on women and medicine and who now return to the field to take a closer, more textured look at their subjects. Some are younger scholars whose research reveals previously unexplored aspects of the lives and careers of women physicians or revises earlier judgments. The essays address mostly new questions. Recognizing that the category "woman physician"—no less than the category "woman"—comprehends widely differing experiences, we have included work on physicians as activists, sex educators, medical "sectarians," college health physicians, and medical missionaries. We also analyze the health status of women physicians, their relationship to their own bodies, and their relationships to the women patients with whom they share concerns about sexuality, sexual orientation, childbirth, and menopause. Finally, we include essays that broach the vexed relationship of women physicians to laywomen in the women's health movement in the nineteenth century and during feminism's "second wave." Unquestionably, women physicians have been an oppressed group within the medical profession, but some were—and are still—privileged as women. Through their position in class and racial hierarchies, they may have participated in the oppression of women patients and less powerful colleagues; certainly their relationships with lay activists were not uncomplicated.

Although these essays, therefore, complicate existing narratives of women physicians' role in medicine in the United States, they also reaffirm much of the basic structure of that narrative. They demonstrate, for example, the central role of biological essentialism in constructing and deconstructing women's place in medicine, reaffirm the centrality of what has often been called the "culture of professionalism" to many women doctors' career aspirations, and reveal, yet again, the power of social class and perceived biological "difference" to deflect promising careers into professional marginality. They reconfirm, too, that women physicians have suffered from both overt and covert gender discrimination. They reaffirm our longstanding assessment that women's rights and women themselves—their values and culture—have always counted for a lot to women who practice medicine. Throughout their history, many women physicians have worked to improve conditions for women and children and for entire communities, readily acknowledging the gendered architecture of their professional and personal choices even if, as one of these chapters suggests, they did not cluster in all-women's medical schools to the extent previously believed.[3]

This book is arranged thematically and, within each section, chronologically. Part I, "Performing Gender, *Being* a Woman Physician," highlights the historical struggles of women physicians to adapt to the masculine culture of medicine and respond to the challenges of *being* a woman physician. These largely biographical essays reveal the techniques by which women physicians internalized and managed complicated and conflicting identities. For example, Mary Putnam Jacobi was renowned for pursuing "science" in medicine, dispelling the myth that women were too emotional and unsystematic to be excellent physicians. Her performance as a *scientific* physician was a crucial buttress for one who was married to an eminent pediatrician with whom she had children, while building an ambitious medical career of her own. However, Carla Bittel shows us a completely new facet of her character, demonstrating that Jacobi consciously exploited her reputation for scientific objectivity by adapting her research to a politicized defense of women's rights.

Arleen Marcia Tuchman shows us a new face of Marie Zakrzewska, founder of the New England Hospital for Women and Children. Tuchman portrays Zakrzewska, after a life lived in opposition to sentimental and stereotypically feminine gender norms, laying claim in her old age to an idealized femininity constructed around women's capacity to bear children. Regina Morantz-Sanchez describes how the gender-challenging surgical activism of nineteenth-century gynecologist Mary Dixon Jones empowered women to read and to speak about their bodies in new ways. Dixon Jones's aggressive surgical practices and interpersonal style put her at odds with fellow physicians, but her openness about gynecological procedures helped to open a permissible discourse for women patients. Judy Tzu-Chun Wu describes the conflicts of a Chinese American woman doctor, Margaret "Mom" Chung, who may have been a lesbian at a time of increasing surveillance of "deviant" sexuality. Ellen S. More analyzes the interplay of personal and professional performance pressures on the controversial physician and sex educator Mary Steichen Calderone, particularly her ability to draw upon childhood sexual trauma for social and professional ends.

In part II, "Challenging the Culture of Professionalism," the authors cast a fresh light on the personal challenges of women physicians as they tried to adapt to or challenge the traditional model of health care professionalism. As Robert A. Nye's essay demonstrates, early women professionals confronted a culture infused with the threat of shame, humiliation, and even violence against those who would infiltrate these highly masculine inner sanctums. Sandra Morgen not only describes the collaboration but also the conflicting expectations and values that sometimes arose between women physicians and laywomen activists during the women's health movement of the 1970s in women's health centers. Susan Wells also examines these issues by analyzing the rhetorical strategies employed by the Boston Women's Health Book Collective in their work *Our Bodies, Ourselves* to appeal to both lay and professional readers. Naomi Rogers shows how, in the political crucible of the 1970s, women medical students were subjected to harassment, innuendo, and sexist humor. Their willingness to "take a joke" or to dish one out was used to measure their capacity to succeed in the masculine culture of medicine.

In part III, "Expanding the Boundaries," we follow women physicians out into the world through recent historical work that broadens our capacity to imagine women physicians in new settings and with wider horizons. Eve Fine describes the flourishing community of sectarian women physicians in nineteenth-century Chicago and their generally congenial relations with women regulars similarly engaged in civic reform. She also challenges us to acknowledge that many more were products of coeducational schools than has been realized. Virginia A. Metaxas demonstrates the complexities of women physicians' attempts to carry their Victorian maternalism overseas as missionaries, another new context for research on women in medicine, a reminder to contextualize our Anglo-American perspective. In the last essay of this section, Heather Munro Prescott analyzes and compares the role of black and white women physicians who directed college health services and the ways their conflicting experiences shaped different goals for themselves and their campus programs.

The conclusion, "Opportunities and Obstacles for Women Physicians in the Twenty-First Century," analyzes women physicians' current situation and suggests avenues for future research and activism. Erica Frank's research on the health of contemporary women doctors investigates the myths and realities of the lives and careers of women physicians. Frank argues that women doctors enjoy a high level of physical and mental health; yet, recent research by health care analysts, sociologists, and women physicians indicates that genuine obstacles persist, even as women make strides in achieving fairer distribution in the ranks of medical leadership and health care policy decision makers.[4] It may be some time before the philosophies and ways of being crafted by women physicians as individuals will be translated into a transformed health care system.

These essays portray a rich and variegated population of women physicians, whose complex and distinct achievements add new depth to our knowledge of the history of the American medical profession and women's struggles to change its internal and external contours. We hope and intend that this book will enhance the excitement of a field that has grown larger and more diverse over the past decade. Just as feminists have learned not to presume a unitary vision when they evoke the term "woman," historians of medicine are learning to complicate and, thereby, enrich their portrayals of women in American medicine.

NOTES

1. The exhibition opened in October 2003 and ran for more than two years. In 2005, a traveling version, developed in collaboration with the American Library Association, began a five-year tour to sixty-one sites in the United States. Papers were presented at

the National Library of Medicine symposium, "Women Physicians, Women's Politics, Women's Health: Emerging Narratives," in 2005, by Carla Bittel, Charlotte Borst, Georgina Feldberg, Eve Fine, Rebecca Flemming, Erica Frank, Vanessa Northington Gamble, Martha Gardner, Monica Green, Judy Houck, Anne Kirschmann, Virginia Metaxas, Regina Morantz-Sanchez, Ellen More, Sandra Morgen, Robert Nye, Heather Munro Prescott, Leslie Reagan, Naomi Rogers, Diane Shrier, Lydia Shrier, Arleen Tuchman, Frederick Wegener, Susan Wells, and Judy Tzu-Chun Wu.

2. Quoted in Virginia G. Drachman, *Sisters in Law: Women Lawyers in Modern American History* (Cambridge, MA: Harvard University Press, 1998), 3. Also see Mona Harrington, *Women Lawyers: Rewriting the Rules* (New York: Alfred A. Knopf, 1994), and Karen Berger Morella, *The Invisible Bar: The Woman Lawyer in America, 1638 to the Present* (Boston: Beacon Press, 1986). As another example, consider the cultural challenges faced by women physicians in the military, a subject researchers have taken up only recently.

3. For a reevaluation of the role of gender-separatist versus integrated medical schools for women medical students before 1890, see Eve Fine, "Women Physicians and Medical Sects in Nineteenth-Century Chicago," Chapter 10 in this volume.

4. Arlene S. Ash, Phyllis L. Carr, Richard Goldstein, and Robert H. Friedman, "Compensation and Advancement of Women in Academic Medicine: Is There Equity?" *Annals of Internal Medicine*, 2004, *141*: 205–214.

Acknowledgments

We would like to thank Jacqueline Wehmueller for her careful shepherding of this project, and the rest of the staff at the Johns Hopkins University Press, especially Andre Barnett and Ashleigh McKown, for their work preparing the manuscript for publication. Sarah Tracy provided detailed and insightful comments that greatly improved the final version. These essays were first presented at a symposium held at the National Library of Medicine in conjunction with the exhibition *Changing the Face of Medicine: Celebrating America's Women Physicians*. Special thanks are due to the exhibition program staff for their invaluable assistance from the beginning of these activities and to Patti Tuohy for her encouragement and guidance. The editors also extend their gratitude to everyone involved in making the symposium such a richly rewarding event. The program participants included Carla Bittel, Charlotte Borst, Georgina Feldberg, Eve Fine, Rebecca Flemming, Erica Frank, Vanessa Northington Gamble, Martha Gardner, Monica Green, Judy Houck, Anne Taylor Kirschmann, Judith Leavitt, Virginia Metaxas, Regina Morantz-Sanchez, Sandra Morgen, Robert A. Nye, Steve Peitzman, Heather Munro Prescott, Leslie Reagan, Susan Rishworth, Naomi Rogers, Diane K. Shrier, Lydia A. Shrier, Susan L. Smith, Nancy M. Theriot, Arleen Tuchman, Frederick Wegener, Susan Wells, and Judy Tzu-Chun Wu.

Ellen S. More

I would like to thank Micha Hofri and Elizabeth More—as always!—for their love, support, wisdom, and humor. I also express gratitude to the Institute for the Medical Humanities faculty and staff at University of Texas Medical Branch, Galveston, my previous academic home, and my new colleagues at the University of Massachusetts Medical School, especially Elaine Martin, director of the Lamar Soutter Library, for their assistance and encouragement. Support for my portion of this work was also provided by fellowships from the Radcliffe Institute for Advanced Study, the National Endowment for the Humanities, and a UTMB Presidential Faculty Development Award, for all of which I am extremely grateful. Finally, I owe a debt of gratitude to the many archivists and librarians whose research assistance makes this work possible. You are real pros.

Elizabeth Fee

I thank all my friends and colleagues at the National Library of Medicine and the energetic and talented staff of the History of Medicine Division for making my professional life so enjoyable; I thank especially Mary E. Garofalo for making life at home so warm, funny, and delightful.

Manon Parry

I thank all the women who inspire me, especially my mother, Helen Parry, and my grandmother, Margo Jones. I am also grateful to Izzy and Pete Gaskarth for providing a jovial and nurturing home away from home and to my extended U.S. and U.K. families for their love and support. My scholarly pursuits have been greatly enriched at the National Library of Medicine and the University of Maryland, and I thank my friends and advisors for welcoming me to both.

Women Physicians and the Cultures of Medicine

New Perspectives on Women Physicians and Medicine in the United States, 1849 to the Present

Ellen S. More, Elizabeth Fee, and Manon Parry

What are the general contours of the history of women physicians in the United States? Despite an important role in Western medicine as doctors, surgeons, midwives, and healers, a legacy extending back to ancient Greece, not until 1849 did the first Anglo-American woman, Elizabeth Blackwell (1821–1910), graduate from a college of medicine in the English-speaking world. However, the act of cultural demarcation implied in this small phrase, "the English-speaking world," indicates a large change in our frame of reference as historians over the past two decades. We now know that women were physicians in ancient times and that their official standing in the profession was gradually rescinded across Europe through the Middle Ages and early modern period; nevertheless, women continued to function as general practitioners, as healers, as midwives, and even as licensed and unlicensed surgeons. One exceptional woman, Dorothea Erxleben Leporin, received a medical degree in 1754 and did carry on a general medical practice in the German city of Halle until her death in 1762.[1] We now know, too, that in Bologna, for a brief period from the end of the eighteenth through the beginning of the nineteenth century, a few women received formal, university-based medical degrees, although few, if any, were able to practice beyond a gender-circumscribed sphere. It is not clear whether these women medical graduates in Bologna are better understood as the last of the early modern era or, at least in the European context, as the first of a new wave of women graduates. But for Blackwell and her peers, a gulf of nearly two centuries stretched from the suppression of women as surgical operators in England to Blackwell's success. We know of no evidence that they were aware of Bolognese precursors.[2]

Thus, Blackwell's graduation from Geneva Medical College in upstate New York signaled a turning point for women in England and in the United States, a bellwether of their entering the modern medical profession.[3] Indeed, despite Geneva's decision not to repeat its dangerous experiment, within the next few years, a sprinkling of medical schools cautiously opened their doors to women.[4] The creation of all-women's and coeducational schools, regular and sectarian, became the engine behind a noticeable increase in the number of women medical graduates in the United States by the end of the nineteenth century, with their numbers increasing from a few hundred in 1860 to approximately three thousand five hundred by 1900.[5] Counting homeopathic and other "sectarian" schools, the majority of these women graduated from coeducational institutions. In addition, woman-run hospitals, such as Marie Zakrzewska's New England Hospital for Women and Children in Boston, and woman-run dispensaries, or clinics, such as the Provident Dispensary of Rochester, New York, began to multiply from the late 1850s onward.[6]

These first two generations of women medical graduates not only faced discrimination, exclusion, and marginalization within the profession but also a profound challenge to their sense of identity as Victorian women. Many Victorian male physicians (and men in general) considered them un-sexed. Their sordid participation in the study of anatomy, surgery, or even obstetrics and gynecology simply disqualified them from that class of women known as "ladies." Male physicians' objections to women entering the profession revealed a complex economy of fear: that menstruation and menopause would render women unsuited to hard work and steady intellectual concentration; that women doctors would appeal more strongly to women and children than would men, costing men many of their patients; and that, given women's presumed delicacy, their moral sensibilities would be destroyed by exposure to the coarseness of medical practice. As one male physician wrote, "a female could scarce pass through the course of education requisite to prepare her, as she ought to be prepared . . . without destroying those moral qualities of character, which are essential to the office [of physician]."[7] Finally, some men worried that their own moral self-control (read: *sexual* self-control) would be gravely shaken by close contact with women students and physicians during the intimacies of the dissecting room, the surgical theater, the hospital ward, or the examining room.[8]

Men's moral qualities surely were no model for emulation when, on November 6, 1869, Dean Ann Preston of the Woman's Medical College of Pennsylvania arranged for the first group of women students to attend clinical lectures, also attended by male students, at the Pennsylvania Hospital in Philadelphia. After an uneasy hour, the men erupted in furious objections to the women. Apparently, the demonstration of a patient's poorly healed, fractured femur—necessitating momentary exposure of his thigh—unleashed a torrent of excited, sexually charged abuse: "Missiles of paper, tinfoil, tobacco quids, etc. were thrown upon the ladies, while some of these men defiled the dresses of the ladies near them with tobacco juice," all accompanied by hissing, clapping, and foot stomping. News of the fracas reached newspapers and periodicals across the United States. One male student described the women for readers of the *New Republic* as a "shameless herd of sexless beings who dishonor the garb of ladies—this beardless set of non-blushers." However, in the weeks that followed, most editorials faulted the men for a lack of chivalry and self-control.[9]

Such rites of passage, refined in the slow-burning embers of memory, provided "defining legends," in Steven Peitzman's fine phrase, for the pioneer generation of women medical graduates and for their successors.[10] Many women had such stories to tell. Elizabeth Blackwell, for example, remembered a surge of shame at hearing that the American consul in Paris, with whom she had had a pleasant dinner, later admitted he couldn't look at her "long, slender fingers without thinking of the anatomical work in which they had been engaged." A few years after the fuss at the Pennsylvania Hospital, Mary Putnam (Jacobi), soon to become the chief exemplar of the woman physician-scientist, was advised to dress as a man to gain entrance to clinical lectures at the École de Médecine in Paris, a suggestion she flatly refused but long remembered. And public health pioneer S. Josephine Baker actually ordered clothes from her dressmaker that would closely mimic male fashion (although she paired her jacket, shirt, and tie with a long skirt), the better to fit in with her all-male staff at the New York City Bureau of Child Hygiene.[11] This matter of embodiment—how to *embody* the dual identity of woman *and* physician—was not taken lightly. From the 1850s to today, women have confronted the challenge of how to *be* a woman physician, scientist, professional: what values to internalize, how to externalize each woman's unique solution to the puzzle of gender identity and role through her choice of clothing, outward demeanor, and lifestyle.

Jacobi summarized the sources of male physicians' resistance to women as colleagues with the scientific precision for which her writing is known:

> Women have in the mass never been publicly and officially regarded as individuals, with individual rights, tastes, liberties, privileges, duties, and capacities, but rather as symbols, with collective class functions, of which not the least was to embody the ideals of decorum of the existing generation, whatever these might happen to be . . . Any symptom of change in the status of women seems, therefore, always to have excited a certain terror . . . It is perfectly evident from the records, that the opposition to women physicians has rarely been based upon any sincere conviction that women could not be instructed in medicine, but upon an intense dislike to the idea that they should be so capable.[12]

Attitudes did not change easily. By 1960, women accounted for less than 7 percent of all physicians in the United States. After more than a century, how could that be? In 1915, women represented 3.6 percent of the profession and increased in proportion until 1920 when they had reached 5 percent. Then progress came to a halt. A multiplicity of factors kept women at the margins. The pressure to "improve" and professionalize medical education by closing sectarian, African American, and all-women's schools (excepting the Woman's Medical College of Pennsylvania, which resisted coeducation until 1969); shrinking class sizes at surviving coeducational medical schools; and increasing the number of years of training required before licensure, along with the cost—all conspired to thwart the burgeoning women's medical movement. By the 1910s and 1920s, a college education and a year of internship had been tacked onto a basic medical education (which had been lengthened from two to three to four years' duration). As a result, few prospective women doctors could raise the needed funds for such a lengthy course of study. Moreover, most medical schools forbade their students to marry before graduation; some women were unwilling or unable to postpone marriage and childbearing until their late twenties. The professionalization of medicine decisively clashed with child-rearing, which was then the exclusive domain of women.[13]

A final note regarding the declining rate of women's entry into the profession in the early twentieth century: On occasion, it is suggested that the increasing "reductionism" and specialization of medicine were uncongenial to women,

Table I.1 American Women Medical Graduates and Practicing Physicians, 1920–2005

Year	Graduates (%)	In Practice (%)
1920	4.0	5.0
1930	4.5	4.4
1940	5.0	4.6
1950	10.7	6.1
1960	5.7	6.8
1970	8.4	7.6
1980	23.1	11.6
1990	33.9	16.9
2000	42.6	24
2005	46.8	27

Sources: Ellen S. More, *Restoring the Balance: Women Physicians and the Profession of Medicine, 1850–1995* (Cambridge, MA: Harvard University Press, 1999), 97, 98, 221, 225; Diane Magrane, Jonathan Lang, and Hershel Alexander, *Women in U.S. Academic Medicine: Statistics and Medical Benchmarking, 2004–2005* (Washington, D.C.: AAMC, 2005), table 1, p. 11; *Physician Characteristics and Distribution in the U.S., 2006*, "Physicians by Gender," AMA Women Physicians Congress web page, www.ama-assn.org/ama/pub/category/12912.html, table accessed February 14, 2006.

who turned away from medicine in favor of other fields after 1900. We have found little evidence this is true. The more closely one looks at women's careers in medicine, and the more one knows about the intensity and uniformity with which they were excluded from internships and residencies up to the 1960s, the less persuasive this hypothesis seems. They did not willingly back away from the new scientific rigor and specialization of medicine nor, in general, equate scientific medicine with dehumanized medicine. Typically, they were prevented from attempting to synthesize the old and the new by their exclusion from hospital internships, residencies, and specialty societies in which new values were disseminated. However, once they found ways to pursue scientific careers, they excelled.[14]

Women's representation in medicine declined until World War II. The war produced an urgent, but temporary, need to increase the number of physicians overseas, allowing a short-lived, dramatic inflation of the admission rates for women medical students. After the war, women were exhorted to abandon their jobs, go home, and take up household duties. By 1950, women made up only 6.1 percent of practicing physicians.

The proportion of women in the profession began to increase again when a perceived shortage of physicians in the postwar years produced national support for opening forty-one new medical schools between 1960 and 1980. Efforts by the Josiah Macy, Jr. Foundation and the American Medical Women's Association also encouraged women doctors to return to practice. These trends and a reawakened feminist movement inspired more women to apply to medical school; during the 1960s, applications from women tripled compared with men's applications (which doubled). On average, women students represented an atypically high proportion of the entering classes of new schools. Soon, existing schools increased their class sizes, and the proportion of women accepted into them rose as the number of applications increased.

Still, gender discrimination in admissions continued. In 1970, all American medical schools found themselves threatened with a lawsuit under Title VII of the Equal Rights Act, an action brought by the Women's Equity Action League. Responding to these pressures, admissions committees gradually grew more open to women applicants. Between 1970 and 1980, the number and proportion of medical school applications from women rose more steeply than applications from men; likewise, the number of women graduates increased fivefold and has continued to rise. In 2005, women made up 46.8 percent of medical graduates.[15]

With the exception of the new admissions policies, however, the profession was slow to catch up with the implications of its new gender profile. The impact of women medical students and physicians as well as their numerical representation within the profession remained limited for another generation. By 1980, they made up only 11.6 percent of practicing physicians. In 1973, Mary Howell, a new associate dean at Harvard Medical School, published a scorching survey of the experiences of the recent cohort of women students. Her findings were so inflammatory that she published under a pseudonym for fear of losing her job. By quoting extensively from women at many U.S. medical schools, her study voiced the outrage and frustration of women fighting discouragement, disparagement, and outright harassment. Summing up the prevailing climate, one student ruefully recalled the many occasions when she was asked, "Why would a *girl* go into medicine?"[16]

This is the climate in which Mary Roth Walsh wrote the first full-length, critical history of women in American medicine. Walsh's 1976 book *"Doctors Wanted: No Women Need Apply": Sexual Barriers in the Medical Profession, 1835–1975* represented something new in the historiography of medicine: an avowedly feminist interpretation of an important aspect of that history. Walsh made no effort to hide her anger and amazement at the evident sexism her research uncovered. Nor did she doubt that such attitudes persisted. Her opening paragraph described an incident that became iconic for feminist historians of medicine: "In the fall of 1974," she wrote, "a doctor in a Boston suburb introduced herself to a patient, a boy of four. The surprised child stepped back and asserted with authority: 'You can't be a doctor, you're a girl.'" (Fourteen years later, pediatrician Perri Klass invoked almost the same scene—in reverse—when she reported her four-year-old son Benjamin's assumption that all doctors *must* be "girls.")[17]

Walsh's book was infused with the spirit of the times, simultaneously a provocation and a needed stimulus to further research. Provocatively, she often conflated male-physician opposition to women doctors with an antipathy to women, a common tendency of the first generation of feminist medical histories.[18] More accurately—and much more significantly—Walsh's work highlighted not only the relative invisibility of women in the medical profession but also their absence from the history and sociology of medicine. A few historians had written articles about women physicians, but, except for a Josiah Macy, Jr. Foundation report published as a book on the obstacles women currently faced in the profession (Carol Lopate's 1968 book *Women in Medicine*), no scholarly book-length account of women's entry into the profession had ever been published.[19]

Walsh's second most important contribution, beyond her success in revitalizing interest in the subject, was to link the history of women physicians to professionalization. Her book was explicitly intended as a call of alarm to the women's movement not to feel overconfident. Women were making huge gains, she conceded, but this was not the first time they had been poised for a takeoff that soon fizzled. Take, for example, the disappointments of the post-suffrage decades. Her subtext suggested a fundamental paradox. Throughout their history, discriminatory practices had derailed women. But, the imposition of professionalizing measures in the late nineteenth century, including more rigorous admissions requirements, upgraded medical school curricula, and postgraduate specialty training (measures partly intended to weed out undesirable candidates such as women or persons of color of either sex), actually benefited women in the long run. Walsh wrote, "Women practitioners needed the advantages of professionalization more than men. Female physicians, already suspect because of their sex, required corroboration of their expertise to meet a disbelieving public." In fact, the public has generally been a big supporter of women doctors, whether regulars or sectarians, as Regina Morantz-Sanchez, Anne Taylor

Kirschmann, and Eve Fine all have demonstrated. But perhaps professionalization helped, over time, to offset opposition by male physicians who, from the outset, were both the principal gatekeepers to, and the most determined opponents of, women's admission to the medical profession. Walsh's analysis astutely pointed the way toward later works by Margaret Rossiter and others that explicitly examined the tensions between the discourse of nineteenth-century (white, middle-class) womanhood and what Burton Bledstein famously termed "the culture of professionalism."[20]

Within a decade of its publication, a cluster of books on women physicians succeeded Walsh's lively but polemical work, demonstrating the maturation of the history of women and gender despite the heightened theoretical tensions emerging within second-wave feminism. While Walsh had written from within the heat of political struggles to bring more women into the professions, her successors' work responded to the increasing complexity of the politics and scholarship of second-wave feminism. Whereas historians of American women during the 1960s began by depicting an idealized experience of womanhood as a white, middle-class, "separate" sphere, by the mid-1970s, they had begun to deconstruct "woman" as a subject into particularities of class, ethnicity, race, sexuality, and individual context.[21] Indeed, the emergence of gender as an analytical category challenged historians of women in health care to also consider women's relationships to men. Thus, raising questions of gender separatism versus integration, entering into the intense but fragmented discourse of "difference" feminism, North American historians of women physicians turned from stories of outright discrimination against women to the complexities of woman-run medical schools, medical societies, and dispensaries. These newer studies deployed the dual discourses of gender and professionalization but only sparingly considered mixed-gender experiences.[22]

Historians of medicine registered the changing political and scholarly imperatives of the era. In the historiographical shifts, which began in the 1970s, we can see the effect of the maturation of departments of history of medicine with mentors such as Charles Rosenberg and Barbara Rosenkrantz, to name only two, and later their students, modeling a history of medicine that contextualized health and disease far more broadly than had been typical in the past.[23] Women historians of medicine also related their work to the women's health movement with a heightened awareness of issues such as reproductive rights and the role of women as patients. This work has been greatly enhanced as scholars turned to matters of race, ethnicity, radical politics, sexuality, or gender performance.[24]

Of works on women physicians published in the decade after Walsh, by far the most influential was Regina Morantz-Sanchez's insightful and expansive book, *Sympathy and Science: Women Physicians in American Medicine*, published in 1985. By this time, historians had been exploring the complexities of women's experiences and sense of identity for two decades.[25] Feminist philosophers of science, such as Evelyn Fox Keller, Sandra Harding, and Helen Longino, were debating the role of gender in science.[26] Historians who had initially characterized nineteenth-century women within the framework of "separate spheres" now found ample materials to document that such women were in fact fully engaged in the social, cultural, and political lives of their communities, routinely subverting the boundaries of the so-called woman's sphere.[27] Within the history of medicine, historians had begun to discover that women patients and physicians were not simply victims of male physicians' malfeasance and ill will. Rather, historians such as Carroll Smith-Rosenberg, Judith Walzer Leavitt, and Linda Gordon documented and interpreted a rich world of women's involvement in the politics of health care. Clearly, women physicians must now be set in a context of Victorian reform as well as female social networks.[28]

In this context, Morantz-Sanchez located the history of women physicians within the social and cultural networks of middle-class American women, not merely as an aspect of medical professionalization. She asked whether women doctors' practices were different from men's and, if so, in what ways. Moreover, she examined how they differed among themselves. Although she detailed systematic discrimination against women in medical school admissions, Morantz-Sanchez focused on nineteenth-century women's social and professional universe, the role of all-women's medical colleges, women physicians' diverse understanding of medical "science," and how that understanding informed their practices. The resulting account revealed the ways women physicians' cultural expectations were built on a substrate of feminine, middle-class cultural norms, a set of expectations to which women doctors could—and did—efficiently adapt. After all, except for women who practiced in state hospitals for the insane, until the 1880s, most women, like men, practiced out of their own homes. By the 1890s, women physicians were more likely to marry than other working women of that era, although most did not marry at all. In *Sympathy and Science*, too, women physicians were more prone to empathize with the needs of women and children, the "sympathy" of the book's title. But they were also, and without contradiction, enthusiastic believers in the value of science—even if they held widely differing ideas of what constituted "scientific" medicine.

Morantz-Sanchez wrote revealingly about the significance of attending an all-women's medical college, but she had relatively little to say about women physicians who were educated at one of the dozens of alternative medical schools populating the nineteenth-century educational landscape, coeducational or otherwise. Most women physicians in practice before 1870 were graduates of homeopathic, eclectic, botanical, or hydropathic schools. As late as the 1890s, sectarian schools still accounted for a sizable proportion of women graduates. According to figures compiled by Anne Kirschmann, for example, women "regulars" began to outnumber sectarians in major East Coast cities only after 1890. Before 1886, they were the majority only in Philadelphia, home of the Woman's Medical College. Works by Thomas Bonner, Edward Atwater, Anne Kirschmann, and Eve Fine also indicate that "sectarian" schools, as they were pejoratively labeled, accounted for most of the coeducation experienced by women doctors in the nineteenth century.[29] Despite women physicians' strong attachment to women's medical institutions, social networks, and political causes, their education frequently occurred within coeducational settings.[30]

Undeniably, the high proportion of women whose medical degrees were awarded by "sectarian" schools over an approximately thirty-year period contributed to the disrepute in which many male "regulars" held all women graduates. As the proponents of these alternative systems battled with the medical "regulars" (particularly the American Medical Association and state medical societies), both for patients and for professional legitimacy, the regulars painted them as ill educated and unscientific. We should not necessarily take the regulars' word for it. The qualitative differences between sectarian and orthodox medical schools before the 1880s were minor. Thus, the eagerness with which the first generations of women matriculated at alternative medical colleges did not mean that they were less committed to medical "science" than most regulars. Women doctors of the early generations presented case reports at medical society meetings, debated therapeutic innovations, and read the medical journals, activities that paralleled men physicians' involvements during that era.

Emblematic figures such as Elizabeth Blackwell and Mary Putnam Jacobi permitted Morantz-Sanchez to demonstrate that "scientific" medicine had as many different meanings and applications for women physicians as it did for men. Recent research by Arleen Tuchman on Marie Zakrzewska and Carla Bittel on Mary Putnam Jacobi, represented by chapters in this volume, reinforces this judgment.[31] By World War I, when the meanings of "science" for medical research and practice had converged around the importance of precise clinical measurement, laboratory analysis, experimentation, and when possible, therapeutic intervention, the majority of women and men had abandoned sectarianism in favor of mainstream approaches to diagnosis and therapeutics.[32]

Writing more than twenty years ago, Morantz-Sanchez held open the possibility that women rejected careers as physicians in the early 1900s because they were dissatisfied with the increasing specialization and scientism of modern medicine. This explanation seems less likely now than twenty years ago, given what we now know about the amount of time required to transform the profession into a field dominated by subspecialties, a point demonstrated in Morantz-Sanchez's recent book on the gynecological surgeon, Mary Dixon Jones. Further, what counted as "science" at the turn of the century was not as sharply defined as it is today. We recognize the outright discrimination leveled at twentieth-century women medical graduates who sought to keep up with the new professional standards by applying for internships at first-rate hospitals. Their exclusion from such opportunities inhibited their ability to penetrate the prestigious enclaves of specialized medical practice for at least a generation.[33]

Ellen S. More's *Restoring the Balance: Women Physicians and the Profession of Medicine, 1850–1995* (1999) is one of several recent works to extend and modify the picture drawn by Walsh and Morantz-Sanchez and to devote considerable attention to the twentieth century.[34] More set out to understand the values and choices that informed the varying career tracks of mainstream women physicians. After medical school, how did they finesse the apparently competing demands of femininity and professionalism? Moving beyond the great (*women*) doctors such as Blackwell, Putnam Jacobi, or Marie Zakrzewska, More considered how *most* women physicians conceptualized and performed the role of medical professional. She showed that throughout the modern history of women physicians in American medicine, the metaphor of "balance" has been integral to their sense of professional identity and their choices of how to structure their lives and careers. In More's account, balance inheres in at least three domains: "first, the process by which American women physicians fought for professional equality in medicine; second, their steadfast resistance to a one-dimensional conception of professionalism by pursuing a judicious balance of personal, community, and professional interests; and third, the attentiveness of many women physicians to interactions among the psychological, social, and physiological dimensions of their patients' lives." "Balance," then, is not meant to suggest a static equipoise or full resolution of the conflict between professional and personal values. Rather, it represents an ongoing struggle to *keep* one's balance, a constant oscillation between poles of personal values that sometimes are complementary but often are in conflict. More also described the complex roles played by men in the career choices and progress of successful women physicians.[35]

Restoring the Balance addressed the resistance of some women doctors to post-Victorian medicine's increasing remoteness from their patients' communities, what she and others termed its "civic professionalism."[36] In some cases, however, their tendency to cling to older "clubwoman" networks when the modern vehicles of professional advance, such as hospitals, medical school faculties, and specialty societies were largely closed to them, resulted in its own class-based remoteness. Finally, she undertook to delineate and acknowledge the multiplicity contained within the category "woman physician," an effort continued by the authors of the chapters contained here. Thus, for example, she tried to incorporate the narrative of African American women physicians into the overall narrative of women in American medicine.[37] She did not trace other threads, however, such as the narratives of Hispanics, Asians, lesbians, and women physicians who were biomedical scientists; only recently have major works, like Judy Tzu-Chun Wu's study of Chinese physician Margaret "Mom" Chung, begun to address these diverse experiences. Full-length studies of women physician-scientists are greatly needed.[38]

As *Restoring the Balance* showed, women physicians are becoming central to the profession. In ways that differ with the perspective of each individual woman, women retain a set of values that, even now, distinguishes them from many male colleagues. Quite a few women physicians aspire to leadership in their profession; yet, their vision of leadership differs subtly from the traditions of past (mostly male) generations. Literary and cultural historians such as Susan Wells and Nancy Theriot, analyzing women physicians' narrative style and rhetorical "voice," also reinforce the idea that gender, as a phenomenon of culture, role, and performance, continues to shape medical practice.[39] Building on the insights of feminist scholarship, many historians of women physicians, indeed, many in this volume, deploy the reinvigorated genre of biography—the *new* biography—to bring to light noteworthy figures who may have been overlooked in their time. The new biography addresses more than the external events of a life; it tries to address the private world and interior perspective of its subject. Of special importance to historians of women and other less visible groups is biography's capacity to reanimate the telling detail or the powerful example embodied in even the least well-known historical subject. Combining the art of historical narrative, the discipline of social history, and the insight of empathic reading, the new biography can illuminate the world of its subjects.[40]

Since the 1990s the professional and social landscape for women physicians has become more fluid, thus making it more difficult to characterize through a single narrative arc. Today's news media abound with images of successful women physicians and scientists: large-scale clinical investigations such as the Women's Health Initiative are directed by women physicians and funded by the National Institutes of Health (NIH). Over the past decade, women have been named directors of the NIH and the Centers for Disease Control and Prevention. They have served as surgeons general, deans of major medical schools, department chairs, president of the American Medical Association, and editor-in-chief of the *Journal of the American Medical Association*. In 2006, women made up more than 27 percent of practicing physicians.[41]

The resulting expectation of imminent gender equality is powerful but may be misleading. Although women have made great strides in the profession, serious inequalities remain. They are no longer a minority among medical students, but as women entering medical school today eventually realize, gender-related obstacles to career advancement merely have been pushed forward into residency and beyond. Women physicians experience gender inequities throughout most of their careers, though they are less blatant than in the past. Like other professional women, they are judged both for the work they do and for the image they present—their particular blend of femininity and professionalism might be scrutinized in the height of their heels, the cut of their suits, the strength of their handshake, or the length of their hair.[42]

Today's women medical students and many younger physicians, both men and women, are trying to understand and embody a tense duality of professionalism and personal fulfillment. The so-called millennial generation, as noted by both educators and employers, evince unprecedented concern over this precarious balancing act, and they are right to do so. Expectations that women will be the primary caregivers and that men will devote most of their energies to their work have changed relatively little. Nor has the profession demonstrated much willingness to accommodate nonstandard career trajectories and family structures. Questions of personal identity and professional performance still speak urgently to women physicians and, of late, to many of their male colleagues.

Nevertheless, some commentators seem convinced that most of the problems of discrimination against women have been solved. In 2005, the *New York*

Times reported that significant numbers of women now *remove themselves* from competitive career tracks to become "full-time mothers." In fact, the reporter's sample was far too small to allow for any valid conclusions. This was only one of several stories purporting to show that women professionals were painlessly choosing to "opt out" of the workforce, stories easily refuted by larger sampling techniques.[43] Nevertheless, even for those privileged women who can make the choice to abandon the workforce, the reported incomprehension of persistent inequities and the problematic choices facing professional women is troubling. The seemingly rapid success of women in American medicine is a complex and unfinished story, with important parallels and a few lessons for women in other professions. A 2003 article in *Discourse and Society* replicated an earlier study by demonstrating that women physicians routinely received letters of recommendation that convey less confidence and enthusiasm than letters for men in comparable situations.[44] Women still are underrepresented in the leadership of American medicine and science, just as they are in politics, business, academia, and the law.

In academic medicine in 2007—nearly forty years since discrimination in medical school admissions began to decline—women represented only 17 percent of full professors in medical schools, 11 percent of department chairs, and 12 percent of deans. These are big gains compared with forty years ago, but they demonstrate a leadership lag that troubles academic medical leaders. Other statistics are even more discouraging. According to the Association of American Medical Colleges, only 1.6 percent of full professors are women of Hispanic, African American, Native American, Hawaiian, or other Pacific Islander origin, reflecting the low representation of women of color overall—6.5 percent of all faculty.[45] Women accept a disproportionate share of the responsibility for balancing family and profession and frequently do not receive the same "fast-track" mentoring for leadership roles that men receive. How these two phenomena are related is open to conjecture.[46]

In response to the current historical moment, historians of women in medicine are addressing a new set of questions regarding gender performance, sexuality, race, ethnicity, political activism among physicians, and the role of science in women's practices and research.[47] The history of women's successful (but incomplete) struggle to take their place in the American medical profession is part of the larger story of women's changing roles in American society. This book is intended to bring to light the story of that struggle and to demonstrate the inventiveness that underlies women physicians' achievements, a bravura gender performance each woman must craft for her own intended sphere of activities, whether clinical care, surgery, public health, basic research, or any other career focus. Only thirty years ago, the woman who chose to become a physician, especially in a traditionally male field such as surgery, was "viewed as if she were performing an unnatural act." Today, the artifacts of past prejudices that still subtly hinder women's progress are the only unnatural things about their place in the medical profession.[48]

NOTES

1. Rebecca Flemming, "New Perspectives on Ancient Physicians," presented at "Women Physicians, Women's Politics, Women's Health: Emerging Narratives," a symposium at the National Library of Medicine, Bethesda, Maryland, March 10–11, 2005. For the medieval period, the authors profited greatly from Monica Green's paper, "Rethinking the Mistress Narrative: Riding the Third Wave to a New History of Women in Medicine," also presented at "Women Physicians, Women's Politics, Women's Health," for its persuasive discussion of the difficulties in reading back into medieval history the existence of recognizable prototypes for the modern-day "woman physician." For Dorothea Erxleben, see Londa Schiebinger, *The Mind Has No Sex? Women in the Origins of Modern Science* (Cambridge, MA: Harvard University Press, 1989), 250–257. Also see Monica Green, *Making Women's Medicine Masculine: The Rise of Male Authority in Premodern Gynaecology* (New York: Oxford University Press, 2008), published too late to be consulted for this volume.

2. Cf. Gabriella Berti Logan, "Women and the Practice and Teaching of Medicine in Bologna in the Eighteenth and Early Nineteenth Centuries," *Bulletin of the History of Medicine*, 2003, 77: 506–535, for a discussion of the few women medical graduates of late eighteenth- and early nineteenth-century Bologna.

3. A small number of women practiced as apprentice-trained botanical and homeopathic practitioners before Blackwell graduated from Geneva Medical College. But from the 1840s or so, they, too, set their sights on admission to medical colleges. See Mary Roth Walsh, *"Doctors Wanted: No Women Need Apply": Sexual Barriers in the Medical Profession, 1835–1975* (New Haven, CT: Yale University Press, 1976), 1–34, for a portrait of the most visible of these women physicians, Harriot Hunt.

4. Regina Morantz-Sanchez, *Sympathy and Science: Women Physicians and American Medicine* (New York: Oxford University Press, 1985). Also see Eve Fine, "Women Physicians and Medical Sects in Nineteenth-Century Chicago," Chapter 10 in this volume, and Anne Taylor Kirschmann, *A Vital Force: Women in American Homeopathy* (New Brunswick, NJ: Rutgers University Press, 2004).

5. No one has yet accurately counted women physicians in practice during the nineteenth century. See Thomas Neville Bonner, *To the Ends of the Earth: Women's Search for Education in Medicine* (Cambridge, MA: Harvard University Press, 1992), 203 n. 92, for one (avowedly conservative) estimate. Eve Fine points out that Bonner's estimate of women practitioners ca. 1900, approximately three thousand four hundred, is likely an

undercount since it did not include women who had died before that year (Fine, personal communication, February 15, 2006). For other sources, see Ellen More, *Restoring the Balance: Women Physicians and the Profession of Medicine, 1850–1995* (Cambridge, MA: Harvard University Press, 1999), 288 n. 16. Also see Kirschmann, *A Vital Force*, 7, who gives an estimate of 1,690 women graduates of homeopathic medical schools between 1852 and 1900.

6. For the question of coeducation, also see Fine, "Women Physicians and Medical Sects," Chapter 10 in this volume; More, *Restoring the Balance*, 42–69; Gloria M. Moldow, *Women Doctors in Gilded-Age Washington: Race, Gender, and Professionalization* (Urbana: University of Illinois Press, 1987); Arleen Marcia Tuchman, *Science Has No Sex: The Life of Marie Zakrzewska, M.D.* (Chapel Hill: University of North Carolina Press, 2006); Virginia G. Drachman, *Hospital with a Heart* (Ithaca, NY: Cornell University Press, 1984).

7. Morantz-Sanchez, *Sympathy and Science*, 49–56; quotation p. 27. Of course, they also worried about male sensibilities but in relation to the coarsening effects of vivisection. Susan E. Lederer, "Moral Sensibility and Medical Science," in *The Empathic Practitioner: Empathy, Gender, and Medicine*, ed. Ellen Singer More and Maureen A. Milligan (New Brunswick, NJ: Rutgers University Press, 1994).

8. Steven J. Peitzman, *A New and Untried Course: Woman's Medical College of Pennsylvania, 1850–1998* (New Brunswick, NJ: Rutgers University Press, 2000), 33–37.

9. Ibid., 9, 34–35.

10. Ibid., 6.

11. S. Josephine Baker, *Fighting for Life* (New York: Macmillan, 1939), 90–95.

12. Mary Putnam Jacobi, "Woman in Medicine," in *Women's Work in America*, ed. Annie Nathan Meyer (New York: Henry Holt, 1991), 139–205; quotation pp. 195–196.

13. Morantz-Sanchez, *Sympathy and Science*, 234–241.

14. See Carla Bittel, "Mary Putnam Jacobi and the Nineteenth-Century Politics of Women's Health Research," Chapter 1 in this volume; More, *Restoring the Balance*, 106–111; Morantz-Sanchez, *Conduct Unbecoming a Woman: Medicine on Trial in Turn-of-the-Century Brooklyn* (New York: Oxford University Press, 1999). For numerous examples of women physicians who were and are fully engaged in research, see www.changingthe faceofmedicine.nlm.nih.gov.

15. More, *Restoring the Balance*, 186–194, 216–225; Diane Magrane, Jonathan Lang, and Hershel Alexander, *Women in U.S. Academic Medicine: Statistics and Medical School Benchmarking 2004–2005* (Washington, DC: AAMC, 2005), table 1, p. 11.

16. Italics added. Margaret A. Campbell, M.D. [pseud. Mary Howell, M.D.], *Why Would a Girl Go into Medicine?* (New York: Feminist Press, 1973), 38. And see Naomi Rogers, "Feminists Fight the Culture of Exclusion in Medical Education, 1970–1990," Chapter 9 in this volume.

17. Walsh, "*Doctors Wanted: No Women Need Apply*," ix; Perri Klass, "Are Women Better Doctors," in *Baby Doctor: A Pediatrician's Training* (New York: Ballantine Books, 1992), 282–283. The original article was the cover story for the *New York Times Magazine*, April 19, 1988.

18. For examples of early feminist polemic, see the two pamphlets by Barbara Ehrenreich and Deirdre English, *Witches, Midwives, and Nurses* (New York: Feminist Press, 1973), and *Complaints and Disorders* (New York: Feminist Press, 1977). Also see the exchange between Ann Douglas Wood ("The Fashionable Diseases") and Regina Morantz

("The Lady and Her Physician") in *Clio's Consciousness Raised: New Perspectives on the History of Women*, ed. Mary S. Hartman and Lois Banner (New York: Harper Colophon, 1974).

19. Walsh, "*Doctors Wanted*," x–xi, n. 4. For examples of scholarly writing from the fifties and sixties, see Richard H. Shryock, "Women in American Medicine" (1950), reprinted in Richard H. Shryock, *Medicine in America: Historical Essays* (Baltimore: Johns Hopkins Press, 1966), 177–199; John B. Blake, "Women and Medicine in Ante-Bellum America," *Bulletin of the History of Medicine*, 1965, *39*: 99–123; and Carol Lopate, *Women in Medicine* (Baltimore: Johns Hopkins Press for the Josiah Macy, Jr. Foundation, 1968). This is not to say that their struggles went unnoticed before World War II. In 1891, Mary Putnam Jacobi's "Woman in Medicine" provided an excellent, chapter-length history; subsequent physicians, such as Kate Hurd-Mead and Esther Pohl Lovejoy, wrote groundbreaking, if uncritical, catalogues of women physicians. Cf. Hurd-Mead, *A History of Women in Medicine: From the Earliest Times to the Present* (Haddam, CT: Haddam Press, 1938); and Lovejoy, *Women Physicians and Surgeons, National and International Organizations* (Livingston, NY: Livingston Press, 1939). Cf. Morantz-Sanchez, *Sympathy and Science*, 371–378.

20. Walsh, "*Doctors Wanted*," 14–15. See Morantz-Sanchez, *Sympathy and Science* and Kirschmann, *A Vital Force*; and Fine, "Women Physicians and Medical Sects," Chapter 10 in this volume. For early histories of professional culture in nineteenth-century America, see Robert J. Wiebe, *The Search for Order* (New York: Hill and Wang, 1967); Burton J. Bledstein, *The Culture of Professionalism* (New York: Norton, 1976); and critiques by Magali Sarfatti Larson, *The Rise of Professionalism* (Berkeley: University of California Press, 1977), and Andrew Abbott, *The System of Professions* (Chicago: University of Chicago Press, 1988). This literature is vast; the foregoing samples merely touch the surface of it. On the effect of scientific professionalism on women before World War II, see Margaret W. Rossiter, *Women Scientists in America: Struggles and Strategies to 1940* (Baltimore: Johns Hopkins University Press, 1982).

21. For an informed analysis of the historiography of women's and gender history (albeit one that fails to deal with the history of medicine and health care), see Kathleen Canning, *Gender History in Practice: Historical Perspectives on Bodies, Class, and Citizenship* (Ithaca, NY: Cornell University Press, 2006), 5–28, 168–180.

22. Patricia Hummer, *Decade of Elusive Promise* (Ann Arbor, MI: University Microfilm Press, 1979); Drachman, *Hospital with a Heart*; idem, "Female Solidarity and Professional Success: The Dilemma of Women Doctors in Late Nineteenth-Century America," *Journal of Social History*, 1982, *15*: 607–619; idem, "The Limits of Progress: The Professional Lives of Women Doctors, 1881–1926, *Bulletin of the History of Medicine*, 1986, *60*: 58–72; Judith Lorber, *Women Physicians: Career, Status, and Power* (New York: Tavistock, 1984); Ruth J. Abram, ed., *"Send Us a Lady Physician": Women Doctors in America, 1835–1920* (New York: Norton, 1985); Penina Migdal Glazer and Miriam Slater, *Unequal Colleagues: The Entrance of Women into the Professions, 1890–1940* (New Brunswick, NJ: Rutgers University Press, 1987); and Moldow, *Women Doctors in Gilded-Age Washington*; Thomas Bonner's book, *To the Ends of the Earth*, focused on the campaign to win admission to medical schools in Europe and the United States. All these works contributed to the understanding of women physicians' dynamic relationship to the forces of separatism, professionalization, and institutionalization of health care. None attempted quite as broad conceptual

or chronological coverage as either Walsh or Morantz-Sanchez, but all contributed to the emerging consensus that a multifactorial approach to women physicians' history is essential to fully understanding its development.

23. A strikingly early example is Charles Rosenberg, *The Cholera Years: The United States in 1832, 1849, and 1866* (Chicago: University of Chicago Press, 1962); as well as even earlier work by Henry Sigerist, Richard Shryock, and John Duffy. Recent discussions of the historiography of medicine include, among others, Frank Huisman and John Harley Warner, eds., *Locating Medical History: The Stories and Their Meanings* (Baltimore: Johns Hopkins University Press, 2004), especially the essays by Frank Huisman and John Harley Warner, "Medical Histories," 1–32, and Susan M. Reverby and David Rosner, "'Beyond the Great Doctors' Revisited: A Generation of the 'New' Social History of Medicine," 167–193; and John C. Burnham, *How the Idea of Profession Changed the Writing of Medical History*, Medical History Supplement No. 18 (London: Wellcome Institute for the History of Medicine, 1998).

24. Two examples of groundbreaking work from this period are Elizabeth Fee, ed., *Women and Health: The Politics of Sex in Medicine* (Farmingdale, NY: Baywood, 1983), and Susan Cayleff, *Wash and Be Healed: The Water-Cure Movement and Women's Health* (Philadelphia: Temple University Press, 1987).

25. Barbara Welter, "The Cult of True Womanhood, 1820–1860," *American Quarterly*, 1966, *18*: 151–174; Carroll Smith-Rosenberg and Charles Rosenberg, "The Female Animal: Medical and Biological Views of Woman and her Role in Nineteenth-Century America," *Journal of American History*, 1973, *60*: 332–356.

26. Evelyn Fox Keller, *Reflections on Gender and Science* (New Haven, CT: Yale University Press, 1985); Sandra Harding, *The Science Question in Feminism* (Ithaca, NY: Cornell University Press, 1986); Helen E. Longino, *Science as Social Knowledge* (Princeton, NJ: Princeton University Press, 1990).

27. Examples of this generation of work are Nancy Cott, *The Bonds of Womanhood* (New Haven, CT: Yale University Press, 1977); Mary Beth Norton, *Liberty's Daughters: The Revolutionary Experience of American Women* (Boston: Little, Brown, 1980); and Linda Kerber, *Women of the Republic: Intellect and Ideology in Revolutionary America* (Chapel Hill: University of North Carolina Press, 1980).

28. An early example is Catherine M. Scholten, "'On the Importance of the Obstetrick Art': Changing Customs of Childbirth in America," *William and Mary Quarterly*, 1977, *34*: 429–431. See essays of Carroll Smith-Rosenberg, most from the 1970s, reprinted in *Disorderly Conduct: Visions of Gender in Victorian America* (New York: Oxford University Press, 1985); Linda Gordon, *Woman's Body, Woman's Right: Birth Control in America* (New York: Penguin Books, 1974); Judith W. Leavitt, "Science Enters the Birthing Room: Obstetrics in America since the Eighteenth Century," *Journal of American History*, 1983, *70*: 281–304.

29. Bonner, *To the Ends of the Earth*, chaps. 2 and 8; Edward Atwater, "Women Who Became Doctors before the Civil War," given as the Radbill Lecture, College of Physicians of Philadelphia, April 18, 1989; More, *Restoring the Balance*, 19–21, 268 n. 36; Kirschmann, *A Vital Force*, 175–176; and Fine, "Women Physicians and Medical Sects," Chapter 10 in this volume.

30. See Fine, "Women Physicians and Medical Sects," Chapter 10 in this volume.

31. Morantz-Sanchez, *Sympathy and Science*, 184–202; Tuchman, "Maternity and the Female Body in the Writings of Dr. Marie Zakrzewska, 1829–1902," Chapter 2 in this

volume; and Bittel, "Mary Putnam Jacobi," Chapter 1 in this volume; and, John Harley Warner, "Ideals of Science and Their Discontents in Late Nineteenth-Century American Medicine," *Isis*, 1991, *82*: 454–478.

32. Kirschmann, *A Vital Force*, however, shows that a remnant of homeopathic physicians maintained a strong allegiance to homeopathic therapeutics all through the twentieth century. See Chapter 7.

33. Morantz-Sanchez, *Conduct Unbecoming a Woman*. For the cascade of exclusionary factors that severely affected women in medicine in the early twentieth century, see More, *Restoring the Balance*, 95–99, 105–112; Morantz-Sanchez, *Sympathy and Science*, 232–255. For the parallel phenomena within the sciences, see Rossiter, *Women Scientists in America*.

34. Also see Steven J. Peitzman, *A New and Untried Course*, the first modern history of the Woman's Medical College of Pennsylvania, which takes into account much recent work on the role of gender in American medicine; and Anne Taylor Kirschmann, *A Vital Force*.

35. More, *Restoring the Balance*, 3.

36. Ibid., 8–9, 16, 42–44, 67–68.

37. Susan Smith, *Sick and Tired of Being Sick and Tired: Black Women's Health Activism in Modern America, 1890–1950* (Philadelphia: University of Pennsylvania Press, 1995). Like Smith, More devoted attention to capturing some of the statistics, general narrative, and personal experiences of African American women physicians: *Restoring the Balance*, 4–6, 165–168, 236–247. For other works, see Vanessa Northington Gamble, "On Becoming a Physician: A Dream Not Deferred," in *The Black Women's Health Book: Speaking for Ourselves*, ed. Evelyn C. White (Seattle: Seal Press, 1990), and idem, *Making a Place for Ourselves: The Black Hospital Movement, 1920–1945* (New York: Oxford University Press, 1995), which explore the effects of sexism and racism on black women doctors and on all black physicians. Finally, see Sharla Fett, *Working Cures: Healing, Health, and Power on Southern Slave Plantations* (Chapel Hill: University of North Carolina Press, 2002), 111–141, which describes the slave women "doctresses" of Southern plantations, whose medical skills were redefined as "menial work" by slave owners and overseers.

38. See Judy Tzu-Chun Wu's "A Chinese Woman Doctor in Progressive Era Chicago," Chapter 4 in this volume. Carla Bittel's forthcoming work on Mary Putnam Jacobi will provide a much-needed examination of the role of science in women's medical careers.

39. Susan Wells, *Out of the Dead House: Nineteenth-Century Women Physicians and the Writing of Medicine* (Madison: University of Wisconsin Press, 2001); Nancy M. Theriot, "Women's Voices in Nineteenth-Century Medical Discourse: A Step toward Deconstructing Science," *Signs*, 1993, *19*: 1–31.

40. Linda Wagner-Martin, *Telling Women's Lives: The New Biography* (New Brunswick, NJ: Rutgers University Press, 1994), discusses the renewed acceptability of biography as a form of historical writing, esp. pp. 1–56. Joyce Antler, *Lucy Sprague Mitchell: The Making of a Modern Woman* (New Haven, CT: Yale University Press, 1987), xiv, who uses her subject to represent the concept "feminism as life process," is an early example of such biographical history.

41. American Medical Association web page, "Member Groups, Women Physicians Congress," tables 1 and 3, www.ama-assn.org/ama/pub/category/12912.html, accessed February 14, 2006.

42. For example, Georgina Feldberg's paper, "Doctors as Debutants: Images of Canadian Medical Women," presented at the symposium, "Women Physicians, Women's Politics, Women's Health: Emerging Narratives," featured a series of Canadian women physicians, each one bedecked with strands of cultured pearls—evidently laying claim to their place in Canadian society as women physicians and *feminine* women, too.

43. Louise Story, "Many Women at Elite Colleges Set Career Path to Motherhood," *New York Times*, September 20, 2005, pp. 1, 18. This story elicited scathing rebuttals, especially for its assumption that its subjects represented more than a small minority, both in the *New York Times* and on Slate.com. See, for example, Jack Schafer, "A Trend So New It's Old," http://slate.msn.com/id/2126760/, posted September 23, 2005. More generally, see E. J. Graff, "The Opt-Out Myth," *Columbia Journalism Review*, 2007 (January/February), accessed at www.cjr.org/issues/2007/2/Graff.asp.

44. Frances Trix and Carolyn Psenka, "Exploring the Color of Glass: Letters of Recommendation for Female and Male Medical Faculty," *Discourse and Society*, 2003, *14*: 191–220.

45. Diane Magrane, Valarie Clark, et al., "Women in U.S. Academic Medicine Statistics, 2006–2007," accessed at www.aamc.org/members/wim/statistics/stats07/start.htm. Figures for women of color exclude women of Asian ethnicity, who are not included under the rubric of "underrepresented minorities" by the AAMC but compose 8.2 percent of women full professors and 1.3% of all full professors.

46. Phyllis M. Carr, Janet Bickel, and Thomas S. Inui, eds., *Taking Root in a Forest Clearing: A Resource Guide for Medical Faculty* (Boston: Boston University School of Medicine for the W. K. Kellogg Foundation, 2003); David Leonhardt, "Scant Progress on Closing Gap in Women's Pay," *New York Times*, December 24, 2006, p. 1.

47. See Morantz-Sanchez, *Conduct Unbecoming a Woman*; Judy Tzu-Chun Wu, *Doctor Mom Chung of the Fair-Haired Bastards: The Life of a Wartime Celebrity* (Berkeley: University of California Press, 2005).

48. Carol Nadelson and Malkah Notman, "Success or Failure: Women as Medical School Applicants," *Journal of the American Medical Women's Association*, 1974, *29*: 167–72, quoted in Regina Markell Morantz, Cynthia Stodola Pomerleau, and Carol Hansen Fenichel, eds., *In Her Own Words* (New Haven, CT: Yale University Press, 1982), 35.

Part I / Performing Gender, *Being* a Woman Physician

CHAPTER ONE

Mary Putnam Jacobi and the Nineteenth-Century Politics of Women's Health Research

Carla Bittel

In 1873, Mary Putnam Jacobi contemplated the subordinate status of American women. Women held secondary positions because of a false but popular "public sentiment": reproduction superseded all other activities of the female sex. She told a group of women's rights activists, "It is so far from true that the bearing and rearing of children suffices to absorb the energies of the whole female sex, that a large surplus of feminine activity has always remained to be absorbed in other than these primitive directions."[1] Jacobi strongly took issue with the idea that reproduction constituted women's sole purpose and that their bodies limited them to domesticity and motherhood. She demonstrated this in her own life, by combining a career in medicine with marriage and children. But most often, she made her case with physiological evidence, not with personal examples. According to Jacobi, women had a surplus of unused energy that could be diverted from the "primitive" to the productive and divided between the private world of home and family and the public world of higher education and professional work. They needed to be contributing members of society, intellectually engaged, and devoted to the greater good. This was not a choice, Jacobi asserted; science showed that the health of women depended on it.

Some vocal physicians and critics disagreed and fueled this "public sentiment," arguing that mental and physical activity could lead to women's physiological downfall and jeopardize their ability to reproduce. As some middle-class women pursued college, the professions, and political rights, they faced resistance from critics who accused them of ignoring their duties as mothers and defying their natural reproductive roles. In the 1870s, as historians know well, the "woman question" was not simply a social debate but a battle over the meanings of biology.[2]

Jacobi opened her New York office in 1871, but she had already embarked on a career that would make her one of the foremost physicians of nineteenth-century America. As the daughter of New York publisher George Palmer Putnam, her family and social class afforded her great educational opportunities. Although primed by her father to be a writer of fiction and didactic literature, she chose a life of science and acquired the highest level of medical training. Her education surpassed that of most medical women and many men. She received three degrees, the first from the New York College of Pharmacy in 1863, and the second from the Female (later Woman's) Medical College of Pennsylvania, where she received her first M.D. in 1864. Then, like numerous American physicians in the nineteenth century, she traveled to Paris to further her education and became the first woman admitted to the École de Médecine, receiving her second M.D. in 1871 and graduating with high honors. Shortly after returning to New York, she met Abraham Jacobi, a German-Jewish émigré from the Revolutions of 1848, who was one of the city's leading physicians and would become known as the "father of pediatrics."[3] After finding her intellectual comrade, someone who shared her interest in science and politics, she married Abraham Jacobi in 1873. For the next three decades, Mary Putnam Jacobi practiced medicine in New York, teaching and working as an attending physician at a number of schools and hospitals, including the New York Infirmary and its medical college. In these years, she combined medicine and activism, becoming a leading advocate for the rights of women.[4]

Jacobi set out to revise common wisdom about women's health and produced studies that aimed to disprove notions of women's biological inferiority, demonstrating that women were capable of education, paid work, and political activity. She believed many of the recent claims about female biology had been produced by medical men whose research did not live up to scientific standards and who allowed their prejudices to taint their investigations. Critiquing their scientific failings, Jacobi often contrasted her work against the research of her theoretical

Mary Putnam Jacobi. Reproduced courtesy of the May Wright Sewall Collection, Library of Congress.

rivals, asserting her own expertise and accusing them of doing "bad science." These medical men were biased against women, did not have statistical support for their theories, and did not back their ideas with experimental methods, she charged. Drawing boundaries between legitimate and illegitimate knowledge, she tried to demarcate scientific from unscientific approaches as she presented more favorable interpretations of female physiology.[5] However, the meaning of scientific medicine was not fixed, as physicians debated the relationship between the clinic and laboratory, argued about the value of particular experiments, and diverged on definitions of positivism.

Jacobi tried to draw clear boundaries around scientific medicine by purposefully displaying and demonstrating her own expertise. As some critics claimed women were incapable of scientific work, she turned the tables on the debate, publicly holding men to higher standards and exhibiting her abilities. In this case, performing gender was not so much about acting masculine but about acting scientific as a woman and acting more scientific than medical men. She tried to show that a woman could be an expert on women's health, not because of her maternalism or physical experience as a woman but because of her mastery of physiology and chemistry. Thus, Jacobi's performance lies in her overt demonstrations of expertise in the midst of gender conflicts and her vocal denunciation of what she saw as mediocre medical work.[6] Her story also further problematizes the idea of "masculine science" because she claimed to have the same or better knowledge and skills as her male counterparts.[7]

Certainly, Jacobi was an accomplished physician with a rigorous and impressive career that combined clinical work with research and experimentation. Although it is tempting to romanticize her as a "real" scientist who did more "objective" research than men on women's health, her knowledge was socially and politically contingent, shaped by her medical training, her philosophy of science, her perceptions of gender, and her desire to engage forcefully in a critical debate over the role of science in medicine.[8] Her work also reflected efforts to establish a place for herself in the profession and to be welcomed in circles of medical men. Although Jacobi claimed to conduct her research "in the interest of truth," her work was far from disinterested. Her medical writings and investigations served as her main platform for political activism as she worked for the expansion of women's education and access to the professions. In the 1870s, while other activists pursued women's rights in the form of legal change and suffrage, she chose to revise women's social status via biological knowledge.

This chapter examines how Jacobi's notions of gender and politics were embedded in her earliest studies of female physiology. It describes how she used concepts of "nutrition" to argue for a more favorable view of female reproductive functions, specifically menstruation. In her efforts to "correct" medical interpretations of female physiology, science became a women's rights activity, and this may have been her most political act of all.

Women, Nutrition, and the Social Organism

Jacobi's writings on female physiology reveal the interrelationship and inseparability of her science and political thought. Her studies on women's health reflect the main tenets of late nineteenth-century physiology and experimentalism, particular notions of positivism, and an expanded conception of women's social roles. Although Jacobi believed her findings were "value neutral," she produced a conceptualization of the female body situated within both the currents of scientific debate and the gender role contestations of her time.

Jacobi based much of her scientific research and writing about women's health on the concept of "nutrition." Nutrition did not refer to the consumption of food or dietary regimens, as it most commonly does today. Rather, in late nineteenth-century physiology and histology, it referred to the organic changes of substances in the body and the building and depletion of body tissues. It described the process by which cells and tissues were continuously destroyed and then regenerated to sustain life. Cells were nourished through the circulation of blood, and blood delivered the chemical substances needed for growth. Nourishment was regulated internally by "force," or vital energy.[9] Jacobi believed that female health rested on normal nutrition and that illness was often a manifestation of failed nutrition. She used this as a conceptual thread to explain various conditions associated with women, including hysteria and anemia, as well as the differences between the sexes. Moreover, concepts of nutrition were central to her interpretation of menstruation.

Whereas some physicians viewed menstruation as a symbol of female weakness, difference, and incapacity, Jacobi asserted a different view, describing menstruation as a "nutritive" process that sustained women's health. Rather than see menstruation as a drain of body force that taxed the system, she argued that menstruation was a sign of vitality. The concept of nutrition allowed her to shift the attention away from pathology, as she tried to normalize menstruation by defining it as a nutritional process, inseparable from other biological functions.[10]

Women and men differed physiologically, she admitted, because nutrition in women differed from nutrition in men, not because women menstruated and men did not. The first and most obvious difference between the two sexes, she asserted, was their muscular mass. In puberty, men developed more motor force than girls because girls' "muscles begin to refuse to assimilate a certain proportion out of each group of nutritive molecules which is brought to them by . . . circulation." Nutrients were diverted from muscles and an "afflux of blood to the uterus" occurred to increase the functional capacity of the uterus, and prepare the body for menstruation or pregnancy. This process of nutrition in women was neither temporary nor periodic but a continuous process that recurred in the course of each month. According to her, women experienced "rhythmic wave[s] of nutrition," or rising levels of vital force, that gradually rose from a minimum point just after menstruation to a maximum point just before the flow; she used the concept of "waves" to visually represent how nutrition and vitality increased in her subjects before and during menstruation.[11] With this strategy, she tried to overcome the notion that menstruation was tantamount to female weakness.

Jacobi's focus on nutrition and histology can be traced back to her medical studies in Paris, where she became an assistant in the microbiology lab of Louis-Antoine Ranvier, an assistant to Claude Bernard, and André-Victor Cornil, allowing her to take part in the scientific culture of Paris medicine.[12] Through Ranvier's studies on morbid tissue conditions, Jacobi (then Putnam) focused her attention on the relationship between pathology and failed nutrition and acquired the basis for her approach to female illness. Her French thesis sought to contribute to the understanding of cellular degeneration and how cellular nutrition affected a body's general health and "vital powers."[13] Jacobi's French thesis also argued for an interactive model of both chemical and physiological analysis, a premise that was later central to her menstruation studies.[14]

Jacobi, of course, was not alone in her attention to nutrition. Several nineteenth-century physicians and gynecologists believed illness resulted from a general failure of the body to build and deplete its tissues properly. She shared basic understandings of nutrition with many American physicians, particularly neurologists. For example, the writings of the famous neurologist S. Weir Mitchell relate conditions of hysteria to "defective nutrition of the nerve centres."[15] Another practitioner, Graily Hewitt, agreed and assigned the pale and feeble characteristics of hysterics to nutritive complications.[16] But while Jacobi and these neurologists could agree on the fundamentals of nutrition, they disagreed on how to apply the concept to the treatment of women. Jacobi's focus on nutrition was not novel, but her use of the concept to define menstruation as healthy was indeed significant.

Nutrition was also a useful point of investigation because, according to nineteenth-century physiologists and chemists, it could be visually measured and recorded. Laboratory techniques were used to track pulse rates, analyze urea, and later, to measure hemoglobin, providing visual indicators of the rate and extent to which aliments were absorbed, processed in the body, and excreted.[17] Such techniques allowed Jacobi to incorporate the laboratory into her evaluation of patients, which became central to her definition of scientific medicine.

Ultimately, they served as some of her most important demonstrations of expertise.

Jacobi advertised how she integrated the laboratory and the clinic in her practice at a critical moment when physicians debated the meaning of "scientific medicine" and disagreed over the significance of laboratory studies for clinical care.[18] Some physicians favored empirical models that based treatment on close observations of patients. Other physicians pursued scientific models of medicine, believing therapeutics should be based on laboratory investigations, such as chemical and physiological experiments. Later in the 1880s, the debate intensified, and Jacobi joined forces with leading advocates of experimentation. But until then, she brought the politics of defining scientific medicine to bear on the "woman question" by associating her studies of menstruation with "true" science and laboratory investigations. In defining her work this way, she attracted the attention of medical men and tried to gain legitimacy from segments of the profession most interested in pursuing forms of experimental physiology.

Jacobi's enthusiasm for the laboratory paralleled her philosophical commitment to positivism. Positivism was a popular ideology in late nineteenth-century America for many middle-class intellectuals, reformers, professionals, and activists who experienced "the spiritual crisis of the Gilded Age."[19] Although the period after the Civil War was a time of religious enthusiasm, it was also a time of religious skepticism. Men and women who felt spiritually disillusioned embraced positivism as a substitute faith based on doctrines of science. Positivists believed professionals could eradicate social problems through science, and a new moral order would emerge in which citizens worshiped humanity and directed their spirituality toward social improvement. It was also attractive to many women, who adapted it to a variety of "feminist" and reform causes.[20]

Positivism became the bridge between Jacobi's science of medicine and her gender politics. As a young woman, she abandoned her religious upbringing and chose science as her faith. But it was in Paris that she became a disciple of Auguste Comte and gained command of positivism as both a philosophy and way of life.[21] Positivism became the underpinning of her daily studies of medicine, but she also realized Comte's ideas as she elevated science above all other forms of knowledge, developed critiques of Catholicism and metaphysics, and defined science as the main source of social reform. Jacobi believed that all knowledge should be based on observable phenomena derived from scientific investigation. These tenets became the framework for her earliest medical studies on women, as she insisted that she could portray the female body as it actually was. For

Jacobi, positivism became the intellectual foundation that connected the laboratory to women's rights.

In her positivist writings, Jacobi argued for the primacy of observable facts over metaphysics and unseen phenomena. "The great Positivist doctrine about truth," she argued, "is, that it is always an expression of relations existing between the human mind and the object that is contemplated by it. Things are not known in themselves, but only as they affect us and as we can perceive them. What is true for us is what we see to be true, and any other truth is inconceivable by our minds."[22] With such discourse, positivists forged a link between their method of "truth seeking" and a mandate for social reform. The Comtean creed, rather than religious teachings, they believed, should direct the course of society, since only observable facts could produce material change.

Jacobi's positivism also assumed an inherent value of the laboratory. Laboratory studies, in her view, represented the ultimate positivist exercise. "It is in the life of the laboratory," she argued, "where the secrets of nature are not only divined but reproduced, that the true joy of knowledge can best be learned."[23] Her idealization of the laboratory reflected her alliance with a type of scientific medicine that based therapeutics on knowledge that could be tested, measured, and reproduced.

Jacobi also advanced an organic image of American society. She rejected individualism and viewed citizens as integral parts of a cooperative social organism. She compared people to cells, families to tissues, and larger groups and classes to human organs. It was in this system that the "life of society is immediately carried on." Thus, all people and social groups contributed in some way to the social order. She also argued that human beings, like human cells, required physical and mental nutrition, in the form of education and meaningful work, to reproduce and function.[24] She explained, "The nutrition of a human being must be divided into two kinds: the nutrition of the body and that of the soul, or in other words, satisfaction of physical, and satisfaction of mental or spiritual wants."[25] Just as the human body failed without nutrition, the social organism was disturbed when all of its members did not participate, she posited. Likening the individual person to the human cell, she explained that an individual "cannot be thought of without the correlative of other persons, and of a larger social life whose central events dominate the local events around each individual."[26]

The concept of the social organism was popular among many positivists and socialists at the end of the nineteenth century because it privileged the needs of a unified social order over the individual. While it justified reform projects for the greater good, the organism concept could also reinforce hierarchies. Certainly, positivism, as defined by Comte, did not support true social equality, particularly gender equality. Comte believed women contributed to the social order but that they did so within the confines of the domestic sphere. In the United States, many positivists tried to translate Comte's ideas into a more democratic vision for social change.[27] American feminists, in particular, saw great potential in positivism and reconfigured it into a justification for women's political participation.

Although many women activists, like Jacobi, tailored Comte to their own advantage, they did not account for the different needs of other disempowered groups and were unable to construct immigrants, African Americans, or the laboring classes as equal contributors to the social organism.[28] Class and race could limit the reach of positivism as an egalitarian force because bourgeois professionals appointed themselves as the vanguard of reform and created social programs that reflected their own values. But positivist principles of scientific investigation could be invoked by minority groups, too. For example, African American and Jewish writers countered scientific racism by using the discourse of science to challenge racial hierarchies.[29] The promise that science could reveal the "truth" about biology and society appealed to some people on the margins as well as those with greater power in society.

The social organism served as a useful metaphor for Jacobi because it connected social relations and gender roles to physiological nutrition.[30] Influenced by the utopian socialist thought of St. Simon and Fourier, Jacobi adapted the organism metaphor to the goal of gender equality. The interdependence of cells and organisms allowed her to set up a reciprocal relationship between individual female bodies and the social body, asserting that women needed to play a critical role in society to ensure its survival.

Jacobi's notions of female health ran counter to the cult of female invalidism in late Victorian America. Frustrated by the preponderance of female illness, she spent her career studying what she considered to be the "real" factors behind American women's frailty. She agreed with many physicians that the harsh conditions faced by New England settlers negatively affected their descendants, causing the (white) race to degenerate and young women to be less vigorous than their ancestors. However, she maintained that it was the "excessive luxury" experienced by contemporary New England girls that caused the most damage. She blamed "the refined and delicate ease of life and sensibility in which so many thousands now contrive to live," which could easily escalate into a "dangerous effeminacy." Constant worry and reflection about health had also caused many young women to become "nothing but bundles of nerves."[31] In response, she supported an active model of femininity, buttressed by the physiological knowledge and evidence presented in her studies.[32]

Jacobi's views of gender, therefore, can be situated at the intersection of physiology, positivism, and a vision of women's rights focused on greater female participation in the public sphere. She believed that health was a gateway to the emancipation of women. Women and men could contribute evenly and equally to the social organism if women received the same opportunities for intellectual training, especially in medicine and science. Finally, female pathologies occurred because of women's second-class status, not because of inherent physical characteristics.

But in the late nineteenth century, pathological notions of female biology were gaining strong cultural currency. Women's rights activists recognized the close relationship between the science of womanhood and the political, educational, and legal rights of women. They published moral condemnations of these ideas, but they believed that science would have to be fought with science. They actively called on women in the medical community to refute claims about female biological inferiority. Mary Putnam Jacobi answered that call and entered medical debates on the side of women's rights. Her efforts represent a deliberate decision to come to the aid of activists, and to be an activist herself, vis-à-vis her medical research.

Fighting Science with Science

Mary Putnam Jacobi's menstruation studies best illustrate the interrelationship of her science and politics in the early years of her career. These studies show a direct link between her work in the women's rights community and the orthodox medical community, as she carefully negotiated her way between them. Jacobi earned the respect of medical men from the beginning of her career. Her extensive medical training helped boost her position as did her personal connection to Abraham Jacobi. More importantly, she did not challenge orthodox medicine but was committed to making "regular" medicine scientific while weeding out homeopathic practices. Her political activism in these early years often came in a scientific package, presented in her scholarly questions and conclusions rather than overt feminist agitation, helping her to maintain her warm welcome in the orthodox medical community.

Indeed, Jacobi was determined to carry out her studies in the most accurate way, applying positivist principles, conducting surveys and data collection, and designing experiments to test and prove her theories. So committed to representing the "truth," her menstruation studies best illustrate her performance of scientific expertise, as she consciously asserted her technical achievements and condemned the failures and inadequacies of her rivals. She delineated between good and bad science, as she confronted work she deemed unscientific, especially the work of the infamous Edward Clarke.

Clarke's classic text, *Sex in Education; or, a Fair Chance for the Girls* (1873), pronounced that menstrual functions and coeducation were incompatible for young American women. Too much education and mental exertion threatened the physical development of girls, especially when undertaken during menstruation. He stated, "Girls lose health, strength, blood and nerve, by a regimen that ignores the periodical tides and [the] reproductive apparatus of their organization."[33] Young women should rest during menstruation, Clarke argued, and concentrate their energies on their reproductive health. If women pursued higher education at the same level as men, they risked sterility and obstructing their reproductive functions. According to Clarke, women's defiance of their "nature" threatened to masculinize a generation of young women. Couched in the language of evolutionary theory and ovarian determinism, Clarke's polemical book, published one year after his retirement from Harvard, articulated common ideas about the female body and expressed fears about the expanding roles of women in American society.

Only one year after the publication of *Sex in Education*, Jacobi produced her first paper in defense of female physiology and women's educational rights. She joined Anna C. Brackett and other women's rights activists to challenge Clarke in *The Education of American Girls*, a collection of essays published by her family's company, G. P. Putnam's Sons. Jacobi's piece, "Mental Action and Physical Health," was an all-out refutation of Clarke's major claims.[34] She designed this essay to corroborate the moralistic essays of the Brackett volume with physiological evidence and to communicate medical information to a nonmedical audience. For these reasons, the piece reads more like a lecture than a medical article. And yet, it was technical enough to gain the attention of physicians and win her a prominent position in medical debates about women's health.

In the Brackett volume, Jacobi accused Clarke of overstating the incapacities of girls during menstruation and falsely attributing menstrual diseases, amenorrhea (absence of menstruation) and menorrhagia (excess of menstruation), to mental exertion. Jacobi's own study of twenty women, several of whom attended medical school and coeducational colleges, showed that only six subjects out of the twenty ever experienced menstrual pain. She argued that Clarke overprescribed rest during menstruation because her study deemed rest unnecessary, positing that an entire week of rest was exorbitant, for her subjects never suffered enough to require such a regimen. Women experienced menstrual discomfort, Jacobi admitted, but pathologies did not result from intellectual endeavor. Instead, they arose from sedentary habits, a lack of exercise, or the interference of emotions with the intellect. While the inferiority of women seemed to be "incontrovertible" to Clarke, he had not proved his arguments sufficiently or empirically, Jacobi charged. Attuned to how conventional theories about the nature of women had been used in arguments against coeducation and women's rights, Jacobi insisted that a positivist reading of the body would demonstrate women's ability to engage in intellectual pursuits.

Calling Clarke out by name, she systematically tried to undermine his authority by linking his ideas to his prejudices toward women, claiming that his work "appeals to many interests besides those of scientific truth."[35] Jacobi also tried to undermine his validity by attacking his vast audience of readers who so easily swallowed his "food for the imagination." Claiming that his work was a clear "exaggeration of fact," she compared him to the French writer Jules Michelet, who claimed, "la femme est une malade,"and tried to associate him with the most extreme claims of female inferiority and to disassociate him from science. In "the interests of truth," Jacobi then offered what she saw as the real science of female physiology.[36]

Despite their many differences, Jacobi and Clarke actually worked from a similar premise, reflex theory. They believed in a direct, physiological interconnection between the body and the mind, between physical development, reproduction, and mental activity. Rejecting the Cartesian model of mind/body dualism, they drew relationships between cerebral activity and the body's many functions, including those of the nervous and reproductive systems.[37] The nervous system, as it was understood, consisted of two interrelated sections: the ganglionic, made up of nerve matter, and the cerebro-spinal, which included the brain, spinal cord, and medulla oblongata. Clarke argued that mental labor (a cerebral action) and menstruation (related to ganglionic nerve force) could not occur simultaneously without causing a physical drain in women. The body, he believed, had a fixed amount of energy; all of the body's functions relied and competed for this limited energy. Women's mental activity was dangerous,

Clarke argued, because it diverted a limited storage of nerve force from reproductive growth and functions. Given this argument, intellectual work absorbed precious energy and deprived the reproductive system of vital force. Clarke believed ganglionic nerves dominated a woman's life; this view subordinated her brain and, as Jacobi quipped, "reduced her to the anatomical level of the crustacea."[38] Although Clarke and Jacobi both subscribed to similar premises of reflex theory, they disagreed on the compatibility between female reproductive functions and cerebral activity.

Jacobi repeatedly denied the need for women to remain inactive during menstruation. In opposition to Clarke, she maintained that "there is no absolute incompatibility between the evolution of nerve force at the ganglionic centres and at the cerebro-spinal." Digestion showed this to be true. Digestion and ovulation were both unconscious processes of the ganglionic system. If digestion did not necessitate rest and "torpor of the brain," she asked, why should menstruation? Rest was not necessary after every meal, and indigestion, or dyspepsia, could actually be even more painful than menstruation, she reasoned. Analogizing menstruation with digestion helped Jacobi to normalize menstruation and its associated discomforts. In Jacobi's rendition, menstruation was just another "rhythmic" process of the body, like the beating of the heart or secretions of the stomach.[39] She argued that both sexes experienced physiological rhythms and rejected the assertion that "periodicity is the grand (i.e., exclusive) characteristic of the female sex."[40]

Two other women physicians, Marie Zakrzewska and Mary Dixon Jones, also tried to deny the centrality of reproductive functions and organs in the female body. As Arleen Tuchman shows, years earlier Zakrzewska had linked the uterus to the digestive system and denied the importance of biological sex differences.[41] Dixon Jones, a gynecological surgeon, also rejected the notion that female reproductive organs controlled the female body, as Regina Morantz-Sanchez demonstrates. Like Jacobi, her preceptor, Dixon Jones refused to accept that the uterus and ovaries constituted womanhood.[42]

Jacobi constructed another important anatomical claim, one that distinguished the influence of thoughts and emotions on women. While Clarke claimed that improper education hindered menstrual health by stimulating the emotions, Jacobi insisted that excessive emotions were generated by a lack of brain stimulation that diminished cerebral nerves.[43] Thoughts were localized on cerebral hemispheres; emotions were located in the ganglia. In the ganglia, emotions could cause blood vessels to dilate (leading to pain and cramps), disturbances in circulation, and the obstruction of the body's nutrition. She concluded, emotions, not brain work, caused discomfort during menstruation and the dominance of the ganglionic nerves emerged from an understimulated brain that created an imbalance. In this way, Jacobi set up a strategic relationship between the brain and the nerves so that they were related but functioned separately.[44] Her interpretation described a female system that privileged rational thought over emotion, a mental model that fit her personal feminine identity as well as the identity she prescribed for other women.

Ultimately, this essay was a preliminary exercise that prepared her to compose and conduct a larger study on menstruation in the following years. She directed the next phase of her work to an audience of physicians, trying to convince the medical community that Clarke's work had no merit. In the next phase, she went further, using the tools of her trade to offer an alternative view of menstruation that challenged readers to rethink sexual difference and thus the social roles of women.

To Rest or Not to Rest: That Is the Question

Jacobi's now well-known study, *The Question of Rest for Women during Menstruation*, is both a systematic argument against Clarke's *Sex in Education* and a performance of her expertise. This study incorporates both statistical analysis and experimental methods to illustrate her arguments and establish her credentials on women's health. She presented an alternative view of female physiology and menstruation, not based on her own physical experience as a woman but based on what she observed and recorded empirically about her research subjects. It was a view that was also more favorable to an active and educated model of womanhood and to the physiological equality of men and women.

In 1876, the Boylston Medical Committee at Harvard University presented Jacobi with the perfect opportunity to refute Clarke, who had been a professor at Harvard. As part of the committee's annual essay contest, the following topical question was proposed: "Do women require mental and bodily rest during menstruation; and to what extent?"[45] The committee, staffed by some of Harvard's most important medical men, including the surgeon Henry Jacob Bigelow, obstetrician David Humphreys Storer, and the clinician and surgeon Morrill Wyman, added fuel to the fire that raged over women's health and higher education. This was no coincidence, as the committee knowingly opened up another line of medical debate on the topic, inviting physicians to either challenge or affirm the principles of Clarke's *Sex in Education*. The committee did not endorse the higher education of women, but its members wanted to open debate about Clarke and his popular essay.

The committee questioned Clarke's work as a legitimate piece of medical writing. Morrill Wyman held reservations about the study, finding it "weak, one-sided and not sufficiently supported by facts."[46] In a private conversation at the home of C. Alice Baker, a Boston women's rights activist and educator, he revealed that he did not consider Clarke's book "a dignified or trustworthy exposition of the subject." Baker learned that other Harvard physicians distrusted Clarke's analysis and that they welcomed other dispassionate studies that used "statistics" to support their assertions. She also heard that "one or two members of the Prize Committee had spoken with great cordiality of [Jacobi's] essay in Mrs. Brackett's book as being strong and sound in its physiology." Finding Clarke's work unsound, Wyman and his colleagues were open to alternative views about menstruation.

With the advice of Baker, Jacobi submitted an anonymous essay in "a masculine" handwriting. She signed it only with the Latin phrase, "Veritas poemate verior" (A truth truer than a poem) in reference to the credibility of her study.[47] However, her piece was not so anonymous; at least one member of the committee, Morrill Wyman, may have recognized Jacobi's submission, having previously and purposely read her Brackett essay, showing it "great respect."[48]

Jacobi submitted a study designed to satisfy an audience of Harvard physicians who valued science above sentiment. With the use of statistics, diagnostic laboratory tools, and the science of nutrition, Jacobi hoped that her medical study would be more effective than moral arguments or outright advocacy for the rights of women. Yet, it was clear that *The Question of Rest* was an orchestrated effort by Jacobi and the women's rights community to use science for political purposes and prove Clarke wrong.[49] Jacobi also understood that the menstruation question was not of abstract significance for women and higher education but could have an impact on the admission of women to Harvard Medical School.[50]

Women physicians had long recognized the educational and symbolic rewards that a Harvard education presented, starting with Harriot Hunt in 1847. In 1867, physicians Susan Dimock and Sophia Jex-Blake made a coordinated effort to apply for admission but did not succeed. In the mid- and late 1870s, the Harvard administration, short of funds for building expansion, was forced again to consider coeducation. When Marion Hovey of Boston offered $10,000 to the school on the condition that they admit women, the medical faculty thoroughly debated the matter, deciding ultimately that women would have to wait.[51] With the Harvard question as the backdrop, Jacobi did her part to show that there was no physiological reason women could not be educated like men.

Although the Harvard faculty refused to admit women, the Boylston committee decided to award Jacobi the prize. Their decision was not simply a victory for women and higher education, although it certainly aided the cause. Rather, it reflected a profession and institution grappling with the meanings of "scientific medicine." Clarke had brought a lot of attention to Harvard with a study that was not based on laboratory studies or experimentalism but anecdotes and moral dictums. Although Clarke was trained in a physiological tradition and believed in "rational therapeutics,"[52] the moral science of female physiology presented in *Sex in Education* did not satisfy his colleagues. While his arguments were designed to have an impact on a broad audience, Jacobi directed her essay to specialists, provided data, and composed her work in technical language, matching the academic expectations of the committee. The Boylston committee cast its votes based more on her methods, than on her conclusions about the social roles of women.[53]

In the following year, Jacobi published an extended version of the essay as *The Question of Rest for Women during Menstruation*, again with the help of G. P. Putnam's Sons publishers. Now known as the "winner" of the Boylston Prize, her book was bound to draw attention, and Jacobi took advantage of the moment. Stepping up her evidence and arguments and armed with more extensive statistical and experimental data, Jacobi converted the essay into an all-out refutation of Clarke's book.

Statistical analysis served as the first step in the refutation of Clarke's arguments. To answer the "question of rest," Jacobi asked participants to describe their menstrual experiences and to indicate whether and when they had pain and for how long. She then asked women to evaluate their strength during menstruation and document their own physical activity. She also gathered the women's medical history, age, duration of their education, and occupation. Although coeducation and women's higher education were middle-class causes, Jacobi's study surveyed women from a variety of occupations and backgrounds. While there is no complete profile of her 268 participants, the study does refer to women whose educations range from "common" to "higher" and whose occupations varied from the professional to the industrial to the domestic. The New York Infirmary provided easy access to a pool of subjects in the city, which included mostly immigrant and working women as well as women from the middle classes. Several participants also worked in teaching and medicine. From her questionnaires, Jacobi produced tables and a detailed statistical analysis of her findings.[54]

Although 35 percent of participants had never suffered any pain, the remaining women reported having some type of discomfort during menstruation. But, of those who suffered, two-thirds had inherited "physical defects" that caused forms of uterine disease or weakened their constitutions; others had serious "organic defects" that rest could not cure. Based on her numbers, Jacobi believed that immunity from menstrual suffering did not depend on rest but, instead, on a healthy childhood, a sound family history, marriage at a "suitable" time, a steady occupation, exercise during school life, and "the thoroughness and extension of the mental education."[55] Jacobi concluded: "Rest during menstruation cannot be shown, from our present statistics, to exert any influence in preventing pain, since, when no pain existed, [rest] was rarely taken." In fact, "in a large proportion of cases, [rest] has been quite superfluous."[56] Attuned to the labor conditions of working-class women, Jacobi said rest could be helpful for excessive pain or for overworked "industrial women," like some of the participants in her study. But rest for the most part was injurious rather than helpful for most menstruating women. She concluded: "There is nothing in the nature of menstruation to imply the necessity, or even the desirability, of rest, for women whose nutrition is really normal."[57]

Jacobi then argued that menstruation was a time of increased vitality by explaining that women had a "reserve of nourishment" in their bodies that was used for reproductive functions, especially menstruation. This nourishment was derived mostly from reserves in the voluntary muscles, whereby the body diverted blood to the uterus to "increase its functional capacity." The diversion of forces for menstrual functions did not necessarily lead to weakness and debility because the body was "supplemented" by a reserve of force from elsewhere. "We have a certain number of facts," Jacobi argued, "which indicate that the period of menstruation may be one of increased vital energy and especially of increased mental force."[58] Jacobi used the concept of waves and the idea of nutritional surplus to rethink common understandings of menstruation and to establish her own menstrual theory.[59]

To test her theory and corroborate the surveys, Jacobi set out to map the physiological signs of her subjects in a chapter entitled "Experimental."[60] Emphasizing her use of laboratory techniques, she combined both physiological

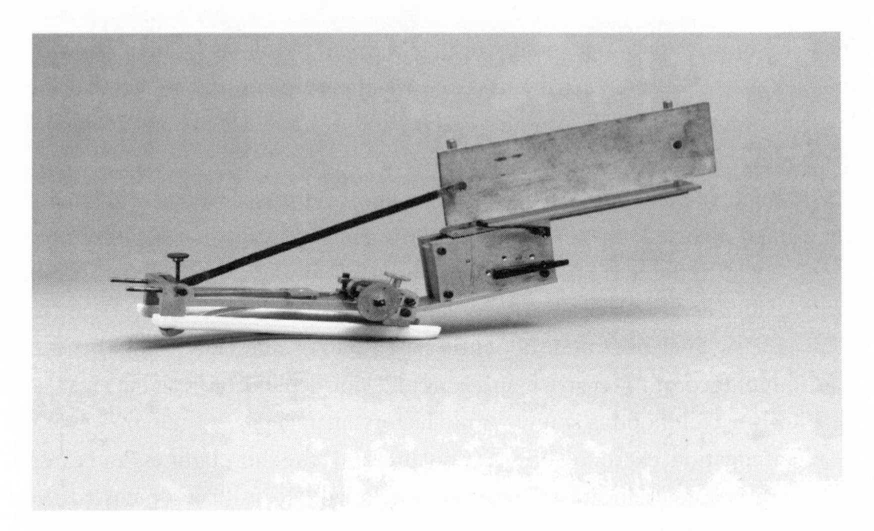

Sphygmograph, ca. 1880. Courtesy of the personal collection of M. Donald Blaufox, M.D., Ph.D.

and chemical methods to measure the levels of nutrition in her subjects. To trace pulse rates, she used a sphygmograph, an early pulse writer developed in European physiology labs in the 1860s. Strapped to the wrists of patients, the sphygmograph created a visual representation of the pulse rate, as a small writing instrument drew wave lines on a rectangular board. The pulse waves illustrated the rise and fall of arterial tension.[61] In healthy patients, arterial tension was at its highest right before menstruation commenced; at menstruation, the tension of healthy patients reached a healthy minimum. A second test showed that patients produced more urea before and during menstruation. Urea, a substance in urine excreted by the kidney, was measured using Liebig's volumetric method to trace the depletion and replacement of the substances in muscular tissue. For Jacobi, chemical analyses indicating an increase in the volume of urea demonstrated an acceleration of the nutritive process and showed the week preceding menstruation was a period of increased vigor and "nervo-muscular strength."[62]

Jacobi tried to demonstrate visually this increase of strength with a dynamometer, a device designed to register muscular pressure, usually through recording the squeezing motion of the hand. She asked a small number of subjects to take part in the experiment: seven women had increased strength, seven had less, and one had the same strength during menstruation. These were not the clear

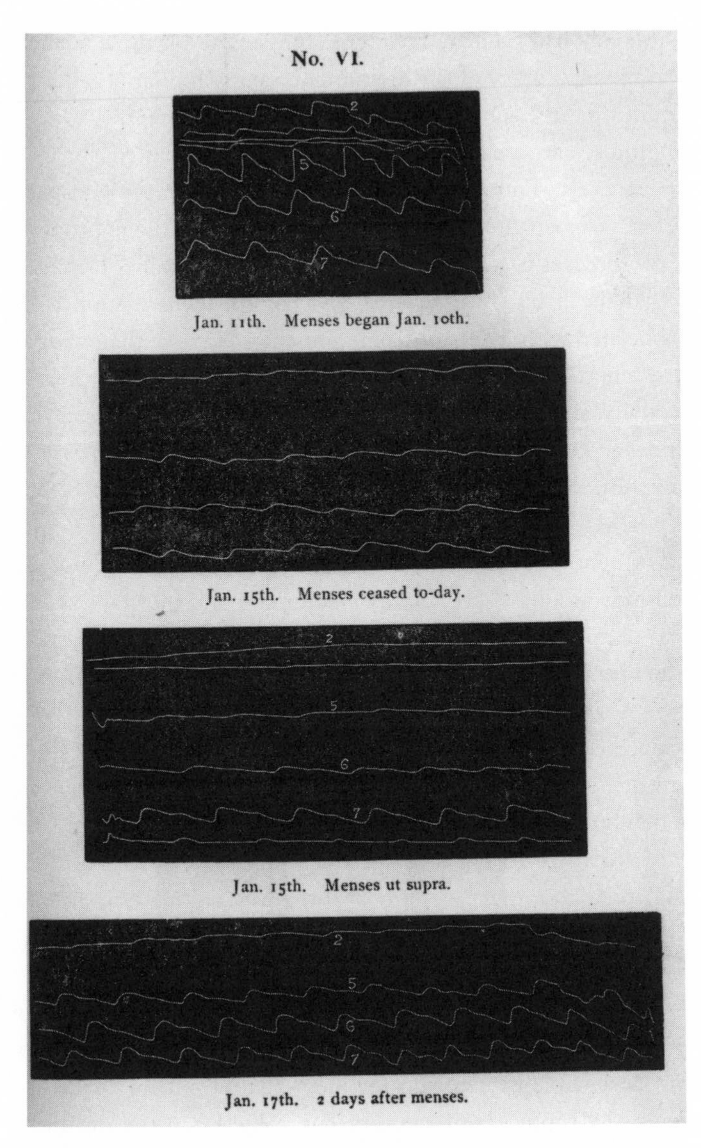

Sphygmograph tracings from Mary Putnam Jacobi's *The Question of Rest for Women during Menstruation*, published 1877.

results that Jacobi had wanted. Trying to compensate for these mixed findings, she blamed the inaccuracy of the instrument, saying it did not reflect muscular strength but rather the subject's skill at using the instrument.[63]

Building muscular strength was important, Jacobi insisted, for maintaining the necessary levels of nutrition needed for normal, painless menstrual experiences. When a woman "breaks down," it was not due to the improper expenditure of nerve force, as Clarke and others argued. "Practically," Jacobi explained, "we find that the habit of disordered and painful menstruation is more frequently associated with habits of feeble muscular exercise than with any other one circumstance."[64] In the years to come, Jacobi stood by her prescription of physical activity, and recommended specific calisthenics, sports, and even exercise machines to keep women healthy.

Mapping physiological signs showed that menstrual functions were not "periodic" but "rhythmic." Jacobi concluded: "Reproduction in the human female is not intermittent, but incessant, not periodical, but rhythmic, not dependent on the volitions of animal life, but as involuntary and inevitable as are all the phenomena of nutritive life."[65] By privileging nutrition, a constant process, she rejected the idea that menstruation was a special crisis. This was important because those physicians who called menstruation an infirmity "base[d] the epithet mainly upon the periodicity of the hemorrhage," she said. Periodicity in the mid-nineteenth century had "come to be considered as a mark of constantly recurring debility, a means of constantly recurring exhaustion demanding rest as decidedly as a fracture or a paralysis."[66] Connecting menstruation to a larger, continuous process allowed Jacobi to shift the focus of medical discourse away from therapeutic remedies designed to rehabilitate women during their week of weakness and toward those that would strengthen them.

In support of her rejection of periodicity, Jacobi also questioned ovulation theory. Ovulation theory was the subject of debate in medical journals but stood as the most acceptable explanation for menstruation in the 1870s. According to physicians, the rupturing of the Graafian follicle and the disintegration of the ovules caused the menstrual flow.[67] Occurring at intervals, many physicians viewed this process as a great disruption, a "perturbation of the economy," and a "morbid circumstance." Ovulation theory, said Jacobi, "isolated menstruation from all other physiological processes, which rendered its ordinary course dangerous, its derangements fatal."[68] To challenge ovulation theory, she looked back to an earlier theory of menstruation, the "plethoric," and applied it to her own research on nutrition. The plethoric theory of menstruation, originating during the time of Hippocrates, equated the menstrual flux with other bodily evacuations. Such evacuations purified and balanced the system, ridding it of superfluous materials, representing excess nutrition, not physical depletion.[69] Equating menstruation with the evacuation of a nutritional surplus, Jacobi, ironically, reached far back to ancient concepts to offer a "new" interpretation of female physiology.

Jacobi rejected a second aspect of ovulation theory: the simultaneity of ovulation and menstruation, and its association with sexuality. Physicians developed ovulation theory from studies of estrus in dogs, not studies of women. Since dogs ovulate and bleed at the same time, many physicians believed that women must ovulate during menstruation. Nineteenth-century physicians associated the new theory of ovulation with estrus, linking nerve force, sexual arousal, and reproductive functions.[70]

The association between human ovulation, heat, and sexual arousal prompted Jacobi's rejection of ovulation theory. She stated, "All processes concerned in menstruation converge, not towards the sexual sphere, but the nutritive, or to one department of it—the reproductive."[71] Jacobi reconceptualized menstruation in humans as a nutritional process, not a sexual one, and rejected new associations between heat and ovulation. To illustrate the centrality of nutrition, she made analogies between women and plants, not animals, such as the ovary and fruit blossoms: "The woman buds as surely and as incessantly as the plant, continually generating not only the reproductive cell, but the nutritive material without which this would be useless." It was women's nutritional reserves that signified their difference from men, not menstruation. By arguing that nutrition "constitutes the essential peculiarity of the female sex," she was able to distinguish between men and women in a way that emphasized their corresponding characteristics rather than their differences.[72] Rejecting an oppositional approach to sex differences, she focused on the physical commonalities between the male and female body, creating a biological model that supported a social model based on symmetrical gender roles.

Meanings of the Debate

As seen in the early writings of Mary Putnam Jacobi, the medical debates on the "woman question" also functioned as disagreements over what constituted legitimate evidence and solid science. She purposefully engaged with this debate by publicly and vocally differentiating between her work and that of Edward

Clarke. Jacobi distanced herself from his sentimental and literary style in both structure and tone. Although Jacobi used clinical anecdotes in some of her other studies, she condemned Clarke for relying on clinical narratives to the exclusion of statistical and experimental data. By contrast, Jacobi transformed her clinic into a laboratory, measuring and evaluating her patients to produce statistics, visual representations, and measurements of physiological processes. Jacobi and Clarke both believed they were scientific practitioners but used different methods of medical analysis and intervention. By constructing her own work as scientific and acting out her expertise, Jacobi tried to undermine the legitimacy of medical men like Clarke. In the process, she drew attention to her research and her alternative views on menstruation.

Beyond the Boylston Prize, her menstruation study won praise from several members of the medical community after appearing in print in 1877. *The Medical Record*, for example, called *The Question of Rest* "masterly" and said, "Every page in the book gives evidence of great erudition."[73] *The Philadelphia Medical and Surgical Reporter* also offered high praise, noting the book's "wide range of observation, experiment, and new statistics."[74] In *The Nation*, Henry Pickering Bowditch of Harvard University said, "Her successful competition for the Boylston prize, rendered her sex a far more important service than if she had directly advocated their claims."[75] Bowditch recognized the meaning of the prize, and how scientific discourse could have more persuasive power than direct calls for women's rights.

Jacobi's physiological theories were also generally well received by physicians who appropriated her methods and sometimes reproduced her results. For example, her challenge to ovulation theory found interested readers, particularly as physicians began to question the notion that menstruation and ovulation occurred simultaneously.[76] Jacobi's work on the menstrual wave also garnered serious attention and was cited and adopted by colleagues. In 1878, physician John Goodman published "The Cyclical Theory of Menstruation," arguing, like Jacobi, that the vital activities of women, body temperature, blood pressure, and pulse, formed a wavelike pattern and correlated to the intermenstrual period. Even George J. Engelmann, president of the American Society of Gynecologists, cited Jacobi for her study of menstrual waves. Like Goodman, he documented changes in the pulse, temperature, and blood pressure of girls during the span of a month. But Engelmann, as late as 1900, still contended that the premenstrual period, the height of the wave, produced "morbid nervous symptoms as characterized by the hystero-neuroses" and claimed that mental influences generated pathology in American girls.[77] Physicians like Engelmann may have accepted her methodologies for "measuring" female physiology, but they still dismissed her arguments about women's education.

Despite these acclamations, Jacobi believed her work had not been fully recognized by the medical community, due mostly to her gender. In 1882, William Stephenson published a medical paper discussing the phenomenon of nutritional waves and referenced Jacobi. Although he rejected her conclusions, he supported her basic concept and called for further study. Thereafter, it became known as "Stephenson's Wave," and Jacobi lost her identification with the wave theory.[78] Years later, she vocally deplored the fact that she had not received "credit where credit [was] due," particularly after winning the Boylston Prize.[79] She was also disappointed that a large body of her work had been ignored, specifically her subsequent articles on the causes of menstruation and the etiology and treatment of uterine disease, entitled "Studies in Endometritis."[80] Jacobi felt slighted by a medical community that respected her skills as a physician but denied her the same intellectual credit awarded to male colleagues.

The Boylston Prize is cited as one of Mary Putnam Jacobi's major achievements, for it illustrated her ability to carry out a complex study, apply laboratory techniques, articulate her findings, and convince an audience of men to rethink concepts of female physiology. Certainly, Jacobi and her feminist allies saw the prize as the triumph of "truth" over prejudice. As more favorable readings of the female body, her studies countered volumes of literature that pathologized women. And yet, like her theoretical rivals, Jacobi could not escape her own views of gender; they were embedded in the very questions, practices, and theories that she employed. Her work represented the intersection of experimental physiology, positivism, and women's rights as she used the tools of orthodox medicine to endorse unorthodox views about women and their bodies.

NOTES

I am grateful for the comments and support I received from colleagues, particularly Ellen More, Arleen Tuchman, Regina Morantz-Sanchez, Naomi Rogers, and Susan Smith. Joan Jacobs Brumberg and Shobita Parthasarathy also offered invaluable suggestions. Research support for this article was provided by Cornell University, the National Science Foundation grant 9910963, and Loyola Marymount University. The Cornell Science Studies Reading Group also provided helpful advice during the earliest stages of this research.

1. Mary E. Putnam Jacobi, "Social Aspects of the Readmission of Women into the

Medical Profession," in *Papers Presented at the First Woman's Congress of the Association for the Advancement of Women* (New York: 1874), 168. Her byline was published incorrectly as "Mary E.," rather than Mary C. Putnam Jacobi.

2. For example, see Cynthia Eagle Russett, *Sexual Science: The Victorian Construction of Womanhood* (Cambridge, MA: Harvard University Press, 1989); Elaine Showalter, *The Female Malady: Women, Madness, and English Culture, 1830–1980* (New York: Penguin Books, 1985).

3. For more on Abraham Jacobi, see Russell Viner, "Abraham Jacobi and the Origins of Scientific Pediatrics in America," in *Formative Years: Children's Health in the United States, 1880–2000*, ed. Alexandra Minna Stern and Howard Markel (Ann Arbor: University of Michigan Press, 2002), 23–46; Viner, "Abraham Jacobi and German Medical Radicalism in Antebellum New York," *Bulletin of the History of Medicine*, 1998, 72: 434–463; Viner, "Healthy Children for a New World: Abraham Jacobi and the Making of American Pediatrics" (Ph.D. diss., University of Cambridge, 1997).

4. For an earlier discussion of Mary Putnam Jacobi and medical science, see the groundbreaking book on women physicians by Regina Morantz-Sanchez, *Sympathy and Science: Women Physicians in American Medicine*, 2nd ed. (Chapel Hill: University of North Carolina Press, 2000), 184–202. For other treatments of Jacobi, see Thomas Neville Bonner, *To the Ends of the Earth: Women's Search for Education in Medicine* (Cambridge, MA: Harvard University Press, 1992), 48–54; Carol B. Gartner, "Fussell's Folly: Academic Standards and the Case of Mary Putnam Jacobi," *Academic Medicine*, 1996, *71*: 470–477; Joy Harvey, "Clanging Eagles: The Marriage and Collaboration between Two Nineteenth-Century Physicians, Mary Putnam Jacobi and Abraham Jacobi," in *Creative Couples in the Sciences*, ed. Helena M. Pycior, Nancy G. Slack, and Pnina G. Abir-Am (New Brunswick, NJ: Rutgers University Press, 1996), 185–195; Joy Harvey, "'Faithful to Its Old Traditions'? Paris Clinical Medicine from the Second Empire to the Third Republic (1848–1872)," *Constructing Paris Medicine*, ed. Caroline Hannaway and Ann La Berge, *Clio Medica* 1998, *50*: 313–335; Joy Harvey, "La Visite: Mary Putnam Jacobi and the Paris Medical Clinics," *Clio Medica*, 1994, *25*: 350–371; Joy Harvey, "Medicine and Politics: Dr. Mary Putnam Jacobi and the Paris Commune," *Dialectical Anthropology*, 1990, *15*: 107–117; Barbara Sicherman, "The Paradox of Prudence: Mental Health in the Gilded Age," *Journal of American History*, 1976, *62*: 890–912; John Harley Warner, *Against the Spirit of System: The French Impulse in Nineteenth-Century American Medicine* (Princeton, NJ: Princeton University Press, 1998), 322–327; Susan Wells, *Out of the Dead House: Nineteenth-Century Women Physicians and the Writing of Medicine* (Madison: University of Wisconsin Press, 2001), 146–192. See also Carla Bittel, "Science, Suffrage, and Experimentation: Mary Putnam Jacobi and the Controversy over Vivisection in Late Nineteenth-Century America," *Bulletin of the History of Medicine*, 2005, *79*: 664–694.

5. Defining science has been described as "boundary-work" by Thomas F. Gieryn: "the attribution of selected characteristics to the institution of science (i.e., to its practitioners, methods, stock of knowledge, values and work organization) for purposes of constructing a social boundary that distinguishes some intellectual activity as non-science." See idem, "Boundaries of Science," in *Handbook of Science and Technology Studies*, ed. Sheila Jasanoff, Gerald E. Markle, James C. Petersen, and Trevor Pinch (Thousand Oaks, CA: Sage Publications, 1995), 405; Gieryn, "Boundary Work and the Demarcation of Science from Non-Science: Strains and Interests in the Professional Ideologies of Scientists," *American Sociological Review*, 1983, *48*: 781–795.

6. Here, I apply the science studies literature on expertise, credibility, and the politics of knowledge. For example, see David Bloor, *Knowledge and Social Imagery* (Chicago: University of Chicago Press, 1991); Susan E. Cozzens and Edward J. Woodhouse, "Science, Government, and the Politics of Knowledge," in *Handbook of Science and Technology Studies*, 540–548; Steven Shapin, *A Social History of Truth: Civility and Science in Seventeenth-Century England* (Chicago: University of Chicago Press, 1994); Steven Shapin, "Cordelia's Love: Credibility and the Social Studies of Science," *Perspectives on Science*, 1995, *3*: 255–275.

7. For another important analysis of Jacobi and the performance of gender, see Wells, *Out of the Dead House*, 146–192. On other women physicians who rejected a masculine monopoly of medical science, see Regina Morantz-Sanchez, *Conduct Unbecoming a Woman: Medicine on Trial in Turn-of-the-Century Brooklyn* (New York: Oxford University Press, 1999); Arleen M. Tuchman, *Science Has No Sex: The Life of Marie Zakrzewska, M.D.* (Chapel Hill: University of North Carolina Press, 2006); Tuchman, "'Only in a Republic Can It Be Proved That Science Has No Sex': Marie Elizabeth Zakrzewska (1829–1902) and the Multiple Meanings of Science in the Nineteenth-Century United States," *Journal of Women's History*, 1999, *11*: 121–142; Tuchman, "Situating Gender: Marie E. Zakrzewska and the Place of Science in Women's Medical Education," *Isis*, 2004, *95*: 34–57. On women who tried to create a balance between science and feminine cultures in this period, see Ellen S. More, *Restoring the Balance: Women Physicians and the Profession of Medicine, 1850–1995* (Cambridge, MA: Harvard University Press, 1999).

8. For more on women physicians and the politics of knowledge, see Nancy Theriot, "Women's Voices in Nineteenth-Century Medical Discourse: A Step toward Deconstructing Science," *Signs*, 1993, *19*: 1–31. For more on situated knowledge and examples of scientific contestations over race (instead of gender), see Nancy Leys Stepan and Sander L. Gilman, "Appropriating the Idioms of Science: The Rejection of Scientific Racism," in *The Bounds of Race: Perspectives on Hegemony and Resistance*, ed. Dominick LaCapra (Ithaca, NY: Cornell University Press, 1991), 72–103.

9. On the history of nutrition, see Harmke Kamminga and Andrew Cunningham, "Introduction: The Science and Culture of Nutrition, 1840–1940," in *Science and Culture of Nutrition, 1840–1940*, ed. Kamminga and Cunningham, *Clio Medica*, 1995, *32*: 1–14. See also several nineteenth-century medical textbooks and dictionaries, for example, Richard Quain, *A Dictionary of Medicine, including General Pathology, General Therapeutics, Hygiene, and the Diseases Peculiar to Women and Children* (New York: Appleton and Co., 1883), 1050; J. C. Dalton, *A Treatise on Human Physiology* (Philadelphia: H. C. Lea, 1875), 31.

10. On the history of menstruation, see Joan Jacobs Brumberg, *The Body Project: An Intimate History of American Girls* (New York: Random House, 1997); Carroll Smith-Rosenberg and Charles Rosenberg, "The Female Animal: Medical and Biological Views of Woman and Her Role in Nineteenth-Century America"; and Vern Bullough and Martha Voght, "Women, Menstruation, and Nineteenth-Century Medicine," in *Women and Health in America*, ed. Judith Walzer Leavitt (Madison: University of Wisconsin Press, 1984), 12–27, 28–37.

11. Mary Putnam Jacobi, *The Question of Rest for Women during Menstruation* (New York: G. P. Putnam's Sons, 1877), 171, 182, 115.

12. V. Cornil and L. Ranvier, *A Manual of Pathological Histology*, trans. E. O. Shakespeare (Philadelphia: Henry C. Lea, 1880). This manual represented the laboratory investigations of the two scientists and their students, who may have included Mary Put-

nam. For other scholarly treatments of (Putnam) Jacobi in Paris, see Bonner, *To the Ends of the Earth*, 48–54; Harvey, "'Faithful to Its Old Traditions'?"; Harvey, "La Visite"; Harvey, "Medicine and Politics"; Warner, *Against the Spirit of System*, 322–327.

13. See Mary C. Putnam, "De la Graisse Neutre et des Acides Gras," Medical Thesis, École de Médecine, 1871.

14. Her thesis referred to ongoing debates between organic chemists and experimental physiologists in Europe. See Frederic Lawrence Holmes, *Claude Bernard and Animal Chemistry: The Emergence of a Scientist* (Cambridge, MA: Harvard University Press, 1974), 1–33.

15. S. Weir Mitchell, *Lectures on Diseases of the Nervous System, Especially in Women* (Philadelphia: Henry C. Lea's Son & Co., 1881), 95–96.

16. Graily Hewitt, *The Pathology, Diagnosis, and Treatment of the Diseases of Women* (New York: Bermingham & Co., 1883), vol. 2, 159. On the physiological approach to neurology and mental disease, see Bonnie Blustein, *Preserve Your Love for Science: Life of William A. Hammond, American Neurologist* (New York: Cambridge University Press, 1991).

17. On these techniques of measuring nutrition, see Terrie M. Romano, *Making Medicine Scientific: John Burdon Sanderson and the Culture of Victorian Science* (Baltimore: Johns Hopkins University Press, 2002); Holmes, *Claude Bernard and Animal Chemistry*. New technologies for measuring hemoglobin, the hemacytometer and hemoglobinometer, emerged in the 1880s. They monitored the number of red blood corpuscles in patients and measured the amount of hemoglobin. See Keith Wailoo, *Drawing Blood: Technology and Disease Identity in Twentieth-Century America* (Baltimore: Johns Hopkins University Press, 1997), 24.

18. Gerald Geison, "Divided We Stand: Physiologists and Clinicians in the American Context," in *The Therapeutic Revolution: Essays in the Social History of American Medicine*, ed. Morris J. Vogel and Charles E. Rosenberg (Philadelphia: University of Pennsylvania Press, 1979), 67–90; John Harley Warner, "The Fall and Rise of Professional Mystery: Epistemology, Authority, and the Emergence of Laboratory Medicine in Nineteenth-Century America," in *The Laboratory Revolution in Medicine*, ed. Andrew Cunningham and Perry Williams (Cambridge: Cambridge University Press, 1992), 110–141; John Harley Warner, "Ideals of Science and Their Discontents in Late Nineteenth-Century American Medicine," *Isis*, 1991, *82*: 454–478; Warner, *Against the Spirit of System*, 3–31.

19. Paul A. Carter, *The Spiritual Crisis of the Gilded Age* (DeKalb: Northern Illinois University Press, 1971).

20. See Gillis J. Harp, *Positivist Republic: Auguste Comte and the Reconstruction of American Liberalism, 1865–1920* (University Park: Pennsylvania State University Press, 1995), 40. On women and positivism, see also William Leach, *True Love and Perfect Union: The Feminist Reform of Sex and Society* (Middletown, CT: Wesleyan University Press, 1989).

21. At the same time, she became enamored with socialism and identified herself with the goals of republicanism in France. As a witness to the Franco-Prussian War and the Paris Commune, Putnam developed sympathies for the idea of social revolution.

22. Mary Putnam Jacobi, *The Value of Life, A Reply to Mr. Mallock's Essay, "Is Life Worth Living?"* (New York: G. P. Putnam's Sons, 1879), 222.

23. Ibid., 89.

24. Ibid., 85–87.

25. Ibid., 85.

26. Ibid., 92.

27. Leach, *True Love and Perfect Union*, 153–157.

28. Harp, *Positivist Republic*, 20, 46–48; Leach, *True Love and Perfect Union*, 182–183; Louise Michele Newman, *White Women's Rights: The Racial Origins of Feminism in the United States* (New York: Oxford University Press, 1999).

29. Stepan and Gilman, "Appropriating the Idioms of Science," 72–103.

30. Jacobi, *Value of Life*, 87–88. On St. Simon and Fourier, see Keith Taylor, *The Political Ideas of the Utopian Socialists* (London: Frank Cass, 1982), 39–67, 111–115.

31. Mary Putnam Jacobi, "Modern Female Invalidism," in *Mary Putnam Jacobi, M.D.: Pathfinder in Medicine*, ed. Women's Medical Association of New York City (New York: G. P. Putnam's Sons, 1925), 478–482, quotations p. 482.

32. On active models of womanhood, see Martha H. Verbrugge, *Able-Bodied Womanhood: Personal Health and Social Change in Nineteenth-Century Boston* (New York: Oxford University Press, 1988). See also, Dio Lewis, *Our Girls* (New York: Harper & Brothers, 1871).

33. Edward H. Clarke, *Sex in Education; or, a Fair Chance for the Girls* (Boston: James R. Osgood and Company, 1873), 126; Sue Zschoche, "Dr. Clarke Revisited: Science, True Womanhood, and Female Collegiate Education," *History of Education Quarterly*, 1989, *29*: 545–569.

34. Mary Putnam Jacobi, "Mental Action and Physical Health," in *The Education of American Girls*, ed. Anna C. Brackett (New York: G. P. Putnam's Sons, 1874), 257–306.

35. Ibid., 258.

36. Ibid., 261, 263, 261, 258.

37. For a history of mind/body dualism, see Merle Curti, *Human Nature in American Thought: A History* (Madison: University of Wisconsin Press, 1980), 187–193. On the mind/body relationship and reflex theory, see Russett, *Sexual Science*, 112–129; Edward Shorter, *From Paralysis to Fatigue: A History of Psychosomatic Illness in the Modern Era* (New York: Free Press, 1992), 40–68; Morantz-Sanchez, *Conduct Unbecoming a Woman*, 116–117.

38. Jacobi, "Mental Action and Physical Health," 272.

39. Ibid., 284. See also 271–272, 274–275.

40. Ibid., 274.

41. Tuchman, *Science Has No Sex*, 75–79; Tuchman, "Maternity and the Female Body in the Writings of Dr. Marie Zakrzewska, 1829–1902," Chapter 2 in this volume.

42. Morantz-Sanchez, *Conduct Unbecoming A Woman*, 130–137; Morantz-Sanchez, "Female Patient Agency and the Trial of Dr. Mary Dixon Jones in Late Nineteenth-Century Brooklyn," Chapter 3 in this volume.

43. Jacobi, "Mental Action and Physical Health," 279–282. She maintained that "no experimental proof at present exists that stimulation of the brain ever does cause such dilatation" of the blood vessels, "that is, ever does become a cause of hemorrhage," or pain. Quotation p. 281.

44. Ibid., 292–293.

45. "Boylston Medical Prize Questions," inserted in *Question of Rest*, 1877.

46. Morrill Wyman quoted in letter, C. Alice Baker to Mary Putnam Jacobi, November 7, 1874, Mary Putnam Jacobi Papers, A-26, Schlesinger Library, Radcliffe Institute for Advanced Study, Harvard University. On C. Alice Baker, see "Charlotte Alice Baker (1833–1909)," in *American Women Historians, 1700s–1990s: A Biographical Dictionary*, ed.

Jennifer Scanlon and Shaaron Cosner (Westport, CT: Greenwood Press, 1996); "Charlotte Alice Baker," in *The Biographical Cyclopaedia of American Women*, ed. Mabel Ward Cameron (New York: Halverd Publishing Co., 1924), vol. 1, 345–347.

47. Minutes of the Boylston Prize Committee, June 5, 1876 (DE10), Harvard Medical Library in the Francis A. Countway Library of Medicine.

48. C. Alice Baker to Mary Putnam Jacobi, November 7, 1874, Mary Putnam Jacobi Papers, A-26, Schlesinger Library, Radcliffe Institute for Advanced Study, Harvard University.

49. Mary F. Eastman offered Baker help in procuring statistics for Jacobi. Mary F. Eastman to "Miss Baker," Mary F. Eastman Letter, E 13, Schlesinger Library, Radcliffe Institute for Advanced Study, Harvard University.

50. Jacobi, "Mental Action and Physical Health," 258.

51. Mary Roth Walsh, *"Doctors Wanted, No Women Need Apply": Sexual Barriers in the Medical Profession, 1835–1975* (New Haven, CT: Yale University Press, 1977), 165–175.

52. Warner, *Against the Spirit of System*, 336; Tuchman, *Science Has No Sex*, 173–174.

53. On the changing discourses of gynecology, see Morantz-Sanchez, "Female Patient Agency and the 1892 Trial of Dr. Mary Dixon Jones in Late Nineteenth-Century Brooklyn," Chapter 3 in this volume.

54. Jacobi produced one thousand questionnaires for distribution but received only 268 answers. Her use of statistics reflected the movement toward a "trust in numbers." On the history of statistics, see Theodore M. Porter, *The Rise of Statistical Thinking, 1820–1900* (Princeton, NJ: Princeton University Press, 1986).

55. Jacobi, *Question of Rest*, 62.

56. Ibid., 62–63.

57. Ibid., 227.

58. Ibid., 109.

59. Jacobi later expanded on her menstruation theory when she studied the etiology of endometritis. See Mary Putnam Jacobi, "Theories of Menstruation—New Theory (Studies in Endometritis)," *American Journal of Obstetrics*, 1885, *18*: 596–606.

60. Jacobi, *Question of Rest*, 115.

61. On the sphygmograph, see Robert A. Frank Jr., "The Telltale Heart: Physiological Instruments, Graphic Methods, and Clinical Hopes, 1854–1914," in *The Investigative Enterprise: Experimental Physiology in Nineteenth-Century Medicine*, ed. William Coleman and Frederic L. Holmes (Berkeley: University of California Press, 1988), 211–290. See also Romano, *Making Medicine Scientific*, 79–86.

62. See Jacobi's sphygmographic tracings (inserts) in *Question of Rest*. See also her laboratory studies on urea, p. 140.

63. Ibid., 146–148.

64. Ibid., 202.

65. Ibid., 165.

66. Ibid., 14–15.

67. Thomas Addis Emmet, *The Principles and Practice of Gynaecology* (Philadelphia: Henry C. Lea, 1879). For a discussion of ovulation and another interpretation of *Question of Rest*, see Thomas Laqueur, *Making Sex: Body and Gender from the Greeks to Freud* (Cambridge, MA: Harvard University Press, 1990), 220–225.

68. Jacobi, *Question of Rest*, 13.

69. Ibid. See pp. 8, 12–13, on the plethoric explanation.

70. Laqueur, *Making Sex*, 214; Margaret Marsh and Wanda Ronner, *The Empty Cradle: Infertility in America from Colonial Times to the Present* (Baltimore: Johns Hopkins University Press, 1996), 84–85.

71. Jacobi, *Question of Rest*, 100. Jacobi said clearly, "Reproduction is essentially a process of nutrition," referencing French physiologist Claude Bernard (ibid., 82).

72. Ibid., 167–168.

73. *The Medical Record*, book review, 1878, *13*: 14.

74. "Book Notices," *Philadelphia Medical and Surgical Reporter*, 1877, *36*: 511.

75. Henry Pickering Bowditch, "The Question of Rest," *The Nation*, September 13, 1877. See W. P. Garrison to Henry Pickering Bowditch, August 3, 1877, inserted in Jacobi, *The Question of Rest*, RG163.J15 c. 2, Boston Medical Library Collection, Harvard Medical Library in the Francis A. Countway Library of Medicine.

76. See the discussion by Paul F. Mundé, "Report on the Progress of Gynecology during the Year 1875," *American Journal of Obstetrics*, 1876, *9*: 127–173.

77. George J. Engelmann, "The American Girl of To-day: The Influence of Modern Education on Functional Development," *Transactions of the American Gynecological Society*, 1900, *25*: 8–45.

78. William Stephenson, "On the Menstrual Wave," *American Journal of Obstetrics*, 1882, *15*: 287–294; John Goodman, M.D., "The Cyclical Theory of Menstruation," *American Journal of Obstetrics*, 1878, *11*: 43, 673–694. Other proponents of the wave idea are noted in Emil Novak, M.D., *Menstruation and Its Disorders* (New York: D. Appleton and Company, 1921), 85.

79. Mary Putnam Jacobi, "The Stephenson Wave," *American Journal of Obstetrics*, 1895, *32*: 92.

80. For example, see Mary Putnam Jacobi, "Studies in Endometritis," *American Journal of Obstetrics*, 1885, *18*: 36–50, 113–128, 262–283, 376–386, 519–537, 596–606, 802–846, 915–932. These studies are beyond the scope of this article; I analyze them elsewhere in the context of the 1880s.

CHAPTER TWO

Maternity and the Female Body in the Writings of Dr. Marie Zakrzewska, 1829–1902

Arleen Marcia Tuchman

In 1854, twenty-five-year-old Marie Zakrzewska, a student at Cleveland Medical College, submitted a medical thesis on "The Organ of Parturition." After discussing the similarities and differences between the organs of parturition in the various classes of the vegetable and animal kingdom, she concluded that the uterus was nothing more than a highly developed differentiation of the intestines. Indeed, she went as far as to equate labor with the "peristaltic motions" of the intestinal system, and to contend that the uterus was to the ovaries what the bladder was to the kidneys.[1] Zakrzewska's purpose was to dethrone the quintessential female organ—to challenge, in her words, one physician's mystification of "this portion of the human frame" as the "Wonder of Nature."[2] This marked, in fact, the first of her many attacks on the kind of biological argument that was often used to justify women's confinement to the home. Thus, in contrast to those who claimed that women were sickly by nature because they possessed a womb, Zakrzewska denied that any meaningful sexual differences existed at all.

More than thirty years later, Zakrzewska, then director of the New England Hospital for Women and Children in Boston, was using quite different language to describe women's bodies and their organs of reproduction. She now heralded childbirth as "the great event" of a woman's life; and she lamented the "lack of sanctity" women showed toward their bodies when they demanded ovariotomies. Women, she insisted, who chose to undergo such a "mutilation" of their bodies, gave "cause for lowering the respect and sacredness of nature."[3]

This essay seeks to make sense of the transformation Zakrzewska underwent in her thoughts and writings on both the female body and motherhood. The analysis provided here aims to deepen our understanding of Zakrzewska—one of the most prominent women physicians in the nineteenth-century United States—while also drawing our attention to the possibilities for, and the constraints upon, women living in a way that challenged entrenched cultural stereotypes about proper feminine behavior.

Marie Elizabeth Zakrzewska was born in Berlin, Germany, in 1829 and trained there as a midwife before immigrating to the United States to study medicine. She earned an M.D. in 1856 from Cleveland Medical College, one of only six women who received a medical degree during the four years it kept its doors open to women. While she was in Cleveland, she not only received a medical education but also a political one. She boarded with Caroline Severance, a leading figure in the women's rights movement and an ardent abolitionist. Through Severance, Zakrzewska came to know Harriot Hunt, William Lloyd Garrison, the Grimké sisters, and other radical reformers, forming friendships that would last her entire life.[4]

After graduation, Zakrzewska moved to New York City, where she spent three years helping Elizabeth and Emily Blackwell establish and run the New York Infirmary for Women and Children. During this time, she also became involved with the community of radical German émigrés, developing a particularly close friendship with Karl Heinzen, a political journalist who had fled his homeland following the failed Revolutions of 1848. This friendship was not romantic; in Zakrzewska's words, it was "based . . . on principle, making the object for which we spoke, our real life's cause."[5] That object was the creation of a truly democratic society, and she shared with these German radicals the conviction that such a society would only come about when religion (especially as preached and practiced by the Catholic Church) was replaced by materialism, atheism, and a scientific approach to the acquisition of knowledge.

Zakrzewska moved to Boston in 1859 in part to follow Heinzen and in part to assume a position as professor of Obstetrics and Diseases of Women and Children, and as resident physician at the New England Female Medical College. She remained at the college barely three years. She had too many differences with the school's director, Samuel Gregory, whom she believed kept the standards too low.[6] Thus, in 1862, at age thirty-three, she founded the New England Hospital for Women and Children with the support of Boston's leading liberal and radical reformers. It was here that Zakrzewska established herself as a leading medical practitioner and educator of women physicians. She retired in 1899 after thirty-seven years of running the hospital and died three years later at age seventy-three.[7]

To understand how and why Zakrzewska's thoughts and writings on the female body changed over the course of her lifetime, we need to begin with a closer reading of her initial position. Indeed, in these early years, Zakrzewska presented a noticeably coherent view. In contrast to most of her peers—those who practiced medicine and those who did not—who insisted in one way or another that women and men were anatomically, physiologically, psychologically, and morally different, Zakrzewska countered that women and men did not differ in any significant way. She argued this not only in her medical thesis, in which she presented a highly mechanistic view of women's bodies, but also in other writings she penned early in her career. Thus, in an 1859 essay about women physicians published in *Der Pionier*, a radical German newspaper that Heinzen edited, Zakrzewska ridiculed the idea that women were physiologically unfit to practice medicine. To the argument that women lacked the necessary intelligence because their brains were too small, she replied that the logical conclusion would be to "regret . . . the death of the 600 pound man who died . . . in New York and in whom we have certainly lost one of the greatest medical geniuses." And to the claim that pregnancy kept a woman from working for weeks at a time, she argued that this was no different than "what happens to male [physicians] who are held up for weeks or months through illness, etc." To Zakrzewska, pregnancy received so much attention only "because it is the most obvious" disturbance, not because it is the most serious. Of course, in rejecting these differences, Zakrzewska articulated a position that most advocates of women's entry into the medical profession upheld. Yet when asked whether women had a special contribution to make to medicine—something to which many women answered "yes"—she suggested that the question made as little sense as asking whether women wrote books because their books were better and more interesting than those men authored.[8]

Zakrzewska's embrace of a mechanistic view of women's bodies was part of her move to erase gender. Whenever "a woman claimed the right of gaining intellectual power," Zakrzewska once complained, "it appeared she stepped out of her sphere. And this claim, so simple and natural, was perverted by a hostile spirit into the claim that she wished 'to become a man.'"[9] Zakrzewska was convinced that the best way to counter such accusations was to challenge constantly the cultural stereotypes in circulation that distinguished sharply between men's and women's nature and between their bodies. She not only articulated this challenge, but she also embodied it: displaying her wit publicly, challenging and ridiculing opponents, even assuming the directorship of a hospital. All were performances that repeatedly placed her in opposition to the stereotype of proper womanly behavior. Her physical appearance, moreover (judging from the few photographs and paintings that have survived), indicates a woman who shunned traditional feminine markers such as ribbons and lace, preferring instead an unadorned and austere look.

Zakrzewska's determination to paint a different picture of what it meant to be a woman also played out in her rejection of both marriage and motherhood. The former she described at age thirty as "an institution to relieve parents from embarrassment. When troubled about the future of a son," she elaborated, "parents are ready to give him to the army; when in fears of the destiny of a daughter, they induce her to become the slave of the marriage bond."[10] As far as children were concerned, she said nothing until later in life; that other unmarried women physicians, such as Elizabeth Blackwell, adopted children, suggests that remaining childless was a choice on her part.

Zakrzewska challenged cultural stereotypes about women in at least one more way. Purchasing her own home in 1860, she became the female head of a complex household that included at times two of her younger sisters; Heinzen and his wife, who shared her home for twenty years; and her close friend and companion, Julia A. Sprague, author of *The History of the New England Women's Club from 1868 to 1893*, who joined the household in 1862 and remained until the doctor's death forty years later.[11] Sprague and Zakrzewska's relationship, especially after Heinzen's death in 1880, may best be characterized as a "Boston marriage," a label that was first used in the late nineteenth century to refer to two women, usually both middle-class professionals, sharing a household.[12] It is unclear why Sprague first joined Zakrzewska's household, but what may have begun as a convenient arrangement seems to have grown slowly into a committed and caring partnership. Although there is little evidence of any passion between them, letters suggest that they viewed themselves as companions and expected others to view them this way as well.[13]

Thus, from her public pronouncements to her private life, Zakrzewska presented a remarkably coherent picture. With a singular determination to challenge gender stereotypes, she presented a mechanical view of women's bodies that functioned to downplay any differences between the sexes; she occupied spaces traditionally reserved for men, including medical school, the directorship of a hospital, and the head of a household; and she claimed for herself such masculine-coded traits as rationality and scientific acumen. Given all this, one might be tempted to say that she lived her life like a man, even going so far as to "simulate" a marriage with a woman. However, such a reading of her life would misrepresent Zakrzewska, who identified as a woman and perceived both her personal and professional battles to be about the liberation of women from the constraints imposed on them.[14] True, she displayed traits and mannerisms labeled masculine in her day, but she did so consciously, trying in her own way to disrupt the boundaries that had been erected to distinguish women from men.

Zakrzewska was not alone among women physicians in arguing that the sexes did not differ in significant ways. From the work of Carla Bittel and Regina Morantz-Sanchez, we know that this was true, at the very least, of Mary Putnam Jacobi and Mary Dixon Jones.[15] We also know that this was a minority view. Whether women physicians believed they were uniquely qualified to bridge what they perceived to be a gap between "sympathy" and "science," or whether they sought to find "balance" between their professional and personal lives, most female practitioners embraced and then redefined, for their own purposes, their culture's emphasis on sexual difference.[16] Zakrzewska thus stood outside the mainstream, although unlike Dixon Jones—and much like Mary Putnam Jacobi—she was not attacked for her transgressions. On the contrary, she garnered much respect from both her male and female peers. She did not threaten the status quo to the same degree as Dixon Jones. Indeed, her life suggests that mid-nineteenth-century middle-class women may have had some freedom to live against the grain, challenge cultural stereotypes about their sex, and achieve a position of respect and admiration despite their unusual ways. This does not imply that cultural constraints did not impose restrictions—Zakrzewska was well aware that her male peers considered her "exceptional"[17]—but for those women willing and able to accept exceptional status, considerable latitude may have existed.

By the late 1880s, however, when Zakrzewska was entering the sixth decade of her life, she was promoting a different view of women's bodies. Thus, she now spoke out against instrumental deliveries, painting a picture of childbirth as a natural event, and describing a woman's "first babe" as the "great event of her life."[18] In a letter to Blackwell, written in 1891, she also lamented the "lack of sanctity for their own body growing up in girls and women" that was sending them to the hospital "on the slightest cause" in order to "urge upon us operations." Well aware of Blackwell's own opinion of ovariotomies as "permanent mutilation," she was letting her friend know that she shared this position.[19]

At the time Zakrzewska wrote this letter, a spirited debate was under way in the medical literature over the use of surgical solutions to cure women's health problems.[20] In this debate, Blackwell represented one extreme. On the other side were women like Dixon Jones, who specialized in ovariotomies, and Putnam Jacobi, whose direct response to Blackwell was to counter that "there is not such special sanctity about the ovary!"[21] Zakrzewska had once shared the latter women's view, but that was no longer the case. She may never have gone as far as to embrace Blackwell's notion of the "spiritual power of maternity," which supposedly taught "the subordination of self to the welfare of others," but she was moving away from her highly mechanistic view of women's bodies and her absolute denial of difference.[22]

Zakrzewska was also modifying her position on the value of educating women apart from men. Although seemingly unrelated to her view of women's bodies, in fact both represented Zakrzewska's willingness to align herself more closely with those who believed in a woman's distinct sphere of influence. When Zakrzewska first founded the New England Hospital for Women and Children in 1862, she had considered it a temporary measure, necessary only until all-male institutions opened their doors and guaranteed women an education equal to men. However, in the 1880s, she started defending separate institutions on principle, arguing the importance "for women to learn *from women*. The value of seeing *women* doing skilful [*sic*] medical work," she told her interns in 1883, "cannot be over-estimated in its inspiring effect upon the young woman practitioner."[23] To be sure, Zakrzewska did not claim that women practiced differently than men; in fact, she seemed more interested in the way female role models could empower young women, stimulating self-confidence by demonstrating to women their ability to perform the same skills as men. Nevertheless, Zakrzewska's embrace of separatism, combined with her view of women's bodies as sacred, marked a radical break from her earlier attempts to erode any boundaries—biological, ideological, or institutional—that sought to distinguish women from men.

Zakrzewska altered her views in part because of changes in her personal life, including a growing sadness over not bearing children. But we cannot understand that lament—to which we shall return later—without first exploring changes in her professional life. When Zakrzewska first founded the New England Hospital for Women and Children in 1862, women were denied access to most medical institutions in the country. Indeed, her hospital was only the third in the country where women could learn at the bedside.[24] She was excited about helping to chart a new path; the small number of women who practiced and studied medicine at the New England Hospital in its first years formed a close-knit group, often socializing together following a long day at the hospital. Sophia Jex-Blake, who interned at the hospital in 1865, wrote letters home to her mother describing the evenings spent at the theater, visiting ice-cream shops, or spending time together in the hospital, singing songs, playing card games, and losing themselves in "roars of laughter."[25] To Zakrzewska, the first few decades marked a time when everyone in her hospital "assumed more of the condition of a family circle."[26] She even assumed a parental role toward some of the interns, writing to Lucy Sewall, who eventually helped Zakrzewska run the hospital, that "you are my child."[27]

By the 1880s, when Zakrzewska's attitude toward women's bodies and motherhood had begun to change, her relationship with her interns also showed signs of undergoing a marked transformation. Indeed, rather than making up a small "family circle," the interns and senior staff became embroiled in a series of conflicts. To be sure, the hospital's growth had something to do with this shift. The year Sophia Jex-Blake wrote her enthusiastic letters home, describing the strong sense of camaraderie they all felt, only five physicians, three interns, three consultants, and twenty directors were associated with the hospital.[28] Two decades later, the numbers had increased to sixteen physicians, six interns, fourteen consultants, and forty directors, making it difficult to remain a close-knit group.[29] But size alone did not alter the culture of the institution; the young interns also viewed the New England Hospital differently. Women now had many more opportunities for clinical internships, including at least half a dozen previously all-male hospitals that had begun accepting female interns.[30] Their assessment of the necessity of all-female institutions thus differed from that of Zakrzewska's generation. Some of the interns, moreover, were graduates of coeducational medical schools, and many were beneficiaries of sweeping reforms that had begun to change medical education, at least among the nation's elite institutions. Notably, schools such as the Woman's Medical College of Pennsylvania and the University of Michigan, where many of the New England Hospital's interns had received their degrees, had lengthened the course of study, established a graded curriculum, and abandoned didactic teaching in favor of "practical" instruction.[31] The graduates of these institutions arrived at the New England Hospital with more advanced skills—and a greater sense of entitlement—than previous generations of interns.

Thus where Sophia Jex-Blake and Lucy Sewall had expressed deep gratitude toward Zakrzewska and accepted willingly the senior staff's parental authority, the interns in the 1880s and 1890s complained of being treated "as mere children" who lacked any "professional standing." What they wanted was more responsibilities for patient care. "We have had abundant opportunities during our college years, to *see*," they complained in 1891, "we come here to *do*." But instead, they were "allowed to *do* absolutely nothing," forbidden from performing "even the simplest operations" or from handling any difficult maternity cases.[32] Most damning were the criticisms of Alice Hamilton, who eventually became an expert in industrial medicine and the first woman to join the faculty at Harvard University (1919). During a brief internship at the New England Hospital in 1893, she branded the institution a "narrow, petty, squabbly [*sic*], idiotic place," voicing in particular her frustration that she was being "treated like a raw school-girl," even being chastised before her patients. She had arrived in Boston excited about the prospect of spending an entire year developing her clinical skills; instead, she found herself either "sitting around and reading text books," or filling her days with boring, menial tasks. Feeling as though she was wasting precious time, Hamilton resented the fact that "not a man medical graduate in the country . . . would accept so inferior a position as this." Although concerned that to quit would mean reneging on her promise to stay an entire year, Hamilton was so indignant that the staff had "not fulfilled their part of the contract," she ended up resigning her position before the year was out.[33]

Zakrzewska did not respond well to these criticisms. She became defensive, accusing the interns of preferring to advance their own careers at the expense of the institution and, in fact, of jeopardizing the cause of all women. As the gap grew between Zakrzewska and the younger generations, the aging director turned increasingly to the past, trying to convince the interns to see the New England Hospital as a "large family" that required them to place its needs before their own. She even spoke passionately about the need for "a good deal of self-sacrifice" to bring to completion "one of the greatest historical reforms," thus embracing a language of self-sacrifice that she had so adamantly rejected in her youth because of her conviction that it kept women from implementing their own plans and satisfying their own interests.[34]

The generational tensions evident at the New England Hospital were not peculiar to this institution. Emily Blackwell also felt compelled to remind the 1899 graduating class of the New York Infirmary and Medical College not simply to work for their own advancement but to recognize "that the work of every woman physician, her character and influence, her success or failure, tells upon all, and helps or hinders those who work around her or come after her."[35] Indeed, historians who have studied late nineteenth-century women's organizations of all types have described the emergence of a "self-conscious professional culture" that seemed to have little in common with the professionalism of their predecessors.[36] The older form, referred to by some historians, as "civic professionalism," was grounded in a sense of community. Professional responsibilities and civic duty were woven seamlessly together, one's place in the community defined by both.[37] The newer professionalism, in contrast, replaced the notion of a community with the image of a group of autonomous individuals joining together to protect their own interests. Even the women's rights movement showed signs of such tensions. Among the older generation (Zakrzewska's cohort), personal experiences of discrimination fueled a political critique that condemned the unjust concentration of power in the hands of white men. To their younger colleagues, the battle for women's rights was a battle for the rights of individuals to pursue their own interests.[38]

A similar dynamic was at play at the New England Hospital in the last decades of the nineteenth century. The young interns did not share Zakrzewska's view that the fate of the women's movement for medical education depended on their allegiance to the institution (and by extension to her). They were not, however, unaware of or indifferent to the plight of women physicians. Indeed, the interns' reminder to the medical staff that they were failing to fulfill their obligation to "assist educated women in the practical study of medicine," indicates that it was the thought of receiving a mediocre medical education *as women* that troubled them so much. Hamilton, we must remember, had been disturbed by the thought that she was being asked to tolerate a situation that "no man medical graduate in the country" would ever be expected to endure. Committed suffragist that she was she understood only too well that her actions and experiences had consequences for the advancement of women at large.[39] Zakrzewska's language, communitarian in spirit, no longer made sense to the younger generation. Thus, where she spoke of sacrifice, family, and stewardship, Hamilton spoke of contractual obligations. Both women were trying to characterize the relationship between two parties, but Zakrzewska assumed the primacy of the group and the existence of a clear hierarchy, where a "protector" assumed moral responsibility for those in her charge. Hamilton, however, assumed the primacy of the individual; in her view, relationships occurred between equal partners bound together by a written document that ensured both parties would fulfill their obligations.[40]

Zakrzewska's growing alienation from her own hospital in the last decades of the nineteenth century clearly was tied to this clash of professional cultures, but it also reflected her inability to divorce herself from her creation or to articulate an alternative identity for the hospital, other than separatism, as all-women's institutions struggled to find a new sense of meaning and purpose. Zakrzewska had good reason to question whether coeducational institutions were committed to granting women an education *equal* to that of men, but her new commitment to the principle of separatism also stemmed from her total immersion in the life of her hospital. It was one thing to imagine that some day the New England Hospital would no longer exist; it was another thing entirely to hear suggestions that it may already have outlived its usefulness. In the end, Zakrzewska's caution proved prescient—women did not fare as well under coeducation as they had when all-women's institutions had flourished.[41] However wise her position may have turned out to be, it was also fueled by a personal identification with the institution that made it difficult for her to contemplate a different path or, for that matter, to hear much criticism at all.

In 1887, just a few years shy of her sixtieth birthday, she stepped down as attending physician, thus withdrawing from the day-to-day management of the New England Hospital. Pulling back from an institution that had been a home away from home for her had already brought some sadness; the loss of two dear friends just a few years later made things even more difficult. Her close friend Mary Booth passed away on March 5, 1889, at age fifty-eight. They had known each other since the 1850s, when Zakrzewska had lived in New York City. They had remained close through the years and had shared, in Zakrzewska's words, an "intimacy [that] was only broken by death." Then, less than a year later, Lucy Sewall passed away at the age of fifty-two. The year, she wrote to Severance on the day of Sewall's death, had been a "sad" and "a cruel one," and she "felt almost tired of life" herself.[42]

These changes in Zakrzewska's life led her to ponder both her own mortality and the meaning of her life's work. Zakrzewska derived considerable satisfaction from the hospital she had founded and her work to further the cause of women physicians, but on several occasions, she implied that these were not enough. Indeed, in a particularly moving letter to her friend Caroline Severance in 1889, she expressed regret that she had not brought any children into her life. Commenting on Severance's son's decision to remain a bachelor, she wrote: "I am sorry for him, because I know he will feel the penalty for doing so, just as keenly as I do, now that I completed my 60th year and have no young life, which belongs to me. . . Now with mature thought, I see that life is selfish when the dread of sorrow prevents us to fulfill our natural mission, and whatever this sorrow may be, it is never as great, as not to be ten times balanced by the joy, which precedes such grief & anxiety."[43]

In the midst of this contemplation and sadness, Zakrzewska found herself drawn to the German biologist and philosopher Ernst Haeckel, who had recently written of his view of the material world as both fundamentally mechanical and capable of generating soul-like properties.[44] Haeckel, who had set out both to deny the existence of "immaterial forces" and to avoid a "soulless materialism," was arguing at the time that the first life forms, or monera, had generated spontaneously from inorganic matter, emerging from that process endowed with the property of irritability. This property eventually evolved, moreover, into consciousness and the human soul. "I am a monist," Zakrzewska wrote Severance. "The whole universe is one great power or material & evolves the spirit."[45]

Zakrzewska appears to have found some comfort in Haeckel's ideas about the material world. She knew she could not, nor did she wish to, believe in an afterlife. She had no desire, she once explained, to join her friends once again "in a form, which is either by virtue of surroundings or an advanced development, so altered for the better or the worse, that they have become estranged to my comprehension or feeling."[46] But perhaps Haeckel's belief that matter could give rise to emergent properties, thus leaving room for something akin to a spiritual element, satisfied a need she had for something to persist after she died. That such thoughts were on Zakrzewska's mind is suggested by a comment she made to Elizabeth Blackwell a few years before her death: "I fully agree with you," she wrote, "and beleive [*sic*] that the spirit cannot die, is indestructible and lives forever, although individual consciousness is lost in Nirvana."[47]

I am suggesting that Haeckel's blurring of the boundary between the natural and the spiritual not only offered Zakrzewska comfort but also contributed to her vision of the body as material but nevertheless sacred. At the very least, it offered her a language and a philosophical foundation, distinct from the religious doctrines she continued to hold suspect, in which to ground her increased reverence toward the human form. Still, this picture is too neat because among the many emotions Zakrzewska expressed at this time of her life, she also showed considerable anger toward the younger generation. As we have seen, she did not hesitate to accuse the interns in her hospital of extreme selfishness, putting their own needs before the interests of women physicians as a whole. But her truly vituperative words she saved for the young women who sought ovariotomies in her hospital. In the letter she wrote to Blackwell in 1891, she accused these women of "indulgence in luxurious living, dislike to work and of self abnegation . . . Yes," she added, "they rather die, than bring up a family of children and work and practise [*sic*] self-denials."[48]

Zakrzewska was clearly struggling later in life to make sense of the decisions she had made at critical moments in the past. I do not, however, wish to present a distorted picture. On several occasions, she expressed contentment with her life and showed pride in the work she had accomplished.[49] Nevertheless, at a difficult point in her life, when she was troubled by tensions within her hospital and overwhelmed by the loss of dear friends, she directed her pain and anger at women who—at least so she felt at the time—were choosing not to bring into the world the young lives she now lamented not having had herself.

Zakrzewska's attitude toward women's bodies had changed considerably since her early adulthood. When she first began formulating her views in the 1850s and 1860s, she was trying to counter beliefs that women were physically and intellectually incapable of practicing medicine, or that by their nature they were inclined to practice a gentler, more humane kind of medicine. She liked neither because, in her estimation, they both confined women to a limited sphere of influence: the former by identifying women with the private sphere of the home, denying them access to positions of power in the public arena, the latter by relegating them to what she perceived to be a subordinate place in the medical hierarchy. In challenging these beliefs, Zakrzewska sought to redefine women's nature, women's bodies, and women's purpose in life. She did this by minimizing the biological differences between men's and women's bodies, by promoting and performing women's ability to engage in rational thought, and by living her life against the grain.

By the last decade of her life, however, Zakrzewska viewed things differently. Still a materialist and an atheist, she now endowed matter with some sentience, and she considered motherhood woman's "natural mission." Some might be tempted to ascribe Zakrzewska's change of view to an elusive "maternal instinct" that finally surfaced toward the end of her life, but I choose to see this shift as evidence of the constraints under which women lived at the time and the difficulty of being an atheist and childless in a society that accepted neither. As a younger woman, Zakrzewska had derived considerable satisfaction from her work, building friendships with her staff and interns that had turned her hospital into a home away from home. As the hospital grew, though, and as the community she had formed began to dissolve, Zakrzewska found it increasingly difficult to understand what her role should be. As she wrote when she retired: "If I had children or grandchildren to work for or a new hospital which must be provided for, I would consider then my work satisfactory and joyful."[50] She had neither, however, and so she stepped back from hospital life to spend more time with old friends. At times this was enough, but as these friends died, she thought increasingly about her legacy. Troubled by the changes taking place around her, and struggling to endow her own life with meaning, Zakrzewska occasionally embraced some of the conventions she had once so vehemently opposed.

NOTES

1. Marie Elizabeth Zakrzewska, "Thesis: The Organ of Parturition," Cleveland Medical Thesis 1855–1856. Howard Dittrick Museum for the History of Medicine, Allen Memorial Medical Library, Case Western Reserve University, Cleveland, Ohio. I am grateful to the staff of the Archives for sending me a transcription of Zakrzewska's thesis. The transcription was prepared by Linda Lehmann Goldstein.

2. The physician was the famous seventeenth-century Dutch physician Jan Swammerdam, as cited in Zakrzewska, ibid., 10.

3. The comment about a woman's "first babe" is from Marie Zakrzewska, "Report of One Hundred and Eighty-Seven Cases of Midwifery in Private Practice," *Boston Medical and Surgical Journal*, 1889, *121*: 557–560. The rest of the comment is from Zakrzewska to Blackwell, March 21, 1891, Blackwell Family Papers, Schlesinger Library, Cambridge, Massachusetts.

4. Biographical information is from Agnes Vietor, ed., *A Woman's Quest: The Life*

of Marie E. Zakrzewska, M.D. (New York: Arno Press, 1972; reprint of New York: D. Appleton & Co., 1924), and Virginia G. Drachman, *Hospital with a Heart: Women Doctors and the Paradox of Separatism at the New England Hospital, 1862–1969* (Ithaca, NY: Cornell University Press, 1984). For Zakrzewska's years at Cleveland Medical College, see Linda Lehmann Goldstein, "Roses Bloomed in Winter: Women Medical Graduates of Western Reserve College, 1852–1856" (PhD diss., Case Western Reserve University, 1989).

5. Zakrzewska to Paulina Pope, October 28, 1901, NEH Collection, Box I, Sophia Smith Collection, Northampton, Massachusetts. I am grateful to Regina Morantz-Sanchez for sending me a copy of this letter. I discuss Zakrzewska's involvement with the community of German radical émigrés in greater detail in Arleen Marcia Tuchman, "'Only in a Republic Can It Be Proved That Science Has No Sex': Marie E. Zakrzewska (1829–1902) and the Multiple Meanings of Science," *Journal of Women's History*, 1999, *11*: 121–142. On the German radical community, see Bruce Levine, *The Spirit of 1848: German Immigrants, Labor Conflict, and the Coming of the Civil War* (Urbana: University of Illinois Press, 1992); Stanley Nadel, *Little Germany: Ethnicity, Religion, and Class in New York City, 1845–80* (Urbana: University of Illinois Press, 1990); and Charlotte L. Brancaforte, ed., *The German Forty-Eighters in the United States* (New York: Peter Lang, 1989).

6. See Martha N. Gardner, "Midwife, Doctor, or Doctress? The New England Female Medical College and Women's Place in Nineteenth-Century Medicine and Society," Ph.D. diss., Brandeis University, 2002; also Arleen Marcia Tuchman, "Situating Gender: Marie E. Zakrzewska and the Place of Science in Women's Medical Education," *Isis*, 2004, *95*: 34–57.

7. On the New England Hospital, see Drachman, *Hospital with a Heart.*

8. "'Eine Aerztinn' [a female physician]," "Weibliche Aerzte," *Der Pionier*, March 19, 1859, *6*: 6. On women physicians' embrace of gender differences, see Regina Morantz-Sanchez, *Sympathy and Science: Women Physicians in American Medicine* (New York: Oxford University Press, 1985). On nineteenth-century women physicians' use of satire in their battle to gain entry into the medical profession, see Susan Wells, *Out of the Dead House: Nineteenth-Century Women Physicians and the Writing of Medicine* (Madison: University of Wisconsin Press, 2001), 92–99.

9. Vietor, *A Woman's Quest*, 156.

10. Caroline H. Dall, ed., *A Practical Illustration of "Woman's Right to Labor;" or, A Letter from Marie E. Zakrzewska, M.D.* (Boston: Walker, Wise, and Company, 1860), 86. See also her comment on p. 72.

11. I have pieced together Zakrzewska's home life from the following: Vietor, *A Woman's Quest*, 311, 477; Carl Wittke, *Against the Current: The Life of Karl Heinzen (1809–80)* (Chicago: University of Chicago Press, 1945), 22–23; Zakrzewska to Dall, February 13, 1867, in C. H. Dall Papers, Box 4, Folder 12, Massachusetts Historical Society, Boston, Massachusetts; and Zakrzewska to Paulina Pope, October 28, 1901. Cf. Julia A. Sprague, *History of the New England Women's Club from 1868 to 1893* (Boston: Lee and Shepard, 1894).

12. On Boston marriages, see Micaela di Leonardo, "Warrior Virgins and Boston Marriages: Spinsterhood in History and Culture," *Feminist Issues*, 1985, 2 (5): 47–68; and Estelle B. Freedman, *Maternal Justice: Miriam Van Waters and The Female Reform Tradition* (Chicago: University of Chicago Press, 1996), esp. 107, 178, 242.

13. There is a collection of about fifty letters from Zakrzewska and Sprague to Caroline H. Severance in the Severance Collection at the Huntington Library. I am grateful to Virginia Elwood for alerting me to this correspondence.

14. See, for example, her comment to her interns that she found it "far more agreeable to be considered first, a woman, and secondarily a 'Dr.'" Zakrzewska to the interns, April 1, 1876, p. 1, NEH Collection, Box 27, Folder 1173, Sophia Smith Collection, Northampton, Massachusetts.

15. See Carla Bittel, "Mary Putnam Jacobi and the Nineteenth-Century Politics of Women's Health Research," Chapter 1 in this volume, and Regina Morantz-Sanchez, "Female Patient Agency and the 1892 Trial of Dr. Mary Dixon Jones in Late Nineteenth-Century Brooklyn," Chapter 3 in this volume. See also Bittel, "The Science of Women's Rights: The Medical and Political Worlds of Mary Putnam Jacobi" (Ph.D. diss., Cornell University, 2003); and Morantz-Sanchez, *Conduct Unbecoming a Woman: Medicine on Trial in Turn-of-the-Century Brooklyn* (New York: Oxford University Press, 1999).

16. Morantz-Sanchez, *Sympathy and Science*; Ellen S. More, *Restoring the Balance: Women Physicians and the Profession of Medicine, 1850–1995* (Cambridge, MA: Harvard University Press, 1999).

17. Vietor, *A Woman's Quest*, 254.

18. Zakrzewska, "Report of One Hundred and Eighty-Seven Cases," 557-560.

19. Zakrzewska to Blackwell, March 21, 1891. Although this may have been a veiled reference to abortions as well, Zakrzewska's focus in this letter was clearly on ovariotomies, Blackwell Family Papers.

20. See Morantz-Sanchez, *Conduct Unbecoming a Woman*, chap. 4; Nancy M. Theriot, "Women's Voices in Nineteenth-Century Medical Discourse: A Step toward Deconstructing Science," *Signs*, 1993, *19*: 1–31; Ornella Moscucci, *The Science of Woman: Gynecology and Gender in England, 1800–1929* (Cambridge: Cambridge University Press, 1990); Deborah Kuhn McGregor, *Sexual Surgery and the Origins of Gynecology: J. Marion Sims, His Hospital, and His Patients* (New York: Garland, 1989).

21. Morantz-Sanchez, *Sympathy and Science*, 195. See also Bittel, "Mary Putnam Jacobi," Chapter 1 in this volume, and Morantz-Sanchez, "Female Patient Agency," Chapter 3 in this volume.

22. Elizabeth Blackwell, "The Influence of Women in the Profession of Medicine," in *Essays in Medical Sociology*, 2 vols. (New York: Arno Press, 1972; reprint of London: Ernest Bell, 1902), vol. 2, p. 9. See also See Regina Morantz-Sanchez, "Feminist Theory and Historical Practice: Rereading Elizabeth Blackwell," *History and Theory*, 1992, *31*: 59–60.

23. Zakrzewska to the interns, March 30, 1883, pp. 3–4, NEH Collection, Box 27, Folder 1173, Sophia Smith Collection, Northampton, Massachusetts.

24. The first two were the New York Infirmary for Women and Children (1857), and the Woman's Hospital of Pennsylvania (1861).

25. Sophia Jex-Blake to her mother, August 18, 1865, reprinted in Margaret Todd, *The Life of Sophia Jex-Blake* (London: Macmillan & Co, 1918), 164–166.

26. Vietor, *A Woman's Quest*, 253.

27. Zakrzewska to Sewall, November 29, 1862, cited in ibid., p. 302. Thanks to Judith Walzer Leavitt for helping me to understand Zakrzewska's maternal feelings toward some of her interns.

28. Front page, *Annual Report of the New-England Hospital for Women and Children for the Year 1865* (Boston: Prentiss & Deland, 1865), 3.

29. Front page, *Annual Report of the New-England Hospital for Women and Children for the Year 1891* (Boston: Prentiss & Deland, 1891), 4. Certainly, compared with an

operation like Massachusetts General, the New England Hospital was still small, but the growth was significant nonetheless.

30. These included New York's Bellevue Hospital, Mt. Sinai Hospital of New York, Cook County in Chicago, Philadelphia's Blockley Hospital, Boston's City Hospital, and Chicago's Wesley Memorial Hospital. See Bonner, *To the Ends of the Earth*, chap.7; Drachman, *Hospital with a Heart*, 108–109; Morantz-Sanchez, *Sympathy and Science*, 78, 164, 166; and Jacobi, "Women in Medicine," 189. Blockley had actually allowed Elizabeth Blackwell to spend a summer as an intern in 1849, and Sarah Dolley to spend eleven months as an intern in 1851, but thereafter it did not hire another woman until 1883. See More, *Restoring the Balance*, 22–23. Bear in mind that some of these hospitals allowed women to walk the wards before this date; however, that is different from offering a clinical internship.

31. Kenneth M. Ludmerer, *Learning to Heal: The Development of American Medical Education* (New York: Basic Books, 1985), 53–57; and Steven J. Peitzman, *A New and Untried Course: Woman's Medical College and Medical College of Pennsylvania, 1850–1998* (New Brunswick, NJ: Rutgers University Press, 2000), 38–44.

32. Interns to the Board of Physicians, October 12, 1891, in NEH Collection, Box 27, Folder 1173, Sophia Smith Collection, Northampton, Massachusetts.

33. Alice Hamilton to Agnes Hamilton, October 29, 1893, and December 6, 1893, in Barbara Sicherman, *Alice Hamilton, A Life in Letters* (Cambridge, MA: Harvard University Press, 1984), 71–72, 75.

34. Zakrzewska to the interns, October 30, 1891, 4. See also Vietor, *A Woman's Quest*, 376, 395, 426.

35. Morantz-Sanchez, *Sympathy and Science*, 60–61.

36. The quotation is from Sarah Deutsch, *Women and the City: Gender, Space, and Power in Boston, 1870–1940* (New York: Oxford University Press, 2000), 148. See also Lori D. Ginzberg, *Women and the Work of Benevolence: Morality, Politics, and Class in the 19th-Century United States* (New Haven, CT: Yale University Press, 1990).

37. Ellen More, "The Blackwell Medical Society and the Professionalization of Women Physicians," *Bulletin of the History of Medicine*, 1987, *61*: 603–628, esp. 605–606, and More, *Restoring the Balance*, 8–9. See also Morantz-Sanchez, *Sympathy and Science*, 144.

38. Kathleen Barry, *Susan B. Anthony: A Biography of a Singular Feminist* (New York: New York University Press, 1988), 200.

39. Barbara Sicherman, *Exploring the Dangerous Trades: The Autobiography of Alice Hamilton, M.D.* (Boston: Northeastern University Press, 1985), 267–270. See also Morantz-Sanchez, *Sympathy and Science*, 178.

40. One should refrain from romanticizing Zakrzewska's position. As Kate Hurd-Mead, who interned at the New England Hospital in 1888, pointed out, Zakrzewska's notion of stewardship "led her to watch even the spare time of the young doctors, and to denounce what she considered harmful contemporary literature, especially the novels of Tolstoi." See Kate Campbell Hurd-Mead, *Medical Women of America* (New York: Froben Press, 1933), 34. I am not, moreover, suggesting that Hamilton lacked a sense of ethical wrongdoing when she decided to leave the hospital. However, she was more concerned about what it would mean to break a contract than what it might mean for the institution.

41. Morantz-Sanchez, *Sympathy and Science*, 159–160, 234, 253–255; and Mary Roth Walsh, *"Doctors Wanted: No Women Need Apply:" Sexual Barriers in the Medical Profession, 1835–1975* (New Haven, CT: Yale University Press, 1977), chap. 6.

42. Zakrzewska's comment about Booth is from Marie Zakrzewska, "Mary L. Booth," *Woman's Journal*, 1889, *20* (April 6): 106. Her comment about the year is from Zakrzewska to Severance, February 14, 1890, Severance Papers, Huntington Library, San Marino, California.

43. Zakrzewska to Caroline Severance, September 8, 1889, Severance Papers.

44. Zakrzewska to Caroline Severance, September 10, 1889, Severance Papers. On Haeckel's monism, see Niles R. Holt, "Ernst Haeckel's Monistic Religion," *Journal of the History of Ideas*, 1971, *32*: 265–280.

45. Zakrzewska to Severance, September 10, 1889, Severance Papers.

46. Zakrzewska to Severance, September 11, no year but probably 1890, Severance Papers.

47. Zakrzewska to Blackwell, undated but probably 1900, National American Women's Suffrage Association Papers, Library of Congress, Washington, D.C.

48. Zakrzewska to Blackwell, March 21, 1891, Blackwell Family Papers.

49. Zakrzewska to Severance, February 27, no year, but probably 1881, Severance Papers. See, as well *Marie Elizabeth Zakrzewska: A Memoir* (Boston, 1903), 26.

50. *Marie Elizabeth Zakrzewska: A Memoir*, 24.

CHAPTER THREE

Female Patient Agency and the 1892 Trial of Dr. Mary Dixon Jones in Late Nineteenth-Century Brooklyn

Regina Morantz-Sanchez

In February and March 1892, Brooklyn, New York, was captivated by a sensational libel trial that pitted its largest newspaper, the *Eagle*, against Mary Amanda Dixon Jones, a gynecological surgeon of no small reputation. At issue was a series of lurid feature stories the newspaper had written about Dixon Jones's medical practice in the spring of 1889. Hinting first at deliberate financial mismanagement of her hospital, the Woman's Hospital of Brooklyn, the *Eagle* went on to paint Dixon Jones as an ambitious and calculating social climber and a knife-happy surgeon who forced unnecessary operations on innocent women and used organs removed from their bodies to advance her reputation in diagnosis and pathology. Journalists gleefully took up the story. Within Brooklyn, the articles set off an avalanche of public criticism, giving rise to two manslaughter charges and eight malpractice suits that made their way through the courts. It took almost two years to clear her name of all criminal and civil charges. When the ordeal was over, she retaliated with a suit against the *Eagle*, seeking $300,000 in damages.

The ensuing trial was the longest libel suit tried in the United States to date. Almost three hundred witnesses testified, including leading physicians, humble craftsmen and seamstresses, immigrants speaking only broken English, tradesmen and their wives, and former patients with babies in their arms. Spectators filled the courtroom day after day, riveted by the display of surgical mannequins and preserved ovaries in specimen jars, which had, seemingly overnight, become common sights in the courtroom. Here, the technical language of surgical praxis and cellular abnormality exposed observers to a newly emerging professional discourse and novel understandings of the human body. The *Brooklyn Citizen*, the *Brooklyn Times*, the *New York Times*, and the *New York World* kept readers informed of the proceedings with dramatic flair. The *Brooklyn Medical Journal* claimed that the event involved the "honor and reputation" of its medical establishment, while the *Philadelphia Ledger* hailed the case as "the most important . . . since the Beecher" scandal, a reference to a salacious adultery trial that pitted the husband of a congregant against a wildly popular Brooklyn-based evangelical minister in the early 1870s.

I have argued elsewhere that this remarkable public event illuminates a complex set of late nineteenth-century social and cultural developments. It represents a chapter in the social history of medicine that confirms the centrality of medical matters to late nineteenth-century American cultural life. It substantiates the emergence of new models of medical specialization, of professional identity, and of relations between doctors and patients. Class tensions and urban growing pains also shaped issues contested daily in the courtroom. Gender themes molded these other phenomena in subtle but specific articulations, gauging the successes and failures of women physicians and enabling frank and open discussions of female health in surprisingly public venues.[1]

Shadowing the controversy at every turn was the enigmatic figure of Mary Dixon Jones. A blatantly ambitious woman, in contrast to the women's networks cultivated by Marie Zakrzewska or Mary Putnam Jacobi's sophisticated social networking, Dixon Jones (deliberately or unknowingly) ignored the gender scripts dictating proper professional behavior for the woman doctor of her time. She was duly punished for her efforts. She lost everything when a male jury found the *Eagle* innocent of libeling her in 1892. The newspaper's campaign left her bereft of influential supporters in Brooklyn, and her hospital's charter was revoked. Defeated, she gave up her practice and moved to New York City, though she continued to publish on gynecological pathology. Exile from Brooklyn marked not only her ouster from a successful medical career built up painstakingly over thirty years but also from the practice of surgery and the perfection of new operative techniques that she had helped pioneer. A person ever

Mary Dixon Jones. Reproduced with permission from the New York Academy of Medicine Library.

desirous of a place in history, Dixon Jones might have considered the eventual elision of her career from the historical record the saddest injury of all.

By 1892, Dixon Jones had become a nationally and internationally known surgeon, a pioneer whose innovative operations and contributions to the cellular pathology of the female reproductive system were well known by male colleagues in the United States and abroad. She was perhaps the only woman physician of her cohort whose published work consistently and over a long period addressed the larger discussions and therapeutic debates taking place in the field of gynecological surgery. She had carefully negotiated this professional identity for herself at a time when boundaries, behaviors, and ideologies of practice were in flux. At the end of the century, fewer than a handful of women physicians had achieved such prominence. Thus, Mary Dixon Jones suffered ignominiously in 1892, partly because she was a woman, but mostly because she was a self-promoting woman who insisted on performing radical gynecological surgery with little inclination to defer to her more conservative male colleagues. Yet, before we conclude that the trial represents just another tired story of male professional power and female physicians' vulnerability, I want to propose a more complicated reading of this event by suggesting that not all women lost when Dixon Jones failed to win her case.[2]

Despite its disappointing outcome, Dixon Jones's libel suit briefly catapulted her remarkable career into the public limelight. For a larger group of women, however—Dixon Jones's real and potential patients—the event achieved much more. The publicity surrounding this drama of medical treatment, played out so graphically in the State Supreme Court of Brooklyn, helped to open a space for evaluating the emerging specialty of gynecological surgery and viewing women and their bodies in a new way. We need only glance at a contemporary newspaper to remind us that scandalous legal proceedings can imbue ordinary social life with dramatic meaning. In engaging a wide public audience, they occasionally lead to moments of cultural transformation and collective self-reflection. I want to suggest that Dixon Jones's libel case was just such an occurrence. On trial, among other things, was surgical therapy for women and its precise, scientific language of pathology and operative procedure. The courtroom became a forum for catalyzing and marking cultural change, while the public nature of its proceedings enabled ordinary women—those who participated in, witnessed, or read about the case—to claim a striking new sense of themselves as manageable bodies and as more effective persons.

Before I examine this claim in more detail, I will briefly review the major strands of medical discourse regarding women in the nineteenth-century Anglo-American world. From the 1830s to roughly the 1890s, sickness and health emerged as ubiquitous concerns of Victorian culture, and a rising faith in medical science prompted many in the middle class to seek medical treatment. Middle-class women became relatively frequent pursuers of health care, for their families as well as for themselves. They suffered from a variety of painful and debilitating ailments, including ovarian cysts; uterine infections; displacements and lacerations following childbirth; menstrual derangements, including profuse bleeding or the absence of periods; and tumors, fibroids, and various forms of cancer. Indeed, their illnesses served to confirm physicians' emphasis on a woman's reproductive apparatus as the driving physiological force of her being, a theory that had been gathering discursive power by virtue of an emerging biological paradigm emphasizing male-female difference that appeared, according to Thomas Laqueur, in the late 1700s.[3]

Increasingly, many practitioners felt justified in the belief that reproduction was fraught with danger and that female ill-health was widespread. When historian Wendy Mitchinson examined the patient charts of a Canadian woman's hospital from the year 1892, she found that attending physicians implicated the sex organs in disease causation even when illness and symptoms appeared totally unconnected, for example, when patients presented with gastritis or tuberculosis.[4] Clearly, for many physicians, one significant hallmark of womanhood in the nineteenth century was to experience "the pathology of femininity."[5] The English obstetrician William Tyler Smith characterized parturition as an event that stood "at the boundary between physiology and pathology." No severe constitutional imbalance, agreed the American gynecologist Arthur Edis, "can long continue in a woman during the predominance of the ovarian function without entailing disturbance . . . And the converse is also true, that disorder of the sexual organs cannot long continue without entailing constitutional disorder, or injuriously affecting the condition of other organs."[6] While the surgeon George Rohé acknowledged: "Among the general public and the medical profession the influence of abnormalities of the sexual organs in producing mental aberrations is . . . believed in to a considerable extent. Indeed, some of the highest authorities in mental diseases, such as Esquirol and Guislain, emphasized the overwhelming influence of the genital organs, especially in women, in the production of insanity."[7]

These examples suggest that many physicians were prone to conceive of female bodies more in the abstract than in the particular. Notions of women's passivity and resignation in the face of fierce biological realities and frequent physical suffering could retain powerful discursive force, especially during the early development of the specialty of gynecology, when practitioners had few treatments to offer. Even medical advertising in the popular press reinforced such views, and there is plenty of evidence that some female patients shared them.[8] Before antisepsis and without any means of alleviating pain, early gynecologists were reluctant to operate on women, especially those of the white middle class, who were assumed to be especially sensitive. Until the 1870s, when surgery became an option for treatment—albeit still a very risky one—most practitioners offered scant recourse, diagnosing serious female illnesses as incurable. They confined themselves to the repeated tapping and draining of fluid-filled sacs, removal of labial growths, minor plastic procedures in the form of perineal repairs, curetting, and topical treatments for a variety of inflammatory indications. Sponge tents and pessaries inserted at the mouth of the cervix were also favorite devices for handling uterine displacements.

Women's bodies remained abstractions to physicians in another sense as well. They were deployed, as Mary Poovey has shown, to perform "the ideological work of gender." The naturalized bourgeois Victorian family, with its model of separate spheres, helped to inscribe a host of new institutions and social relations, from shaping policies toward the poor, regulating prostitution, and rethinking divorce law, to new forms of literary expression and the development of social, political, and economic theory. Images of long-suffering, virtuous women—always conceived as mothers or potential mothers—accompanied by modest and unthreatening constructions of female desire were indispensable to emerging male capitalist social identities in both England and the United States. Morality became highly dependent on the new discourses of gender embedded in the language of domesticity. Virtue, now identified with femininity, sympathy, altruism, and social benevolence, found itself safely ensconced in the private realm of family life. According to Poovey, the new morality allowed Victorian society to have its cake and to eat it too. In linking ethical, empathetic, self-sacrificial behavior to the figure of Woman, "immune to the self-interest and competition integral to economic success," virtue could be preserved "without inhibiting productivity." Social order and confidence in a humane and stable society depended on believing that the images of home-oriented, maternal women represented reality.[9]

This ideology combined with medical discourse about the female body to exercise a powerful role in shaping late nineteenth-century definitions of femininity that constrained women and, as the essays by Bittel and Tuchman in this volume have discussed, limited their opportunities in a variety of ways. Nancy Theriot and others have demonstrated how women's discursive experience—the injunction to submission, self-sacrifice, and suffering—was written on the body. Theriot argues that in the first two-thirds of the nineteenth century, women, plagued with the physical effects of fertility control, abortion (which was widespread), and childbirth, learned to accept physical pain as an idealized symbol of their moral superiority and devotion to others.[10]

Scholars from a number of disciplines have enriched this picture by singling out invasive medical procedures used to treat women—such as cauterization, bloodletting, and clitoridectomies—which allegedly were animated by these ideas. Many have also underscored the orgy of gynecological surgery that took place in the last two decades of the century, arguing that conceptions of female biology sanctioned "transgressive" procedures, and that such operations endangered lives while cynically expanding opportunities for practitioners to try out a range of new therapeutic modalities and techniques.[11] In the past two decades, my own work and that of other medical historians have found these claims to be overdrawn. Though helpful in focusing our attention on the importance of evaluating medical practice through the lens of gender, they often paint too stark a picture of a more complicated interaction between patient care, medical theory, the medical marketplace, scientific and technological innovation, and professionalization.

Evidence from the Mary Dixon Jones libel trial can elaborate and extend these critiques. If we read it carefully, we can see that the emergence of surgical gynecology in the last third of the century gave female patients more agency in their struggle with illness and gradually enabled them to view their bodies differently. Eventually, they challenged the dominant cultural and medical discourses regarding female self-sacrifice and suffering. Surgical gynecology in the last third of the century may well have risked lives, but it also contributed to the fashioning of new representations of sick women, not solely as abstract bodies and moral/ideological constructs but as persons with a wider range of choice for coping with bad health. The publicity and public discussion the proceedings engendered present the historian with an opportunity to gauge this transition and explore its consequences, as forms of self-assertion and the means to validate women's newfound agency in health matters were made more visible. In addition, the event may have been transformative in that, by publicly marking this shift, it not only offered those women who participated as witnesses or observers a new perspective but generated a different image of women for public consumption.

Let me turn first to my claim that gynecological surgery altered the nature of medical discourse on women. It has been established that it was the female body that enabled surgeons to boldly explore the internal body cavity and experiment with a variety of surgical treatments that helped create the specialty. Many surgeons continued to believe in the overarching influence of women's reproductive organs on their general health. However, there are also countervailing developments to be noted. While it was primarily men who first used the authoritative language of science to speak and to write about women's bodies, the careers of Mary Dixon Jones and a handful of other prominent women physicians suggest that the absence of women's voices in medical practice was never as stark in the United States as it was in England or in Europe. Women learned to speak for themselves in America, first as physiological lecturers in the health reform movement, then as sectarian medical practitioners, and, finally, after midcentury, as regularly trained and professionally licensed physicians.[12] Women physicians participated enthusiastically in expanding gynecologists' role as monitors of female health and aided in broadening the specialty's authority among the lay public. There is clear evidence that women doctors had an effect on the views of their male co-specialists. By the end of the century, many male physicians conceded female colleagues a certain expertise where female health was concerned, while others felt it necessary to cite women doctors' opinions on the subject to bolster their own authority.[13]

In addition to the inroads women physicians were making in the profession, and despite the effusion of Comstockian reserve about sexual matters after the Civil War, specialization, technology, and new theories of disease generated notable changes in medical language. Terminology gradually became more "objective" and ostensibly more "neutral." This turn away from morally loaded metaphors facilitated an eventual decline in Victorian reticence. It enabled surgical innovators to reconnect with and reframe in scientific rather than in religious terms the longstanding interest of antebellum physiological reformers in understanding the anatomical body, a desire Michael Sappol has characterized as a ticket to "a more rational and elevated moral order." Sappol's emphasis on the complex relationship between anatomical understanding and emerging middle-class culture suggests that Victorian prudery regarding sexuality always existed in tension with an alternative view that held body knowledge to be emblematic of middle-class "self-mastery." In this trial, we can see these shifts at work.[14]

My examination of the medical literature on female health from the 1870s through the beginning of the twentieth century gauges this shift in content and focus. Medical treatises that used biology to speculate philosophically about proper female roles and behavior decreased markedly in favor of approaches typical of the more reductionist gaze of the new medical science. Debates about technique, diagnosis, and treatment replaced the extended social commentary of midcentury. As Nancy Stepan and Sander Gilman have shown, this shift in language and orientation is easily identified in the larger body of Western medical and scientific literature as a whole. Moral and political argument was marginal-

Mary Dixon Jones. Reproduced with permission from the New York Academy of Medicine Library.

ized as science's "conceptual categories, rhetorical styles, and methodologies were adopted."[15] Surgeons welcomed and perhaps even catalyzed this development, which meant highlighting diseased tissue and inflamed organs more often than pondering the social role of these body parts in diagnosis.

Dixon Jones embraced the use of pathology and the contributions laboratory science was making to clinical practice. Her writings can be linked to a younger group of gynecological surgeons, a cohort that stands in marked contrast to the generation of gynecologists whose careers spanned the beginnings of the specialty in the 1850s and peaked in the 1880s. This elder group of pioneers was aware of pathology and its importance to diagnosis and treatment, but when it came to issues of women's health, they speculated not only on strictly scientific matters, but also on the moral obligations of maternity, the necessity of preserving female sexual purity, and the significance of the incommensurability between male and female bodies. Men such as Thomas Addis Emmet (1828–1919), E. H. Clarke (1820–1877), and A. J. C. Skene (1837–1900) constructed the female body with reverence, carving out a safe and subordinate place for women in the home.[16]

Standing in marked contrast to the morally inflected scientific language characteristic of the early works of Skene and Emmet was the publication of a definitive volume edited by Howard A. Kelly and Charles P. Noble, two active gynecological surgeons in midcareer at the turn of the twentieth century, who helped to refine and solidify developments in the specialty in their early careers and gradually became dominant figures after 1890. Their textbook, *Gynecology and Abdominal Surgery*, published in 1907, opens a broad window onto the changes of the previous decades. Clearly, technique and diagnosis had made great strides. But what is most instructive about the volume is that it is absolutely devoid of the philosophical theorizing endemic to most gynecology texts only two decades before. Skene contributed an essay on "ovariotomy," and, though his death before the work was published evokes a kind of symbolic passage, even *his* language had changed. His chapter, like the other twenty-four, including the two penned by women physicians, is stark in its utter concentration on scientific data and surgical technique.[17]

The maturation of gynecological surgery also meant that surgeons could offer their patients real possibilities for cure. Techniques remained risky, to be sure, and in the 1890s, mortality rates for such operations were still roughly at 15 percent (down from 66 percent in the 1860s). But given the new therapeutic options, doctors needed a different kind of compliance from their patients, one likely to involve more communication and negotiation. This was especially necessary for invasive and dangerous procedures, though it is not clear that all surgeons understood these new imperatives or adopted them with alacrity. Hence, physician testimony at the libel trial reveals a wide range of opinion on the issue of informed consent. How much information should doctors share before surgery, and what was the relationship of new modes of communication to female agency?

Among its many allegations, the *Eagle* had accused Dixon Jones of a poor rapport with patients, though trial testimony was inconclusive. A. J. C. Skene insisted that the effects of surgery should be understood by "all patients except children and insane people." He was "especially careful to make himself understood by an ignorant woman." Several others expressed discomfort with Dixon

Jones's sometimes cavalier attitude toward husbands, and Skene observed that "he didn't think you could find a surgeon who would operate on a woman if her husband objected." Two key witnesses on her behalf, radical surgeons Gill Wylie and Abel M. Phelps, offered unusually candid observations. Their remarks contrasted with the consensus on communicating with patients, which was discussed in medical journals. Phelps admitted that he rarely discussed procedures with "extremely nervous women incapable of understanding a scientific explanation." Nor was it usual for him "to explain beforehand to ignorant patients" because "they might be frightened off the operating table" if told what would happen to them. Gill Wylie agreed. "The danger from laparotomies [has] now become so slight," he declared, "that much less ceremony was observed than there used to be, about obtaining consents, securing assistants, and the like."[18] Although our current understanding of informed consent was not in place until well into the twentieth century, surgeons were among the first to debate it, according to courtroom testimony. They recognized the likelihood that they would be sued for malpractice more often than general practitioners and other specialists.[19]

Surgeons might have been motivated to speak frankly with patients because their theories of disease causation became rapidly oriented toward the local and the specific, in contrast to the systemic and holistic approaches of the general practitioner. It is possible that these gynecologists quickly learned to explain that pain and discomfort were caused by limited infection in particular organs, rather than by the biological and teleological status of womanhood and that such organs could be removed, with positive results. The ability to bear children might well be impaired, but even a discussion of that eventuality made it possible for doctor and patient to negotiate intelligently about the relative merits of wellness over motherhood.[20]

Finally, because in the last third of the nineteenth century, gynecologists did not yet agree on the efficacy of surgery in treating female illness, women patients were compelled to become active agents in choosing a physician; indeed, their choices often turned on which practitioner was willing to explain a therapeutic course of action clearly and most to their liking.

We discern some of these shifts when we examine what Mary Dixon Jones's patients had to say about the adequacy of her communication with them. I have already observed that trial testimony was contradictory. For example, Mary Webster claimed that she wasn't well informed about the nature of the operation she received and learned about it from a fellow patient. Margaret Fisher testified that she never authorized her operation and might not have done so had she realized it would render her incapable of having children.[21] But others were quite satisfied with preoperative explanations and clearly felt the decision for surgery was collaborative. Dixon Jones told Mary Huck that the tumor to be removed was the "size of an orange"; while Clara Hartisch explained that her two pus-filled tumors were the size and shape of a goose egg. "Dr. Jones removed these tumors and cured me entirely. She was very good to me, as she was to all patients." John Bruggeman, whose wife eventually underwent surgery, admitted that he could not remember the details of the first interview with Dixon Jones because "my wife did all the talking." Perhaps most eloquent of all was Mina Emmerich, who testified that when Dixon Jones cautioned her that surgery might not save her, she responded that she wanted anything that would relieve her, "dead or alive."[22]

What these accounts reveal about Dixon Jones's bedside manner is less important for our purposes than that informed conversations between doctors and patients were valued positively by all the women who testified at this trial. Physicians who served as witnesses disagreed about how much information to impart—just as they do today—but the tone of the discussion in the courtroom suggests that patients expected a minimum standard of negotiation and information gathering to occur during the encounter. Many doctors confessed to being wary of such frank discussions, but even the most reluctant understood their usefulness in deterring malpractice suits. Articles in contemporary medical journals echoed these courtroom debates.[23]

There is also much to be gleaned from this evidence about power relations between these patients and their doctors. In much of the historiographical literature on male physicians and female patients about this period, an implicit assumption exists that sick women are brought to the doctor primarily by husbands, family members, or friends. Carroll Smith-Rosenberg's insightful study of hysteria, for example, presents us with the image of a woman ceasing to function within the family, taking to her bed, and forcing others to assume the role of wife, mother, or daughter. "Worry and concern bowed the husband's shoulders," we are told. "His home had suddenly become a hospital and he a nurse."[24] In Joan Brumberg's *Fasting Girls*, parents—most often the mother—seek medical advice and treatment for recalcitrant daughters.[25] Finally, Mary Poovey's exploration of the "silenced female body" and its relationship to professional struggles in midcentury England represents female patients as passive participants. Doctors "enter" the birthing room; they "attend" the hysteric but exactly who authorizes their presence remains a mystery. What seems lacking in Poovey's account is the possibility that physicians might actually have been *invited* to deliver treatment by the sufferers themselves.[26]

In contrast, trial testimony matched up with Dixon Jones's published case records provide some fascinating information about illness complaints and treatment. We hear of the nature of patients' physical symptoms, what they felt about their bodies and their illnesses, their expectations from doctors, patterns of patronage, and how Dixon Jones and several of her male colleagues presented themselves to their clients. We learn that middle-class women were not dependent on their doctors for information about their own bodies. Apparently, they kept carefully abreast of developments in gynecology and were familiar with the possibilities of surgical intervention. In urbanized Brooklyn, where proximity made frequent communication possible, the material is strikingly abundant. Female patients' bodies in this city were far from silenced; on the contrary, sick women attended to physical signs and symptoms and monitored their illness experiences with self-assertive determination. They sought the advice of numerous practitioners to mollify their enduring pain and chronic discomfort. If they deemed treatment ineffective, they readily moved on, demonstrating no enduring loyalty to a specific doctor or a particular mode of therapy.[27]

Patients felt justified in seeking another opinion because consensus on how to treat gynecological ailments, even among specialists in the field, did not exist. Most women sought to mitigate disabling symptoms and restore functionality and competence to their lives. The testimony at Dixon Jones's trial reveals that in some cases women consulted with more than twenty physicians before they appeared at her office. The dictates of consumer culture were exercised by many in their quest for effective medical care. It was, after all, primarily women who composed the clientele of department stores in this period, and shopping was coded a feminine pursuit in the late nineteenth century. Becoming the primary purchasing agents for their families aided women in viewing themselves as responsible household managers, and they took pride in their ability to shop around. Why should not the same rules of consumption apply to medical care? A lively medical marketplace existed in Brooklyn and New York City at the end of the nineteenth century, and the business of gynecology was apparently booming.[28]

Mrs. Alfred Strome, for example, told both the Brooklyn *Eagle* and the *Citizen* in 1889 that she had been sick for fourteen years before she consulted Dixon Jones. She sought the opinion of Dr. Westbook, who had advised against surgery on the grounds that it would kill her. "Not satisfied with this," Strome recalled, "I went to several other physicians in the city and elsewhere, but with the same negative result." "Having abandoned all hope of a radical and permanent cure" she was eventually convinced by a friend to visit Dixon Jones. The doctor removed Strome's abdominal tumors and cured her.[29] Another patient Dixon Jones treated in 1884, about whom she published a case record, had tried twenty-one doctors before she contacted Dixon Jones, and a third patient, whose discomfort was so acute that she threatened suicide, confessed to having appealed to "thirty different physicians."[30]

Some of the best specialists in New York City and Brooklyn had examined Mrs. A. E. Scholtz, including "Dr. Skene, Dr. Thomas, Dr. Fowler and others." None had suggested surgery. Her friends urged her to seek out Dixon Jones, who successfully operated on her for a tumor.[31] Victoria James, an African American patient, confessed to seeing "six doctors before I went to . . . [Dixon Jones's] hospital and none of them did me any good."[32] Mary Vibert had fourteen doctors in as many months come to her house.[33] When Mary Gearon was asked on the witness stand what Dixon Jones had told her about the nature of her surgery, she replied, "Well, I can't remember, I have been through so many doctors since."[34] Similarly, Mrs. Frances Stroble came to Dixon Jones's Madison Avenue dispensary in 1887, after being ill for twenty years. "She had been to all the doctors she could get at," she said, according to the *Eagle*, "but they did her no good."[35] Finally, Mrs. Mina Emmerich, who had pains in her lower abdomen, had seen at least five different practitioners from 1883 to 1887.[36]

The publicity and spectacle surrounding the trial made public these informal networks of women taking an active part in managing their own health care. The paper's account having rendered their agency normative no doubt encouraged others, especially those who displayed a keen interest in the event. From the trial's opening days, newspapers remarked on the unprecedented attendance of women at the proceedings. "The Jones case has begun to exert a morbid fascination upon the women of the city," the Brooklyn *Eagle* observed on February 12, 1892. "Most of them have been present at nearly every session of court since the trial of the case began, and yesterday afternoon the back part of the courtroom was pretty well filled with alert feminine faces, eagerly stretched upward and forward as the women on the witness stand told the pitiful story of their sufferings." The newspaper argued that it was "only natural" that there should be a greater proportion of women in attendance than was usual at "courtroom crowds," noting that the testimony appealed "to the sympathy of women with peculiar force."[37] For the most part, this was a middle-class assemblage, "well-dressed," "refined," and "intelligent."[38] Why were they there?

The answer, at least in part, is that the female spectatorship authorized by this event, along with the changing nature of the doctor-patient encounter enabled by gynecological surgery, affirmed women's new perceptions of their own bodies and offered them an effective forum for vicarious sharing, information gathering, and even for voyeurism. For the middle-class and poor immigrant women who came to court to speak in public about the burdensome experience of illness, the trial provided an unusual public venue to air disappointments and complaints about the female condition, or even recount small triumphs, in a manner that attracted a sympathetic and captive audience.

We have already seen that gynecological surgery took on a special urgency at the end of the nineteenth century, especially as advances in medical technique were increasingly able to turn ill women's "wants," namely, the desire for good health, into "needs," the right to demand it from their doctors. In a secularizing culture that increasingly equated medical science with progress and began to question traditional religious notions of the moral necessity of gendered suffering, physicians with the technological skill to alleviate women's ill health were at a premium.

This was evidenced no more readily than in changing women's attitudes toward their own bodies. The trial affords a rare opportunity to witness such conversations taking place both outside of professional medical journals, to which laywomen had little access, as well as away from the informal encounters of female friendship networks, which were private. In this courtroom, the female body (and its parts) could be held up to full public view without incident. It could be discussed scientifically by medical experts while a decorous but inquisitive audience looked on.

This chapter from the history of medicine and health care, read in combination with several new studies of late nineteenth-century female subjectivity from a variety of disciplines, suggests that, by the close of the Gilded Age, expectations of passivity and the sequestration of Victorian women in the chaste and modest privacy of the domestic sphere was rapidly coming to an end.[39] A confluence of factors came together—social, political, economic, and cultural—to bring about such change, and it would take another seventy-five years for women patients to find ways to institutionalize their self-advocacy in ways that forced physicians to take it for granted.[40]

Sensational public trials always tell at least two stories at once. The first belongs primarily to the characters involved. The second, perhaps even more compelling and instructive than the first, mirrors the wider historical context in which the trial takes place. Fortunately, by providing evidence that women were becoming not just the objects of surgical intervention but conscientious consumers of new health treatments, Mary Dixon Jones's libel suit opens a window onto an important societal transition.[41]

NOTES

1. Readers interested in an extensive analysis of the meaning of this event can consult the author's book, Regina Morantz-Sanchez, *Conduct Unbecoming a Woman: Medicine on Trial in Turn-of-the-Century Brooklyn* (New York: Oxford University Press, 1999).

2. In fact, the gender discrimination that Dixon Jones experienced was complex and contradictory. Although by the end of the century women had made great strides in medicine, they were expected to behave in a ladylike fashion, defer to their male peers, and display a more nurturing and sympathetic bedside manner, even when they used the same treatment modalities. Dixon Jones's style was neither collegial nor deferential, and this rankled not only male but also female colleagues in Brooklyn. Three Brooklyn women physicians testified against her, and there is evidence that several thought she was not setting a good example for proper female professional comportment. Though she was not the only woman physician who performed gynecological surgery and argued in favor of surgical excision as the most effective treatment for pelvic disease, the public still understood surgery to be a "masculine" specialty. Dixon Jones's behavior at the trial reinforced the perception among many that she was not sufficiently feminine, while her unbridled enthusiasm for her surgical work and for laboratory science in general raised the specter of "science gone mad." Male physicians were permitted to demonstrate this interest openly, but listening to a *woman* speak without flinching of diseased ovaries and damaged fallopian tubes apparently troubled many. Dixon Jones's reluctance to enact a generic, socially scripted femininity was not lost on the all-male jury. For a more elaborate discussion of the issues of male and female professionalism and the relationship of gender to medical treatment, see "Doctors and Patients: Gender and Medical Treatment in Nineteenth-Century America," chap. 8 in my book *Sympathy and Science: Women Physicians in American Medicine* (New York: Oxford University Press, 1985; Chapel Hill: University of North Carolina Press, 2000); Morantz-Sanchez, "Negotiating Power at the Bedside: Nineteenth Century Patients and Their Gynecologists," *Feminist Studies*, 2000 (Summer): 287–309, and idem, "The Gendering of Empathy: Late-Nineteenth Century Medical Practice and Ideas of the Good Physician," in *The Empathic Practitioner: Empathy, Gender, and Medicine*, ed. Ellen Singer More and Maureen A. Milligan (New Brunswick, NJ: Rutgers University Press, 1994), 40–58.

3. For emerging theories of male-female incommensurability, see Thomas Laqueur, *Making Sex: Body and Gender from the Greeks to Freud* (Cambridge, MA: Harvard University Press, 1990). For the increased faith in medical science among women, see Regina Morantz, "Making Women Modern: Middle-Class Women and Health Reform in the 19th Century," *Journal of Social History*, 1977, *10*: 490–507.

4. Wendy Mitchinson, *The Nature of Their Bodies: Women and Their Doctors in Victorian Canada* (Toronto: University of Toronto Press, 1991), 47.

5. Ornella Moscucci, *The Science of Woman: Gynecology and Gender in England, 1820–1929* (Cambridge: Cambridge University Press, 1990), 102.

6. W. T. Smith, "Lectures on Parturition, and the Principles and Practice of Obstetricy," *Lancet*, 1848, *2*: 119; Arthur Edis, *Diseases of Women: A Manual for Students and Practitioners* (Philadelphia: 1882), 20; Robert Barnes, "Women, Diseases of," in *A Dictionary of Medicine: Including General Pathology, General Therapeutics, Hygiene, and the Diseases Peculiar to Women and Children*, ed. R. Quain, 2 vols. (London, 1882), II, p. 1790. Quotes from Barnes and Smith are cited in Moscucci, *The Science of Woman*, 102.

7. George H. Rohé, "The Relation of Pelvic Disease and Psychical Disturbance in Women," *Transactions of the American Association of Obstetricians and Gynecologists*, 1892, *5*: 320–348, 321.

8. Sally Shuttleworth has shown that ads for female pills assumed a "direct continuity between the operation of the menstrual cycle and mental health." Shuttleworth, "Female Circulation: Medical Discourse and Popular Advertising in the Mid-Victorian Era," in *Body/Politics: Women and the Discourses of Science*, ed. Mary Jacobus, Evelyn Fox Keller, and Sally Shuttleworth (New York: Routledge, 1990), 47–68. Andrew Scull and Diane Fabreau make this point as well in "'A Chance to Cut Is a Chance to Cure': Sexual Surgery for Psychosis in Three Nineteenth-Century Societies," *Research in Law, Deviance, and Social Control*, 1986, *8*: 3–39, 17. See also Nancy Theriot, "Women's Voice in Nineteenth-Century Medical Discourse: A Step toward Deconstructing Science," *Signs*, 1993, *19* (Autumn): 1–31, and her remarks about the physical roots of the nineteenth-century ideology of suffering in *Mothers and Daughters in Nineteenth-Century America: The Biosocial Construction of Femininity* (Lexington: University Press of Kentucky, 1999), 40–61; Regina Morantz-Sanchez, *Sympathy and Science*, 215–216. Also interesting in terms of patient attitudes is Edward Shorter, *From Paralysis to Fatigue: A History of Psychosomatic Illness in the Modern Era* (New York: Free Press, 1992), chaps. 2, 3. See also Mitchinson, *The Nature of Their Bodies*, 50, who suggests that the medical perception of female ill health gained credibility at least partly because it reinforced what people already believed. Of course medical ideas both reflected as well as helped shape ideas about women.

9. Poovey, *Uneven Developments: The Ideological Work of Gender in Mid-Victorian England* (Chicago: University of Chicago Press, 1988), 1–23, quotation p. 10. See also Ruth Bloch, "American Feminine Ideals in Transition: The Rise of Moral Motherhood, 1785–1815," *Feminist Studies*, 1978, *4*: 101–126. Also helpful in charting the reframing of female sexuality in the eighteenth and nineteenth centuries is Nancy Cott, "Passionlessness: An Interpretation of Victorian Sexual Ideology, 1790–1850," *Signs*, 1978, *4*: 219–236.

10. See Theriot, *Mothers and Daughters*, 65; also see Carla Bittel, "Mary Putnam Jacobi and the Nineteenth-Century Politics of Women's Health Research," Chapter 1 in this volume, and Arleen Marcia Tuchman, "Maternity and the Female Body in the Writings of Dr. Marie Zakrzewska, 1829–1902," Chapter 2 in this volume.

11. See, for example, G. J. Barker-Benfield, *The Horrors of the Half-Known Life: Male Attitudes toward Women and Sexuality in Nineteenth-Century America* (New York: Harper & Row, 1976); Elaine Showalter, *The Female Malady* (New York: Pantheon, 1985); Mary Poovey, "'Scenes of an Indelicate Character': The Medical 'Treatment' of Victorian Women," *Representations*, 1986, *14*: 137–168. For a different view, see Jane E. Sewell, "Bountiful Bodies: Spencer Wells, Lawson Tait, and the Birth of British Gynaecology" (Ph.D. diss., Johns Hopkins University, 1991), 290. The word "transgressive" is Sewell's, but she is much more careful to place developments within a detailed understanding of

medical theory and practice, highlighting exigencies other than the desire to keep women in their place that might have motivated physicians to attempt such procedures. Her work is complex and rewarding.

12. Morantz-Sanchez, *Sympathy and Science.* See also Martha H. Verbrugge, *Able-Bodied Womanhood: Personal Health and Social Change in Nineteenth-Century Boston* (New York: Oxford University Press, 1988); and Susan E. Cayleff, *Wash and Be Healed: The Water-Cure Movement and Women's Health* (Philadelphia: Temple University Press, 1987).

13. Morantz-Sanchez, *Conduct Unbecoming a Woman,* 119–126.

14. See Morantz-Sanchez, *Conduct Unbecoming a Woman,* esp. chaps. 4 and 5, and idem., "Feminist Theory and Historical Practice: Rereading Elizabeth Blackwell," *History and Theory,* 1992, *31*: 51–69. For an excellent summary of Comstockian efforts to suppress various forms of sex radicalism and commercial exploitation, see Helen Lefkowitz Horowitz's *Rereading Sex: Battles over Sexual Knowledge and Suppression in Nineteenth-Century America* (New York: Vintage Books, 2002). Michael Sappol's *A Traffic in Dead Bodies: Anatomy and Embodied Social Identity in Nineteenth-Century America* (Princeton, NJ: Princeton University Press, 2002), 189, provides an interesting counterpoint to Horowitz. On the complicated role of the language of professionalism in securing client loyalty at the end of the century, see Joanne Brown, "Professional Language: Words That Succeed," *Radical History Review,* 1986, *34*: 33–52. For women's role in anatomical instruction, see Regina Morantz-Sanchez, "Making Women Modern: Middle Class Women and Health Reform in 19th-Century America," *Journal of Social History,* 1977, *10*: 490–507.

15. Nancy Stepan and Sander Gilman, "Appropriating the Idioms of Science: The Rejection of Scientific Racism," in *The Bounds of Race,* ed. Dominick LaCapra (Ithaca, NY: Cornell University Press, 1991).

16. See Morantz-Sanchez, *Conduct Unbecoming a Woman,* chaps. 4, 5.

17. Howard Kelly and Charles P. Noble, eds. *Gynecology and Abdominal Surgery* (Philadelphia: W. B. Saunders, 1907). The women physicians who were invited to write for the volume were Anna Fullerton, a surgeon at the Woman's Hospital in Philadelphia for a portion of her career, and Elizabeth Hurdon, a Johns Hopkins M.D. and a protégée of Howard Kelly.

18. For a more extended discussion of the role of informed consent and its role in the trial, see Morantz-Sanchez, *Conduct Unbecoming a Woman,* 187–188. For whether women physicians were indeed better communicators and exhibited more caring, see Morantz-Sanchez, *Sympathy and Science,* 203–231, and "The Gendering of Empathy: Late-Nineteenth Century Medical Practice and Ideas of the Good Physician," in *The Empathic Practitioner,* More and Milligan.

19. See especially James Mohr, *Doctors and the Law* (New York: Oxford University Press, 1993); Kenneth De Ville, *Medical Malpractice in Nineteenth Century America* (New York: New York University Press, 1990); and Regina Morantz-Sanchez, *Conduct Unbecoming a Woman,* 197–203.

20. Morantz-Sanchez, *Conduct Unbecoming a Woman,* chap. 5.

21. Brooklyn *Eagle,* February 11, 10, 1892.

22. Brooklyn *Citizen,* June 10, 1889; *Eagle,* February 12, 1892; *Citizen,* June 10, 1889; *Eagle,* February 24, 1892.

23. Morantz-Sanchez, *Conduct Unbecoming a Woman,* 272 n. 80.

24. "The Hysterical Woman: Sex Roles and Role Conflict in Nineteenth-Century America," *Social Research*, 1972, *39*: 652–678, 208.

25. Joan Jacobs Brumberg, *Fasting Girls: The Emergence of Anorexia Nervosa as a Modern Disease* (Cambridge, MA: Harvard University Press, 1988), 140–163.

26. Mary Poovey, "'Scenes of an Indelicate Character': The Medical Treatment of Victorian Women," *Representations*, 1986, *14*: 137–168, 148, 152, 153. This prevailing view stands in contrast to Judith Walzer Leavitt's revisionist account of women's continued active negotiation in the management of childbirth throughout much of the nineteenth century. See her *Brought to Bed: Childbearing in America, 1750–1950* (New York: Oxford University Press, 1986).

27. Regina Morantz-Sanchez, "Negotiating Power at the Bedside: Historical Perspectives on Nineteenth-Century Patients and Their Gynecologists," *Feminist Studies*, 2000, *26*: 287–309. Nancy Theriot argues that daughters coming of age in the 1880s were less willingly resigned than their mothers' generation to the pain and suffering expected as a matter of course to accompany the female reproductive life cycle. See Theriot, *Mothers and Daughters*, 80–82. Her findings underscore Edward Shorter's argument that, in contrast to the traditional patient who had a high pain threshold and greater tolerance for chronic illness, the modern patient displayed greater sensitivity to the body's vicissitudes and sought advice from experts more quickly. See Edward Shorter, *Bedside Manners: The Troubled History of Doctors and Patients* (New York: Simon and Schuster, 1985), 61. The fact that women moved from doctor to doctor would not surprise modern medical sociologists, who have found that, although women may appear passive in the doctor's office, they take a more active role in managing their health outside the formal clinical setting. Women are still likely to change their doctors more often than men. Some scholars argue that they also turn more readily to "unconventional methods of healing." In short, writes Alexandra Dundas Todd of present-day female attitudes, "Women use medical care more than men, perhaps because of a more help-oriented socialization and their need of health services for reproductive care as well as disease. The literature . . . illustrates that in seeking this care, women report many dissatisfactions. There is no evidence, however, that these dissatisfactions are voiced to doctors. In fact, women's complaints, in general, seem to take the form of silent rebellion, such as noncompliance and changing practitioners, rather than direct confrontation." See Alexandra Dundas Todd, *Intimate Adversaries: Cultural Conflict between Doctors and Women Patients* (Philadelphia: University of Pennsylvania Press, 1989), 40–41.

28. On women, shopping, and consumer culture, see Elaine Abelson, *When Ladies Go A-Thieving: Middle-Class Shoplifters in the Victorian Department Store* (New York: Oxford University Press, 1989); William Leach, *Land of Desire: Merchants, Power, and the Rise of a New American Culture* (New York: Vintage Press, 1993). Two other books in the medical field that suggest women's increasing interest in taking responsibility for health and disease was stimulated by the development of consumer culture and advertising campaigns targeting them are Nancy Tomes, *The Gospel of Germs: Men, Women, and the Microbe in American Life* (Cambridge, MA: Harvard University Press, 1998), and Rachel Maines, *The Technology of Orgasm: "Hysteria," the Vibrator, and Women's Sexual Satisfaction* (Baltimore: Johns Hopkins University Press, 1999), esp. 82–110. By the 1930s, the American Cancer Society had organized the Women's Field Army, which enlisted volunteers from the General Federation of Women's Clubs to mount an early detection campaign for

breast cancer. See Ellen Leopold's chapter on the development of women's breast cancer awareness and its relationship to consumer advertising in Ellen Leopold, *A Darker Ribbon: Breast Cancer, Women, and Their Doctors in the Twentieth Century* (Boston: Beacon Press, 1999), 153–214.

29. Brooklyn *Eagle*, May 11, 14, 1889; *Citizen*, June 20, 1889.

30. See Mary Dixon Jones, "Another Hitherto Undescribed Disease of the Ovaries. Anomalous Menstrual Bodies," *New York Medical Journal*, 1890 (May 10), *51*: 511–551, 542–542; idem, "Diagnosis and Some of the Clinical Aspects of Gyroma and Endothelioma of the Ovary," *Buffalo Medical and Surgical Journal*, 1892 (November), *32*: 197–214, 209.

31. *Eagle*, February 13, 1892. See also May 19, 1889. Though parts of Scholtz's story are contradictory, her incessant search for the perfect doctor is not.

32. *Citizen*, February 24, 1892.

33. *Eagle*, February 26, 1892.

34. *Eagle*, February 9, 1892.

35. *Eagle*, February 25, 1892.

36. *Citizen*, June 10, 1889. See also the following cases: Mrs. Bruggeman, *Eagle*, May 7, 1889, *Citizen*, February 12, 1892; Mrs. Hulten, *Eagle*, May 11, 1889; Mrs. Fisher, *Eagle*, May 14, 1889, February 10, 1892, *Citizen*, February 10, 1892; Mrs. Nash, *Eagle*, May 17, 1889, *Eagle*, February 12, 1892; Miss Olsen, *Eagle*, June 14, 1889; Ida Hunt, *Eagle*, May 31, 1889, February 17, 1892; Mrs. Rettinger, *Eagle*, February 13, 15, 1892; Mrs. Clara Hartisch, *Citizen*, June 10, 1889; Margaret Walsh, *Citizen*, June 10, 1889; Miss Hattie Coulson, *Citizen*, June 20, 1889; Mrs. Maggy Laklbreunar, *Eagle*, February 24, 1892; Charlotte Mason, *Eagle*, Feburary 25, 1892. Dixon Jones's case records tell exactly the same story.

37. *Eagle*, February 12, 1892.

38. *Eagle*, February 3, 4, 1892.

39. Along with Theriot's insights about mothers and daughters, Jane Hunter's marvelous study, beginning in the 1880s, charts the role of coeducational public schools in producing young women with different expectations and more ambitious self-understandings of female possibilities in the world. This is the type of young woman who was likely to own her body more readily. See Jane H. Hunter, *How Young Ladies Became Girls: The Victorian Origins of American Girlhood* (New Haven, CT: Yale University Press, 2002). Also helpful is Kathryn R. Kent, *Making Girls into Women: American Women's Writing and the Rise of Lesbian Identity* (Durham, NC: Duke University Press, 2003).

40. A wonderful example of how self-advocacy evolved over time can be found in comparing breast cancer patient Barbara Mueller's relationship to her doctor, William Stewart Halsted, in the early 1920s, with Rachel Carson's management of her own breast cancer, with the help of Dr. George Crile Jr. in the early 1960s. See Leopold, *A Darker Ribbon*.

41. To follow up on the theme of women's medical consumerism beyond the turn of the century would be another study entirely, which I urge other researchers to take up. In the meantime, Judy Leavitt's wonderful discussion of twilight-sleep delivery in *Brought to Bed* is a good start. Ellen Leopold's *A Darker Ribbon* is also excellent on patient advocacy. See also Sandra Morgen, *Into Our Own Hands: The Women's Health Movement in the United States, 1969–1990* (New Brunswick, NJ: Rutgers University Press, 2002), and Jule Dejager Ward, *La Leche League: At the Crossroads of Medicine, Feminism, and Religion* (Chapel Hill: University of North Carolina Press, 2000).

CHAPTER FOUR

A Chinese Woman Doctor in Progressive Era Chicago

Judy Tzu-Chun Wu

On a gloomy Thanksgiving Day in 1916, twenty-seven-year-old Margaret Jessie Chung arrived in Chicago via the Santa Fe Railroad. Just that summer, she had graduated from the University of Southern California's College of Physicians and Surgeons in Los Angeles. Women from China had obtained medical training in the United States before Chung; in fact, the first Chinese female physician graduated from the Woman's Medical College in New York in 1885, four years before Chung's birth.[1] However, Chung was the first known woman of Chinese ancestry born in the United States to attain her medical degree. At the time of her graduation, however, she received very little recognition for her accomplishment. Mainstream Americans, especially in California, tended to conflate Asian ethnicity with Asian nationality. They regarded individuals of Chinese ancestry as foreigners and not members of the American polity.

Chung, in fact, had never visited her parents' birth land and had never even left her native California. She must have enjoyed her eastward journey, which took her across the majestic Rocky Mountains, the golden mesas of the Southwest, and the expanse of the central plains before depositing her in the burgeoning midwestern metropolis of Chicago. Nevertheless, the brisk autumnal weather in the windy city caught her by surprise. In Chung's unpublished autobiography, written approximately four decades later during and after World War II, she recalled: "When I left California, the sun was shining brightly, the lawns . . . were green, the poinsettias were as high as the eaves on the [Santa Fe Railroad] hospital, and a brilliant red. When I got to Chicago, it had snowed the day before, but a very black coal dust had settled over the snow, and it was frozen hard, icy cold. All I had on was a thin suit and a twelve dollar overcoat, which as you may guess provided very little warmth."[2]

What motivated Chung to leave the Edenic California landscape for the gray harshness of Chicago? In her home state, she had been unsuccessful in obtaining an internship, a position increasingly necessary to launch a career.[3] As modern medicine promoted the development of specialties and hospital-based health care, recently graduated physicians were expected to serve as unpaid interns before being appointed as paid residents in medical institutions. After gaining experience, establishing contacts, and earning the requisite resources, physicians could then establish their own practices. Although Chung attended a coeducational school and crossed the color line in her class cohort, she continued to face both gender and racial barriers to obtaining an increasingly coveted internship.

Most hospitals in the late nineteenth and early twentieth centuries denied female and black doctors the opportunity for postgraduate training.[4] The only other woman in Chung's class did not secure an internship by graduation either, even though she won a prize for achieving the highest scores in the Gynecology and Abdominal Surgery final. Chung's classmate eventually served as the first female intern at the Los Angeles County Hospital, the site of their clinical training in medical school; however, the hospital only accepted its first African and Asian American interns in the 1920s.[5] After all, California boasted not only glorious weather but also a long history of anti-Asian discrimination. Nativist forces in the state had sparked a national campaign that culminated in the Chinese Exclusion Act of 1882, which barred the entry of Chinese laborers and denied all Chinese immigrants the right to American citizenship. Within California, the anti-Chinese movement passed a series of laws that restricted the economic livelihoods, places of residence, and marriage partners of Chinese people and sometimes enforced these restrictive social practices through violent means.[6]

In the Midwest, home to far fewer residents of Chinese ancestry, Chung was offered the opportunity for postgraduate training. She completed her internship and residency at institutions that were engaged with Progressive reform. That is, the medical personnel who instructed her in surgery and gynecology at the Mary Thompson Hospital and in psychiatry at the Illinois Juvenile Psychopathic Institute (JPI) shared a passion for "civic medicine."[7] They sought to use their scientific expertise and worked in conjunction with the state to address pressing social problems, particularly those that resulted from rapid industrial growth and massive immigration during the turn of the twentieth century. Although Chung returned to California after just two years, her brief medical career in Chicago reveals the opportunities available to as well as the constraints facing a woman of color in Progressive medical institutions.

This biographical portrait of Margaret Chung during her internship and residency in the Midwest illuminates broader historical trends regarding the significance of gender, race, religion, and sexuality in the medical profession. First, her career path provides insight into the maternalist nature of Progressive women's reform movements during the early twentieth century and their connection with earlier Christian-inspired forms of female activism at both the local and international levels. On the one hand, scientifically trained women, like Mary Putnam Jacobi, increasingly challenged the assumptions of nineteenth-century separate spheres ideology, which assigned women to the realm of home and family. On the other hand, progressive reformers nevertheless continued to use the beliefs regarding women's innate nurturing and maternal instincts to argue for a uniquely female role in improving the broader society. In fact, as the career of Marie Zakrzewska demonstrates, women physicians evolved and sometimes vacillated in their thinking regarding gender identity and the medical profession. Chung's time in Chicago also suggests continuities as well as contrasts between West Coast and Midwest forms of racialization and reveals the possibilities for bridging racial and religious divides in a period of rising nativism. Strikingly, all of her mentors, male and female, came from immigrant backgrounds and tended to be Jewish. Although they traced their ancestry to Europe rather than Asia, these physicians occupied a similar, but certainly not equivalent, "outsider" status as Chung due to their gender or ethnicity. Finally, Chung's encounters with medicalized forms of discipline demonstrate the often masked but pervasive significance of sexuality for the professional identity of a female doctor. The scant evidence of her private life during this time suggests that her behavior increasingly came under institutionalized and medicalized scrutiny. As a Chinese American female physician who engaged in intimate homosocial relationships, Chung not only served as an agent but also became the object of Progressive reform.

Maternalist Medicine

Long before Margaret Chung sought an internship to further her medical training, her professional aspirations and education had been shaped by maternalist ideologies and institutions. She was born in 1889 to parents who had immigrated to the United States from China and who had both converted to Christianity in the United States. In fact, her mother, Ah Yane ([1869]–1914), was "rescued" from a life of servitude and likely prostitution by white Christian women affiliated with the San Francisco Presbyterian Mission Home.[8] These female reformers subscribed to the Victorian belief of innate gender differences but believed that women's moral and motherly instincts justified their efforts to reform the broader society. Fueled by religious fervor and seeking to "uplift" disadvantaged women, these missionaries sought to inspire their charges to form Christian families and even to pursue professional work, such as medicine, so that the "rescued" could themselves become "rescuers."

Steeped in Christian ideology, Margaret Chung at an early age expressed a desire to travel to China to serve as a "medical missionary"; her "ultimate aim [wa]s to reach the soul by ministering to the body," especially the bodies of other Chinese women.[9] In fact, following graduation from the University of Southern California, Chung applied to serve in China. However, her application was turned down repeatedly by the missionary board. She declined to comment about this episode in her unpublished autobiography, but she gave an interview in 1939 that suggested that the rebuff resulted from her American birth and Chinese ancestry: "Three times her application came up before [missionary] administrative boards and three times it was turned down—not because of any lack in her qualifications, but because there was no provision in the rules and regulations governing funds to send American missionaries to China that covered a case of an American of Chinese descent."[10] The publications of the Presbyterian Occidental Board indirectly support Chung's claim of racial discrimination. Of the forty-one missionaries sponsored in China from 1875 to 1920, none was of Chinese descent.[11] Despite their goal of converting the "natives," the American missionary society in China was largely segregated.[12] Chinese Christians mainly participated as marginal figures, sometimes as domestic servants or Bible women, who were paid slight sums to proselytize. The few Chinese women who received their medical training in the United States were respected figures in the Chinese and white Christian communities. However, unlike Chung, they had been born and raised in China and had the ability to develop and use connections with Chinese political leaders and benefactors. As an American-born Chinese, Chung was both too American and too Chinese for the purposes of white missionaries. The inability to carry out her vocational goal resulted in a religious crisis for Chung. According to her sister Dorothy, Margaret "was never known to have gone to church after she graduated from medical school."[13] The possibility of an internship in Chicago allowed Chung to redefine her professional and life goals in more secular terms. Both the Mary Thompson Hospital and Chung's mentor there, Dr. Bertha Van Hoosen, exemplified Progressive versions of maternalist reform.

Chung's internship at the Mary Thompson Hospital in many ways harkened back to her mother's upbringing at the San Francisco Presbyterian Mission Home. Both institutions trained women to assist other women. Mary Thompson Hospital was initially established in 1865 in a house with fourteen beds to provide medical services for widows and orphans of the Civil War. It became one of many institutions created during the mid- to late nineteenth century that focused on providing specialized care for women and children. By 1922, the hospital expanded into a five-story brick building that accommodated seventy-five to one hundred patients at a time. The staff consisted entirely of female doctors who not only cared for the sick but also trained female interns and nurses. Even as women increasingly attended coeducational medical schools during the early twentieth century, hospitals like the Mary Thompson offered much-needed opportunities for female physicians to obtain postgraduate training.[14] Like the missionaries at the Presbyterian Mission Home, Margaret's mentors used motherhood to explain why women, even unmarried and biologically childless women, should exert a protective and nurturing influence over their patients as well as their trainees. At the San Francisco Mission Home, Margaret Chung's mother, Ah Yane, had become a surrogate daughter for Margaret Culbertson, a single woman who served as the home's "matron." At the Mary Thompson Hospital, Margaret Chung became a "surgical daughter" for Bertha Van Hoosen, an unmarried female doctor.[15]

Both the hospital and mission home also focused on improving the lives of the immigrant poor. Located on the West Side of Chicago and close to the Hull House Settlement Project, the Mary Thompson Hospital was surrounded by

Italian, Greek, Bohemian, Polish, and Jewish communities. These so-called new immigrants from eastern and southern Europe began arriving in Chicago and other urban communities throughout the United States during the late nineteenth and early twentieth centuries. Along with black and white migrants from the rural South and Midwest, the new immigrants contributed to the phenomenal growth of Chicago from a mere 300,000 residents in 1870 to 2.7 million in 1920.[16] Chung was familiar with these new migrant populations, especially those from eastern and southern Europe, for they were the same groups that contributed to the expansion of Los Angeles during the first two decades of the twentieth century. The neighborhoods where she grew up were racially and ethnically diverse and included Italian, Jewish, and African American residents. However, one of the key differences between California and the Midwest was the relative absence of Mexican and Asian Americans.[17] Approximately two thousand Chinese lived in Chicago; the same number resided in Los Angeles, a city more than four times smaller than the Windy City. In the midwestern city, some Chinese clustered around the old Chinatown in the downtown Loop district and increasingly in a new concentration on the South Side, close to the growing African American community.[18] In addition, a few international and American-born Asians attended the various academic institutions in Chicago.[19] It is unclear how much contact Chung had with the Chinese communities in the Midwest. In the predominantly European immigrant neighborhood of the West Side, a few scattered Chinese operated laundries and restaurants that catered to non-Chinese clientele.[20] Almost all immigrant men, they labored for long hours and had limited language skills to interact with their customers and neighbors. Chung, apparently the only Chinese American on the Mary Thompson Hospital staff during her year there, no doubt experienced a similar sense of racial isolation.

However, as an intern at the Mary Thompson Hospital, Chung learned to care for her European immigrant neighbors. The hospital treated an estimated 1,800 patients per year. In addition, it accepted 100 "charity" cases and operated a free dispensary for 12,000 low-income individuals annually.[21] Furthermore, for four months of her internship, Chung served as an "externe, visiting charity patients at their homes"; there, she "administer[ed] . . . medicines" and also taught "hygiene in the everyday life and in the convalescence of the sick."[22] In essence, Chung served as a secular medical missionary in Chicago. However, as a Chinese American physician to her European charges, she reversed the dominant racial pattern of the white rescuers saving nonwhite people.

The female community at the Mary Thompson also differed in important ways from the Chinese Mission Home. Whereas earlier reformers emphasized voluntarism and Christianity, Progressive era activists emphasized professionalism and interfaith cooperation.[23] Chung's mentor, Van Hoosen, serves as an example of this "new" generation of female reformers. Born in 1863 to a farming family in Stony Creek, Michigan, Van Hoosen studied medicine at the University of Michigan, one of the best medical institutions and one of the first coeducational schools in the country. After obtaining postgraduate training at women's hospitals in Detroit and Boston, she opened a practice in Chicago in 1892. Van Hoosen subsequently accomplished a number of firsts for women in the medical field: first woman to be appointed under the Civil Service as chief gynecologist at the Cook County Hospital, first woman to be a professor and chair of Obstetrics in a coeducational medical school, and first woman to present a paper at the International Congress of Medicine.[24] In addition to these personal accomplishments, she also advocated for better medical care and opportunities for other women. In 1915, on the fiftieth anniversary of the Mary Thompson Hospital, she helped found the American Medical Women's Association (AMWA) to advance the professional interests of female physicians.

Van Hoosen's career exemplified Progressive strategies of using scientific expertise to improve society. For example, she gained recognition during the early twentieth century for promoting a new medication, referred to as "twilight sleep," to facilitate childbirth.[25] Hospitals provided the most effective setting for this technique of delivery. Consequently, Van Hoosen's campaign helped remove the experience of childbirth away from the home and the guidance of female family members and midwives. Even mothers were divorced from the process of birthing because the injection of morphine and scopolamine did not eradicate a person's ability to respond to physical discomfort but did erase the memory of pain. In Van Hoosen's eyes, only particular groups of women, with specialized training and access to medical institutions, could offer the best health care.

Van Hoosen's interest in promoting modern medical techniques encouraged her to train her intern "daughters" in the highly prestigious and male-dominated field of surgery.[26] Chung's fascination with this particular branch of medicine began in childhood. She explained: "My first love was surgery. . . As a very young child, having no dolls or toys to play with, I would frequently take banana peels or cabbage stems and make believe I was operating upon them. Living on a farm or a ranch from time to time, I had to cook and prepare chickens, wild ducks, quail, or rabbit, and even while I was preparing the food I would make believe I was operating on the chicken or the rabbit."[27]

Chung's preference for surgery coincided with its rising prominence during the turn of the century. Led by pioneers like Mary Dixon Jones, surgery became safer with new knowledge about asepsis and antisepsis; it also epitomized the new scientific ethos of medical care. Rather than a holistic approach to medical care, surgery emphasized the role of specialized physicians who could treat discrete illnesses and malfunctioning body parts. Surgery increasingly took place in hospitals, which provided the facilities, equipment, and personnel necessary for the procedures. Because male-dominated hospitals not only denied women internships but also operating privileges, Van Hoosen and the Mary Thompson Hospital staff undertook the task of training female surgeons. Chung apparently displayed "a great aptitude for surgery"; Van Hoosen and her colleagues gave her "unaccustomed opportunities to operate."[28]

Chung benefited specifically from her mentor's commitment to assisting female physicians of Chinese ancestry. Van Hoosen not only offered Chung a position but also appointed the first intern from China at the Mary Thompson Hospital. Li Yuin Tsao graduated from the Woman's Medical College of Pennsylvania in the early 1910s but like Chung was unable to obtain an internship. Van Hoosen explained,

> My first acquaintance with Chinese women physicians dates from the time that Dr. Mary McLean came to Chicago to get an internship in one of the hospitals for one of her protegees, Dr. Li Yuin Tsao.
>
> I wagered that she would not meet with success and told her, "When you fail, come back, and we will take Li Yuin at the Women's and Children's Hospital."
>
> Disappointed and resentful, she returned to accept my offer.
>
> Nevertheless, I began to regret my impulsive bravado and to dread the constant defense of a Chinese intern. I could almost hear such remarks as these: "I won't have a 'Chink' doctor me." "Don't let that foreigner come near me." "Why must we have a foreigner when we haven't enough positions for our own women?"[29]

Although Chicago lacked a large population of Chinese and did not have California's vociferous anti-Chinese movement, xenophobia also permeated the Midwest. Despite the fears about patient and professional reactions to a Chinese female doctor, Tsao's medical and interpersonal skills encouraged Van Hoosen to provide opportunities for other physicians of Chinese ancestry. In fact, the year before Chung arrived at the Mary Thompson, another Chinese female physician served as an intern there.[30] Interestingly, these women doctors do not appear to have been recognized by the Chinese American community in

Chicago. A 1926 study of the population notes the presence of fourteen male physicians but makes no mention of the female physicians of Chinese ancestry who interned at the Mary Thompson Hospital.[31]

Van Hoosen's commitment to training Chung and other Chinese women physicians reflected the influence as well as a transformation of female missionary efforts. When Van Hoosen appointed Tsao as her first Chinese intern, she explained that "at the Women and Children's Hospital we do not only profess Christianity, we practice it."[32] In other words, like women physicians in Greece and Armenia described by Virginia A. Metaxas in Chapter 11, she viewed the acceptance of diverse individuals as an expression of Christian values. Van Hoosen not only trained Asian women in the United States, but she also traveled abroad to promote scientific exchange. Retracing the earlier movement of American missionaries overseas, Van Hoosen visited China, Japan, and India in the early 1920s to learn about medical practice in these countries and to offer instruction for Asian female physicians. She was particularly eager to reunite with Tsao and also meet Ida Kahn and Mary Stone, fellow graduates of the University of Michigan and two of the earliest and best-known Chinese medical missionaries.

Van Hoosen's international interests led her to support a secular version of religious medical missions. Shortly after the founding of the AMWA, the organization sponsored health missions abroad to provide wartime, postwar, and emergency medical care. The female missionary movement previously viewed science as a tool for spiritual conversion. Van Hoosen and other members of AMWA may have been motivated by Christian beliefs and used religious networks to assist their work but they viewed the dissemination of scientific knowledge and the provision of medical care as their primary goals.

Chung benefited from the commitment of Van Hoosen and other female medical reformers who assisted those on the margins. Their willingness to help "outsiders" and "foreigners" partly stemmed from their personal backgrounds. Van Hoosen's father emigrated from the Netherlands. May Michael, a Jewish pediatrician, vouched for Chung's "good moral and professional character" on her Illinois State Board of Health Application.[33] Rachelle Yarros, another female Jewish physician, eventually helped Chung obtain a medical residency. All three of her mentors came from immigrant backgrounds and dedicated themselves to providing medical services and training for women, especially women in ethnic communities. Michael emigrated from England and devoted much of her practice and energies to Jewish service organizations in Chicago.[34] Yarros, born near Kiev, Russia, in 1869, worked in a New York sweatshop before becoming a physician. In Chicago, she completed her residency at the Michael Reese, a Jewish hospital. Yarros eventually affiliated with Hull House, a reform community in the heart of poor immigrant neighborhoods on Chicago's West Side, and became known for promoting sex education and birth control.[35] Chung, later in her medical career, would assist women in obtaining access to abortions.[36] No doubt she was influenced both by Yarros and the experiences of her own mother. Ah Yane bore eleven children, seven of whom survived to adulthood. She died at the relatively young age of forty-five from tuberculosis, an illness whose symptoms are aggravated by childbirth.

The ethnic and religious backgrounds of Chung's mentors facilitated their ability to empathize with her status as a Chinese American. Nativist movements targeted immigrants from both Asia and eastern Europe. As far back as the mid-nineteenth century, exclusionists argued that America should be reserved for the "white" Anglo-Saxon race. Whiteness, however, did not necessarily correlate with European ancestry. Along with individuals from Ireland, Italy, and other parts of southern and eastern Europe, Jewish immigrants were largely perceived as less than white.[37] They faced a variety of anti-Semitic restrictions and practices that determined where Jews could live, work, and attend school. In fact, during and after World War I, nativist forces would push for increasingly restrictive immigration laws that reduced the entry of individuals from "undesirable" parts of Europe and almost completely cut off migration from Asia.[38] Being ostracized oneself does not necessarily lead to compassion for other despised groups. During the second half of the nineteenth century, Irish Americans led some of the most ferocious anti-black and anti-Chinese movements throughout the North and West. While some groups and individuals chose to reinforce discriminatory practices to gain mainstream acceptance, others advocated racial equality.[39] Chung fortunately came into contact with female reformers who supported accessible health care and training. The outlook of the Mary Thompson Hospital was expressed in its mission statement: "no person shall be deprived of the benefit of this institution on account of nationality, race, or religious profession."[40]

The Limits of Sisterhood

Despite the Mary Thompson Hospital's ideal of creating a supportive professional environment for women of diverse backgrounds, Chung's memories of her internship largely dwell on the tensions within this female community. Instead of a nurturing workplace, she discovered social isolation and professional hierarchy. Her negative experiences stemmed in part from the restrictiveness of life as an underpaid employee who resided at her place of work. In the early twentieth century, women's institutions were coming under criticism for fostering deviant sexuality. To deflect these sorts of criticisms, female organizations sometimes reinforced divisions between women by monitoring and censoring their private behavior.

Chung's first exposure to the Mary Thompson Hospital created a sense of its impersonal and institutional character. Upon her arrival in Chicago on Thanksgiving, she discovered that: "No one met me at the train. . . . The trees were leafless, the wind was cold, and the welcome in the hospital was hardly warmer. I was shown to my room—a little small cubicle with enough room in a little closet to hang up about four suits, an old iron bed which sagged unmercifully in the middle, one chair with the bottom out."[41] Facing financial difficulties that almost forced closure in 1911, the Mary Thompson offered fare that was little better than the accommodations. Chung recalled that "the meals were exceedingly meagre. For breakfast we had cold toast, coffee, and an egg; for lunch, meat, a vegetable, potatoes, bread and butter, coffee, and perhaps a dessert; for supper, cold toast again, and always, without exception, the same dish of stewed dried apricots and more coffee. . . . The few dollars which I had saved . . . were soon gone buying extra food to keep myself going."[42]

Despite these personal hardships, Chung concluded that the "work was exceedingly interesting" because the Mary Thompson provided numerous opportunities to develop her surgical skills.[43] However, she did not characterize her professional relationships in familial and nurturing terms. Rather, she emphasized the hierarchical differences between the medical personnel:

> The Staff Doctors of the Mary Thompson Hospital were very happy to turn over a great many of the clinic cases which were to be operated on free. Especially during the summer when it gets unbearably hot in Chicago, and of course in those days air conditioning was unheard of . . . and they had enough confidence in me to say to me, "Well, Dr. Chung, there's so many patients in the clinic to be operated on and when work is not too heavy here you can have the office send out postcards to these people and schedule a few operations each day for yourself, and you can clean up the list of operative cases in the clinic." This was welcome news for me indeed, and by the time they came back in September I had cleaned out the whole clinic of all the operative cases which came to over six hundred cases.[44]

Chung acknowledges that the staff doctors had "confidence" in her. However, she also suggests that they took advantage of her status as an unpaid intern, who could not afford to take a summer vacation.

Chung's critical attitude toward the Mary Thompson Hospital may have partly stemmed from the institution's efforts to monitor her private behavior. Despite her initial sense of neglect, Chung apparently became quite popular. In an unpublished draft of Van Hoosen's autobiography, Chung's mentor described her as "a favorite with nurses and interns to the degree that the hospital, for the first time, made a ruling that two people must not sleep in a single bed."[45] The nature of Chung's relationships with these nurses and interns is not clear. Chung's own autobiography makes no mention of them. During the Victorian era, as the discussion of Marie Zakrzewska's "Boston marriage" in Chapter 2 makes clear, women developed romantic friendships with one another that sometimes (but not always) included physical intimacy. In a society that presumed fundamental gender differences, women tended to form their strongest emotional bonds with other women. The passionate sensuality that characterized some of these relationships appeared natural, not abnormal.

In preventing Chung from sharing her bed with other women, however, the Mary Thompson Hospital labeled her behavior as socially unacceptable. The new policy responded to emerging ideas about sexuality that critics used to stigmatize female institutions. As women broke gender barriers in the professional and political arenas, their personal behavior drew increasing scrutiny. At the turn of the century, a branch of medicine pioneered by individuals such as Havelock Ellis and Richard von Krafft-Ebing focused on the topic of "gender inversion," the phenomenon of individuals who refuse their assigned gender roles. Chung was doubly "guilty" of transgressing social norms. She not only entered a traditionally male profession, but she also adopted masculine dress during this period in her life.[46] She, like other nonnormative women, became suspect for harboring "mannish" sexual desires. In essence, the new "sexology" studies created new categories of identity, branding certain people as normal and others as scientifically deviant based on their behavior, appearance, or desires. Although "gender inversion" research did not seek to stigmatize populations, it nevertheless helped fuel criticisms of women's education and professional training. In particular, female institutions, like the Mary Thompson Hospital, faced charges of fostering lesbianism. In response, the hospital, along with women's schools during the 1910s and 1920s, increasingly regulated female social and sexual behavior.[47]

Chung's mentors were not entirely unsympathetic toward her. Van Hoosen held her surgical daughter in high regard and kept in contact with her. In the published edition of Van Hoosen's autobiography, released shortly after World War II, she omitted the anecdote about her mentee's ability to attract bed partners. No doubt she attempted to shield both Chung and the Mary Thompson Hospital from public censure. The female community at the hospital had provided valuable opportunities for Chung to continue her ascent up the professional ladder. However, as she gained scientific expertise, she also became subject to the sexual scrutiny of modern medicine.

Saving Children

In October 1917, as Chung neared the completion of her first year at the Mary Thompson Hospital, she heard about the tragic death of her father in a car accident. According to her youngest sister Dorothy, who was eight at the time, Chung Wong ([1860]–1917), was driving his "first and only auto," when he became "caught between two Pasadena Red Cars . . . going opposite ways."[48] He lost his leg in the accident and died from loss of blood; with proper medical care, he might have survived the accident, but the local hospital denied Chung Wong admission because he was Chinese.[49] Margaret, lacking financial resources to travel, did not return to California for the funeral. Instead, she sent a cable, inviting her siblings to join her in Chicago. Dorothy remembered that "soon after the funeral 'the Doctor' sent them a cable that told them she would bring them to Chicago. They all immediately quit school and it was some time before things were straightened out and back to school some of them went. Margaret could not afford proper winter clothes much less finance ten [*sic*] children. The children stayed in Los Angeles, the older ones worked and took care of the younger ones and Margaret sent a bit of money now and then."[50] With little to offer, not even a place for her six younger sisters and brothers to stay, Chung used her contacts among female medical reformers to seek a paid residency to provide financial assistance to her siblings in California.

Chung's residency at the Illinois JPI exposed her to a new medical specialty and a new set of professional pressures. Her entry into psychiatry partly resulted from the professional obstacles she continued to face. In April 1917, the United States entered the "war to end all wars." As medical professionals left to fulfill their military duty, hospitals faced a personnel shortage.[51] Even under these conditions, Chung could only obtain a position at a psychopathic hospital, not

Margaret Chung appears as (*left*) a "Victorian lady" (courtesy of the personal collection of Judy Tzu-Chun Wu) and (*right*) a masculine young medical professional (courtesy of the Dorothy Siu Collection). The left photo, most likely taken when Chung graduated from the University of Southern California Preparatory School in 1911, evokes the era of gender separatism and maternalist reform that inspired and facilitated Chung's entry into the medical profession. The photo on the right, which Chung presented to her friends under her nickname "Mike," reveals how she transgressed gender norms not only by becoming a physician but also in the way she presented herself.

a particularly desirable posting. In the nineteenth century, the medical community generally held psychiatry in low regard. Practitioners primarily served in state hospitals, detaining and isolating the mentally deranged in asylums. Despite the "low professional status" of institutional work, women doctors "were particularly happy to take such positions. . . Uncertain of the rewards of private practice, women were often attracted by the security of such appointments, and the opportunities they afforded to gain expertise."[52] Chung's career path in fact followed Van Hoosen's. Her mentor had previously interned at the Kalamazoo State Hospital for the Insane.

While the lack of attractive alternatives forced female physicians like Chung to accept institutional work, the changing field of psychiatry in the early twentieth century also offered an appealing opportunity to use scientific expertise to improve society. Instead of detention and isolation, Progressive era practitioners focused on research, education, and reform. For example, the JPI, created in 1909, assisted the work of the Juvenile Court of Chicago by providing psychiatric evaluations for troubled youth and developing recommendations for their rehabilitation. The institute's mission, to understand and save children, might have appealed to Chung personally.[53] After all, Presbyterian missionaries in San Francisco had "rescued" her mother from a potential life of crime.

Rachelle Yarros assisted Chung in obtaining the residency through her connections with female reformers and Jewish physicians. Hull House activist Julia Lathrop, along with members of the Chicago Woman's Club, had successfully lobbied for the creation of the first Juvenile Court in the country in 1899.[54] Seeking to exert a protective maternal influence over adolescents who ran afoul of the law, Lathrop explained that "children should not be treated as criminals, but as delinquent children needing wise direction, care and correction."[55] Ten years after the founding of the Juvenile Court, she also played a crucial role in establishing the JPI. With the support of philanthropist Ethel Sturges Dummer, Lathrop hired medical experts and social workers to study juvenile delinquency and encourage rehabilitation. The privately funded experiment lasted five years before the institute officially became part of the state apparatus. Because the institute came under the purview of the newly founded Illinois Department of Public Welfare, Chung officially qualified for the residency by passing a civil service examination.[56] However, Yarros provided the initial introduction. Her influence stemmed from a history of women's involvement in developing the vision for the JPI and successfully demonstrating the value of state and medical cooperation.

Yarros's contacts with Jewish male physicians also assisted Chung's entry into a new medical field. In November 1917, Chung received a brief appointment as an assistant physician at the Kankakee State Hospital, where she took a course in psychiatry with the superintendent, H. Douglas Singer. Like May Michael, Singer was a Jewish immigrant from England.[57] When Chung joined the JPI in December 1917, she worked under Herman Adler. Adler, the grandson of a prominent Talmudic scholar, helped redefine the mission of psychiatry in the early twentieth century and successfully obtained the newly created position of Illinois state criminologist shortly before Chung's arrival.[58] Just as the initial low status of the specialty facilitated the entry of women, the emerging field provided opportunities for religious and racial minorities to advance their careers as well. In fact, "when psychology was viewed as a branch of medicine," the field "was understood to be potentially appropriate for Jews."[59] Chung recalled Adler and his wife with great fondness. They "did several nice things for her," and even "noticed her inadequate winter clothing and bought her a coat." Her sister Dorothy remembered that Margaret "always had a soft spot for Jews" because of the Adlers' quasi-parental interest in her.[60]

Chung, however, did not enjoy the pressures associated with her new position. She worked and lived in two wards of Chicago's Cook County Hospital, which housed the JPI. Unlike the Mary Thompson Hospital, where Chung worked with an all-female staff to help female patients, her JPI colleagues included male and female "psychiatrists, psychologists, psychoanalysts, [and] social service workers."[61] Men tended to occupy the primary positions of authority, as directors, judges, and physicians. Since male offenders came before the court in greater numbers than females, men even dominated the patient base.[62] Chung's responsibilities included training nurses from psychopathic hospitals throughout the state. However, she primarily assisted in "the examination, observation, and recommendation to the different judges as to the position of borderline cases, criminals who feigned insanity, and unusual court cases in which the judges had difficulty in deciding what to do with the individual before them." Although the institute initially focused on juvenile cases, Adler advocated for an increased role in all court proceedings. As Chung recalled, "More frequently, we had adults rather than children." The large caseload, an average of 100 per month during her residency, overburdened the resources of the institute and allowed little time for reflective research.[63] She found the work "very interesting" but also "nerve wracking for a young woman in her early twenties who would sometimes have to give a recommendation which might result in a man's getting a death sentence."

Although Chung was then in her late, not early, twenties, the coercive aspect of her work as a court psychiatrist disturbed her. Because the JPI served as an extension of the legal system, researchers not only offered treatment but also assisted the court in meting out punishment. Overall, Chung found the experience "too depressing."[64]

Chung's anxiety over her responsibilities as a court psychiatrist partly stemmed from the scientific tools at her disposal to understand criminality and abnormality. The field of psychiatry increasingly assumed cultural importance during the early twentieth century by offering explanations for social behavior. However, unlike most branches of medicine, psychiatry "lacked the critical capacity to inscribe itself on the body. Psychiatrists had few signs with which to identify disease; they had to be content to work with the ephemera commonly known as symptoms."[65] To buttress the scientific claims of their specialty, psychiatrists created categories of mental abilities and personality traits to classify and to diagnose their patients. However, these classifications tended to reveal much about the subjective attitudes of the scientists and the nature of their interactions with the patients.

For example, the Juvenile Court and the JPI kept meticulous statistics on their subjects' "nationality" to examine correlations with criminality and insanity.[66] The concept of "nationality" did not refer to one's national citizenship but rather to so-called race and racial stock. Separate listings existed for "American" and "colored," implying that "American" only referred to white individuals. American-born children of immigrants also did not fall under the category of "American." Instead, researchers listed them under the nationality of their parents. The seemingly strange classification system revealed the influence of prevailing eugenics and nativist ideas that associated race and foreignness with mental defectiveness. The researchers, some of whom were immigrants and minorities themselves, increasingly emphasized environmental factors and individual personality to explain criminal behavior and psychological disorders. However, the statistics on and references to the patients' ancestry and race suggest that heredity lurked as a subtext for explaining mental deficiency. Reflecting the demographics of Chicago as well as the community surrounding Cook County Hospital, the overwhelming majority of Chung's patients at the JPI were southern and eastern European immigrants, along with sizable populations of Irish, native-born whites, and African Americans. During her residency, no Chinese and only one individual of Mexican descent were processed.[67]

Chung's uneasiness with psychiatry perhaps also stemmed from the field's role in classifying deviant and normal forms of sexuality. The study of "sexology" emerged through research on "psychopathology." Physicians defined gender inversion and "sexual perversion" as mental disorders that required psychological and psychiatric treatment. In fact, cases involving cross-dressing and homosexuality appeared before the Juvenile Court and the JPI during Chung's residency.[68] Her hesitation about her professional responsibilities perchance reflected a certain distrust of the field of psychiatry and the way in which it was applied at the JPI. The new sexology research helped liberate some individuals by providing a medical vocabulary, as opposed to a religious/moral framework, to understand their behavior and desires. However, in the context of the JPI, the evaluation of "patients" was directly connected with the determination of whether these individuals might be institutionalized within the judicial system.

Near the completion of her residency, Chung took steps to obtain a new position. Initially seeking to follow in the footsteps of her supervisor, Herman Adler, she requested indefinite leave from her civil service position to offer her skills to the military. During World War I, psychiatrists played an important role in instituting mental testing of soldiers and treating shell shock. However, the armed forces, despite the lobbying efforts of the American Medical Women's Association, refused to admit female physicians into their ranks.[69] In November 1918, however, after two years away from her family and home state, Chung was ready to return to southern California and launch her own medical practice.

Conclusion

Chung's internships and residency in Chicago introduced her to various professional possibilities to engage in Progressive reform. At the Mary Thompson Hospital, she developed her surgical skills, a traditionally male specialty, in a female-staffed institution that catered to female patients. At the JPI, she trained as a psychiatrist, focusing on traditionally "feminine" concerns such as emotional well-being, marital relations, and child-rearing. Ironically, she learned these skills from male experts and focused on treating male patients.[70] Women reformers, seeking to assert a maternal influence over their communities, initiated both institutions. However, while the Mary Thompson remained a privately funded women's establishment, the JPI became part of the governing apparatus. The JPI demonstrated both women's success in shaping the state and the diminution of their status in gender-mixed settings.

Of the two specialties that Chung developed during her time in Chicago, she preferred surgery to psychiatry. As a psychiatrist, she treated indeterminate mental disorders that frequently required lengthy therapy with no guarantee of improvement. As a surgeon, she interacted with her patients' bodies and produced immediate results. Chung favored this more direct, almost heroic style of medical practice. In addition, she appeared to be increasingly comfortable with a predominantly male patient base. When she returned to Los Angeles, Chung worked as a staff physician at the Santa Fe Railroad Hospital and treated the predominantly male workforce. Even as she moved beyond maternalist medical institutions, however, she discovered that her patients still sought to project their maternal fantasies on her. Chung's skill in removing steel filings from workers' eyes, a common injury in the railroad machine shops, was attributed to the "natural" gentleness of a woman doctor. Her suggestion for increasing salt consumption to prevent heat exhaustion sounded like the simple but effective advice that a mother would offer.[71] In other words, Chung's patients perceived her as a good physician, not in spite of, but because of, her sex.

To examine Chung's career in Chicago purely through the lens of gender, however, would overlook the significance of race, ethnicity, religion, and sexuality in her professional career. Chung's status as an American-born woman of Chinese descent was connected in the minds of her mentors and her family's supporters with their interests in China. That is, she was regarded not solely as a racialized person in the United States but was treated as part of a broader effort to foster missionary and scientific outreach to Asia. This internationalization of race defined not only her personal status but also the position of Americans of Asian descent more generally. Furthermore, the immigrant and, in particular, Jewish backgrounds of Chung's mentors underscored the exclusivity of "whiteness" during the early twentieth century. Perhaps an awareness of the ambiguous racial status of Jews encouraged these physicians to foster professional opportunities for others who also experienced social ostracism.

Finally, the censuring of Chung's sexual behavior in her place of work and residence reveals the contradictory significance of modern science that both empowered and disciplined medical practitioners. Some individuals found comfort in the findings of sexology research and adapted the theories to create a positive sense of their own identities. However, Chung's experiences during her internship and residency underscored the regulatory power of science, especially in the realm of sexuality. As the essay on Mary Steichen Calderone demonstrates (Chapter 5 in this volume), these tensions between liberation and discipline were central in shaping the sexuality and professional identity of female physicians.

Chung's early training in Chicago was crucial not only in launching her medical career but also in shaping her public and personal identities. She eventually established one of the first western clinics in San Francisco Chinatown, which boasted the largest Chinese population in the United States. Chung's preference for surgery would place her in an awkward position in a community that overwhelmingly preferred noninvasive forms of traditional Chinese medicine. As a consolation, though, Chung increasingly catered to a predominantly white and mainstream clientele. In fact, during World War II, she became a political celebrity who adopted glamorous feminine clothing and socialized easily and regularly with well-known entertainers, politicians, and military personnel. She elected to remain single, but she assumed a fictive maternal identity by "adopting" over a thousand predominantly white American "sons" to support the Allied war effort. Although she never entered the military, Chung used her connections to lobby for the creation of the WAVES (the women's naval reserve). Finally, while Chung continued to develop homoerotic and romantic relationships with other women, she skirted overt identification with lesbianism and in fact crafted an outwardly asexual persona that harkened back to Victorian gender norms. Chung's personal negotiations through the racial, gender, and sexual terrains of Progressive era Chicago reveal the possibilities as well as the parameters of professional identity formation for a Chinese American woman doctor and set the stage for her emergence as a maternal and seemingly asexual political broker of international politics during World War II.

NOTES

1. Her name was King Ya-Mei (also known as Jin Yunmei). Elizabeth Lee Abbott, "Dr. Hu King Eng, Pioneer," in *The Life, Influence and the Role of the Chinese in the United States, 1776–1960* (San Francisco: Chinese Historical Society of America, 1976), 243–249; Margaret E. Burton, *Women Workers of the Orient* (New York: Woman's Press, 1919); and Weili Ye, "Crossing the Cultures: The Experience of Chinese Students in the U.S.A., 1900–1925" (Ph.D. diss., Yale University, 1989), 263–282.

2. Margaret Chung, "Autobiography," Margaret Chung Collection, Box 1, Folder 1, Asian American Studies Collection, Ethnic Studies Library, University of California, Berkeley. The collection contains a typed manuscript and some handwritten notes to the autobiography. The unpublished work is not paginated. For more biographical information about Margaret Chung, please see Judy Yung, *Unbound Feet: A Social History of Chinese Women in San Francisco* (Berkeley: University of California Press, 1995); idem, *Unbound Voices: A Documentary History of Chinese Women in San Francisco* (Berkeley: University of California Press, 1999); Leila J. Rupp, *A Desired Past: A Short History of Same-Sex Love in America* (Chicago: University of Chicago Press, 1999); and Judy Tzu-Chun

Wu, *Mom Chung of the Fair-Haired Bastards: The Life of a Wartime Celebrity* (Berkeley: University of California Press, 2005).

3. The American Medical Association reported that only 50 percent of medical students obtained hospital training after graduation in 1904. However, by 1914, 75 percent to 80 percent of physicians took internships. Paul Starr, *The Social Transformation of American Medicine: The Rise of a Sovereign Profession and the Making of a Vast Industry* (New York: Basic Books, 1982), 123–124.

4. Darlene Clark Hine, *Hine Sight: Black Women and the Re-construction of American History* (Brooklyn: Carlson Publishing, 1994); Regina Markell Morantz-Sanchez, *Sympathy and Science: Women Physicians in American Medicine* (New York: Oxford University Press, 1985); Ellen S. More, *Restoring the Balance: Women Physicians and the Profession of Medicine, 1850–1995* (Cambridge, MA: Harvard University Press, 1999), 95–121; Mary Roth Walsh, *"Doctors Wanted: No Women Need Apply": Sexual Barriers in the Medical Profession, 1835–1975* (New Haven, CT: Yale University Press, 1977), 219–221.

5. Agnes Scholl was the other female graduate from Chung's medical class. Barbara Bronson Gray, *120 Years of Medicine: Los Angeles County, 1871–1991* (Houston: Pioneer Publications, 1991), 43–44.

6. Ronald Takaki, *Strangers from a Different Shore: A History of Asian Americans* (Boston: Little, Brown, 1989); Sucheng Chan, *Asian Americans: An Interpretive History* (Boston: Twayne Publishers, 1991).

7. David J. Rothman, *Conscience and Convenience: The Asylum and Its Alternatives in Progressive America* (Boston: Little, Brown, 1980), 293–323.

8. Peggy Pascoe, *Relations of Rescue: The Search for Female Moral Authority in the American West, 1874–1939* (New York: Oxford University Press, 1990); Judy Tzu-Chun Wu, "'The Ministering Angel of Chinatown': Missionary Uplift, Modern Medicine, and Asian American Women's Strategies of Liminality," in *Asian/Pacific Islander American Women: A Historical Anthology*, ed. Shirley Hune and Gail Nomura (New York: New York University Press, 2003), 155–171.

9. Chung, "Autobiography"; Mrs. J. B. Stewart, "Medical Missions," in *Occidental Leaves* (San Francisco, 1893), 38.

10. Helen Satterlee, "The Story of a Persevering Chinese Girl Who Reached the Heights of Surgical Fame," *Los Angeles Times Sunday Magazine*, June 25, 1939, p. 20.

11. *Forty-Seventh Occidental Board Report* (1920), 48–54. The Occidental Board sponsored a total of 137 missionaries from 1875 to 1920. Cecilia Tsu has identified second-generation Chinese Americans who served as missionaries in China. However, their names do not appear in the listings of the Presbyterian Missionary Board. Cecilia M. Tsu, "'Winning These Americans for Christ': Protestant Women and the Vision of Chinese American Assimilability in California, 1870–1920" (unpublished paper, Berkshire Conference on the History of Women, June 6–9, 2002).

12. Jane Hunter, *The Gospel of Gentility: American Women Missionaries in Turn-of-the-Century China* (New Haven, CT: Yale University Press, 1984).

13. Mariko Tse, "Made in America," Research Project for East West Players on Chinese in Southern California (unpublished paper, June 1979, Private Collection of Judy Yung), 10.

14. Morantz-Sanchez, *Sympathy and Science*, 168.

15. Rose V. Mendian, "Bertha Van Hoosen: A Surgical Daughter's Impressions," *Journal of the American Medical Women's Association*, 1965, *20*: 349. Also see Regina G.

Kunzel, *Fallen Women, Problem Girls: Unmarried Mothers and the Professionalization of Social Work, 1890–1945* (New Haven, CT: Yale University Press, 1993).

16. Joanne J. Meyerowitz, *Women Adrift: Independent Wage Earners in Chicago, 1880–1930* (Chicago: University of Chicago Press, 1988).

17. George J. Sánchez, *Becoming Mexican American: Ethnicity, Culture and Identity in Chicano Los Angeles, 1900–1945* (New York: Oxford University Press, 1993), 71–75.

18. Tin-Chiu Fan, "Chinese Residents in Chicago" (master's thesis, University of Chicago, 1926); Susan Lee Moy, "The Chinese in Chicago: The First One Hundred Years, 1870–1970" (master's thesis, University of Wisconsin–Madison, 1978); Paul C. P. Siu, *The Chinese Laundryman: A Study of Social Isolation*, ed. John Kuo Wei Tchen (New York: New York University Press, 1987).

19. Henry Yu, *Thinking Orientals: Migration, Contact, and Exoticism in Modern America* (New York: Oxford University Press, 2001).

20. Approximately 30 percent of the Chinese in Chicago operated businesses in non-Chinese neighborhoods. Siu, *The Chinese Laundryman.*

21. Council of the Chicago Medical Society, *History of Medicine and Surgery and Physicians and Surgeons of Chicago* (Chicago: Biographical Publishing Corporation, 1922), 256.

22. Mary Harris Thompson, "The Chicago Hospital for Women and Children," in *In Memoriam, Mary Harris Thompson*, ed. Maria S. Iberne (Chicago: Board of Managers, 1896), 59.

23. Lori D. Ginzberg, *Women and the Work of Benevolence: Morality, Politics, and Class in the 19th-Century United States* (New Haven, CT: Yale University, 1990).

24. Constance M. McGovern, "Bertha Van Hoosen," in "Bertha Van Hoosen Biography File," Chicago Historical Society; Mabel E. Gardner, "Bertha Van Hoosen, M.D.: First President of the American Medical Women's Association," *Journal of the American Medical Women's Association*, 1950, 5: 413–414.

25. Judith Walzer Leavitt, "Birthing and Anesthesia: The Debate over Twilight Sleep," in *Mothers and Motherhood: Readings in American History*, ed. Rima D. Apple and Janet Golden (Columbus: Ohio State University Press, 1997), 242–258.

26. Starr, *The Social Transformation of American Medicine;* Charles E. Rosenberg, *The Care of Strangers: The Rise of America's Hospital System* (New York: Basic Books, 1987).

27. Chung, "Autobiography."

28. Bertha Van Hoosen, *Petticoat Surgeon* (Chicago: Pellegrini & Cudahy, 1947), 219.

29. Van Hoosen, *Petticoat Surgeon*, 218.

30. The 1916 Chicago Medical Directory lists Lin Hie Ding, who graduated from the University of Illinois in 1915, as an affiliate of the Mary Thompson Hospital.

31. Fan, "Chinese Residents in Chicago," 37, 40.

32. "Manuscript," Bertha Van Hoosen Papers, Bentley Historical Library, University of Michigan, Ann Arbor, Box 2, File 1, p. 440.

33. "Margaret Jessie Chung," Application for a Certificate from the Illinois State Board of Health, Illinois State Archives, Department of Registration and Education, Register of Licensed Physicians and Surgeons, p. 208.28.

34. Hyman L. Meites, ed., *History of the Jews of Chicago* (Chicago, 1924; facsimile, Chicago: Chicago Jewish Historical Society and Wellington Publishing, 1990), 409.

35. Marilyn Elizabeth Perry, "Rachelle Yarros," University of Illinois at Chicago, Jane Addams Memorial Collection.

36. Elsa Gidlow, *Elsa: I Come with My Songs* (San Francisco: Booklegger Press, 1986),

207; Percival Dolman, "Letter to Frederick N. Scatena," July 29, 1950, Margaret J. Chung File, Deceased Physicians File.

37. James R. Barrett and David Roediger, "Inbetween Peoples: Race, Nationality, and the 'New Immigrant' Working Class," *Journal of American Ethnic History*, 1997, *16*: 3–44; Karen Brodkin, *How Jews Became White Folks and What That Says about Race in America* (New Brunswick, NJ: Rutgers University Press, 1998); Matthew Frye Jacobson, *Whiteness of a Different Color: European Immigrants and the Alchemy of Race* (Cambridge, MA: Harvard University Press, 1998); David R. Roediger, *The Wages of Whiteness: Race and the Making of the American Working Class* (New York: Verso Press, 1991).

38. Mae N. Ngai, "The Architecture of Race in American Immigration Law: A Reexamination of the Immigration Act of 1924," *Journal of American History*, 1999, *86*: 67–92. The 1924 immigration act did not target Filipinos for complete exclusion because they originated from an American territory.

39. Michael Rogin argues that activism on behalf of racial equality and exploitation of racial stereotypes might be two sides of the same coin. Jewish participation in NAACP and the Civil Rights Movement as well as Jewish adoption of blackface in entertainment both use racial masquerade to achieve their acceptance in mainstream American society. Michael Rogin, *Blackface, White Noise: Jewish Immigrants in the Hollywood Melting Pot* (Berkeley: University of California Press, 1996).

40. Quoted in Martha Douglas Bost, "History of Mary Thompson Hospital, 1865–1973" (unpublished paper, Chicago Historical Society, December 1973), 2.

41. Chung, "Autobiography." This quote incorporates Chung's handwritten as well as the typed versions of her autobiography.

42. Ibid.

43. Ibid.

44. Ibid.

45. "Manuscript of Autobiography," 403, Bertha Van Hoosen Papers, Box 3, Folder 1.

46. Judy Tzu-Chun Wu, "Was Mom Chung a 'Sister Lesbian'?: Asian American Gender Experimentation and Interracial Homoeroticism," *Journal of Women's History*, 2001, *13*: 58–82.

47. Carroll Smith-Rosenberg, *Disorderly Conduct: Visions of Gender in Victorian America* (New York: Oxford University Press, 1985); Bert Hansen, "American Physicians' 'Discovery' of Homosexuals, 1880–1900: A New Diagnosis in a Changing Society," in *Framing Disease: Studies in Cultural History*, ed. Charles E. Rosenberg and Janet Golden (New Brunswick, NJ: Rutgers University Press, 1992): 105–133; Harry Oosterhuis, *Stepchildren of Nature: Krafft-Ebing, Psychiatry, and the Making of Sexual Identity* (Chicago: University of Chicago Press, 2000).

48. Dorothy Siu, "Oral History," interviewed by Jean Wong, January 12, 1979, and November 6, 1980, *Southern California Chinese American Oral History Project*, University of California, Los Angeles.

49. Gilbert Siu, personal communication, June 24, 2001, Los Angeles, California.

50. Tse, "Made in America," 10.

51. Stuart K. Jaffary, *The Mentally Ill and Public Provision for Their Care in Illinois* (Chicago: University of Chicago Press, 1942), 20–21.

52. Morantz-Sanchez, *Sympathy and Science*, 153.

53. Anthony M. Platt, *The Child Savers: The Invention of Delinquency* (Chicago, 1969; revised, Chicago: University of Chicago Press, 1977), 75–100.

54. Victoria Getis, *The Juvenile Court and the Progressives* (Urbana: University of Illinois Press, 2000); Robert M. Mennel, *Thorns & Thistles: Juvenile Delinquents in the United States, 1825–1940* (Hanover, NH: University Press of New England, 1973).

55. Louise de Koven Bowen, *Growing Up with a City* (New York: Macmillan Company, 1926), 103.

56. State of Illinois, Department of Mental Health, "Employment Record for Dr. Margaret J. Chung," No. 84350. Chung was appointed assistant physician at Kankakee on November 1, 1917. She was transferred to the Division of Criminology on December 16, 1917, and appointed as a resident at Cook County Psychopathic Hospital, which housed the Juvenile Psychopathic Institute. "Minutes of Meetings," State of Illinois, Department of Public Welfare, RG 206, December 5, 1917; Herman M. Adler, "The Juvenile Psychopathic Institute and the Work of the Division of the Criminologist," *Institution Quarterly*, 1918, *9*: 6–10.

57. Singer, born in 1875, received his medical training at the University of London and worked in Nebraska and Illinois in the field of psychiatry.

58. Adler received his medical degree from Columbia and served as chief of staff of the Boston Psychopathic Hospital as well as assistant professor of psychiatry at Harvard Medical School. Invited to review the institutions for criminals and delinquents in Illinois, Adler was appointed to succeed William Healy, the first director of the Juvenile Psychiatric Institute, in 1917.

59. David A. Hollinger, *Science, Jews, and Secular Culture: Studies in Mid-Twentieth-Century American Intellectual History* (Princeton, NJ: Princeton University Press, 1996), 25.

60. Tse, "Made in America," 10–11. Adler, born in 1876, was thirteen years older than Chung. He married Frances Porter in March 1917.

61. Chung, "Autobiography."

62. From 1915 to 1919, approximately one in five juvenile delinquency cases involved girls. Helen Rankin Jeter, *The Chicago Juvenile Court*, U.S. Department of Labor, Children's Bureau, Publication No. 104 (Washington, D.C.: Government Printing Office, 1922), 18.

63. From December 17, 1917, to September 30, 1918, the Juvenile Psychiatric Institute processed 939 cases. *Institution Quarterly*, 1918, *9*: 59. For a discussion of Adler's accomplishments at the JPI, see Getis, *The Juvenile Court and the Progressives*, 92–97.

64. All of Chung's quotes in this paragraph come from her "Autobiography."

65. Elizabeth Lunbeck, *The Psychiatric Persuasion: Knowledge, Gender, and Power in Modern America* (Princeton, NJ: Princeton University Press, 1994), 118–119.

66. Getis, *The Juvenile Court and the Progressives*, 122–127; Lunbeck, *The Psychiatric Persuasion*, 121–126.

67. Cook County Hospital, *Annual Reports* (Chicago, 1918), 230.

68. *Twelfth, Thirteenth, and Fourteenth Annual Reports of the Municipal Court of Chicago* (December 2, 1917–December 16, 1920), Chicago Historical Society.

69. Emma Wheat Gillmore, "An Unprecedented Opportunity for Women," *Southern California Practitioner*, 1918, *33*: 133.

70. Lunbeck, *The Psychiatric Persuasion*, 34.

71. Satterlee, "Two Remarkable Women," 5.

CHAPTER FIVE

Professionalism versus Sexuality in the Career of Dr. Mary Steichen Calderone, 1904–1998

Ellen S. More

> Mary Calderone suddenly rose from obscurity to become one of the most controversial figures on the American scene. A Quaker grandmother in her sixties, crusading in the cause of sex—this was something sensational even in the sex-saturated United States.
>
> DAVID MACE, 1971

> Time and trouble will tame an advanced young woman, but an advanced old woman is uncontrollable by any earthly force.
>
> DOROTHY SAYERS, 1926

This essay proposes an alternative approach to the conflict between women physicians' professional identity and feminine gender norms. By highlighting the "intimate relation between identity and shame,"[1] I wish to focus on the competing influences of gender, professionalism, and *sexuality* in the professional identity-formation of women physicians. In the nineteenth century, many male physicians considered women doctors to be un-sexed. Because of their sordid participation in the study of anatomy, surgery, and even obstetrics and gynecology, they were charged with transgressing the bounds of gender propriety and sexual identity, although the latter claim rarely was made explicit.[2] By the end of World War II, explicit attacks on women physicians' sex/gender identity had waned; the presence of women in medical school classes gradually was assimilated into feminine professional and gender norms, while the concept of sexual deviance had achieved diagnostic status and, as such, had become identified not with physicians but with their patients. Yet even into the twenty-first century, professional and gender identity conflicts persist; anxieties still surface in discussions with and about women physicians over such issues as marriage and motherhood.[3] I suggest that this is primarily because of a submerged sexual text underlying modern discussions of gender and professionalism. For many women professionals, even doctors, recognizing one's own body as a site of knowledge and experience remains deeply problematic, evoking embarrassment or even shame. As all the chapters in this section demonstrate, ordinary displays of embodiment, such as one's physical carriage or mode of dress, or rites of passage such as choosing a life partner or to become pregnant, tap into unresolved issues of feminine sexuality and conflict with deeply entrenched structures of professional identity. This unyielding fact of life for American women professionals (and something which is not confined to medicine) shadows their life histories and informs our analyses.[4]

This chapter will map the problem of sexual identity onto the now standard discourse of gender and professionalism by analyzing the strategies adopted by Mary Steichen Calderone (1904–1998), physician and sex educator, to transform her own feelings of suspect sexuality into a source of personal power and cultural authority at a time of pervasive ambivalence toward women in authority. Calderone gained prominence as the medical director of Planned Parenthood Federation of America from 1953 to 1964 and then, from 1964 to 1982, as co-founder and director of the Sex Information and Education Council of the United States (SIECUS), a nonprofit clearinghouse for information about sexuality as a "health entity" and sex education. From the 1940s through the 1980s, she explicitly drew on personal experience to lecture and write on sexual health to popular and to professional audiences. Sexuality, she asserted, was a fundamental component of human identity, constitutive of mental and physical well-being. Calderone thus played a major role in popularizing a new discourse of sexuality expressed in the languages of personal liberation and public health. I will discuss the conflicting performative demands on Calderone, as a woman physician and as an advocate of sexual health.

"I Always Planned to Be a Heroine"

At her death in 1998, the *New York Times* declared Mary Steichen Calderone the "Grande Dame of Sexual Education," and during her lifetime, she received many plaudits for her work.[5] To the John Birch Society and the Moral Majority, however, she was nothing but "an aging libertine." One such opponent stood up after a lecture by Calderone in 1971 to accuse her of "rape of the mind." Reactions to her work on behalf of sex education ranged from "the sexiest grandmother" in the field to charges by the radical right of immorality and Communist sympathies.[6] Mary Calderone was once a household name—famous or infamous, depending on the household—during her heyday from the 1960s to the 1980s. Yet, there is little consensus on the nature of her contribution to twentieth-century sexual health education or her location along the political spectrum.[7] Calderone's extraordinary life, her thirty years in the public eye, and her artfully crafted performance as a national symbol of mainstream sexual morality beckon for deeper analysis. Her life, after all, spanned nearly the entire twentieth century. She embodied many of the contradictions in the American discourse on sexuality—competing visions of sexuality as a wellspring of personal empowerment, as an uncontrollable, psychobiological drive, or as a subject of scientific study and professional management.

Calderone's own upbringing was rife with such contradictions. Her father, renowned photographer Edward Steichen, translated his love of beauty into countless studies of the human form—from fashion layouts for *Vogue* to the sturdy populism of the Museum of Modern Art's exhibition, "The Family of Man." He combined esthetic ambition, technical precision, and journalistic populism. In his older daughter, Mary, these traits issued as an intense, if conflicted, sensuality, a faith in science, a sweeping vision of the role of public health, and a flair for the dramatic. Her mother, Clara Smith Steichen, however, was fully convinced that sexuality was dangerous and displays of the nude form, definitely not "nice." Her husband's own roving eye no doubt reinforced such convictions. Calderone usually credited her father as the chief influence on her life, but her mother also left her mark. Until the end of her life, Clara Steichen's bitterness heightened her daughter's sexual ambivalence but also stiffened her professional resolve.[8] Ostensibly a bohemian in her youth (Calderone was pregnant with her first child before her first marriage in 1926), she was never fully rid of an internalized sense of sexual shame. Even in her earliest public sex education lectures in the 1940s, she insisted on the linkage between early feelings of "fear and fright and of anger, or disgust . . . relating to matters of sexuality" and later struggles for self-acceptance as a sexual being.[9] Thus, although she acknowledged her mother's influence less often than her father's, it surfaced, as we will see, whenever Calderone considered why her ambition and intense drive led her to the field of sexuality. Calderone dedicated her career both at Planned Parenthood and at SIECUS to resisting the repression and stigmatization of

Mary Steichen Calderone, July 1, 1982, her seventy-eighth birthday. Reproduced with permission from the Schlesinger Library, Radcliffe Institute, Harvard University.

human sexuality. Whereas Margaret Sanger, founder of Planned Parenthood, was an unabashed proponent of uninhibited (hetero)sexual passion, Calderone's psychology was far more conflicted.[10]

But Calderone could not have won widespread renown if her work had not resonated with more than her personal struggle. Calderone reached adulthood in the 1920s, during what she saw as the "re-sexualization of women," but which might better be understood as the re-sexualization of everyone. Against a background of profound social and sexual transformation, Mary Calderone's professional persona oscillated between the rhetoric of sexual liberation and sexual responsibility. She embodied a complex character and performed it superbly.[11]

From several months after Mary Steichen's birth in 1904 until her family fled the war zone and returned to New York in 1914, Mary, her younger sister Kate Rodino Steichen (1908–1988), and their parents lived what Mary remembered as a "fairy tale" life in the tiny village of Voulangis, near Crécy-la-Chapelle, about forty miles east of Paris. Edward Steichen brought his family there so he could paint, photograph, and socialize with the French artistic avant-garde. The village lay among gently rolling hills, fields, and forests in a region that was beginning to attract artists because costs were low, housing available, and Paris reasonably close. A photograph from the period shows a modest, two-storied, graceful house, with shuttered "French" windows and an enclosed, densely flowered rear garden with a shed. Tangled wisteria clung to a trellis across the front door.[12] In contrast, for Mary's mother, Clara Steichen, the decade in France was a prelude to despair and divorce. Clara had given up hopes of a career in music to marry the charismatic Steichen, but she was never suited to be the wife of an artist—certainly not one as flamboyant as Edward Steichen. While despairing over her husband's frequent flirtations, she struggled to control her bright, willful, and sexually precocious daughter, Mary, who delighted Edward but tormented her. Clara always saw her older daughter as a reflection of Edward, "so selfishly disposed[,] so DEMANDING," as Clara wrote in 1916.[13]

Life at Voulangis, where Mary rode her own horse, attended the village school, and was enthralled by her father and his artistic circle, challenged and excited her.[14] Late in life she gathered together a few cherished items from her childhood in a folder labeled "Historical." They included a handmade birthday card to her father (addressed to "St. Edward"), a picture postcard of Voulangis, a school recitation notebook (in her good French handwriting, she copied out excerpts from the *Fables* of La Fontaine), photographs of her graduating class from medical school at the University of Rochester, and reprints from two leading psychoanalysts, one of whom she may have seen as a patient in the 1940s.[15] One item stands out from the rest. A childishly bold watercolor with vigorous brush strokes in vivid colors, it portrays a male figure standing on green grass at the center of the frame under a bright yellow sun. He holds an upraised scythe, his legs are planted apart, and he is surrounded by tall clusters of brilliant blue flowers, perhaps the delphiniums for which Steichen's garden was famous.[16]

I begin with this striking watercolor to try to understand the significance of Calderone's career and its distinctive linkage between private and public life, the needs of children, and the responsibilities of adults. What propelled her beyond a moderately progressive career as medical director for the Planned Parenthood Federation into a far riskier, high-profile campaign for sex education and sexual health? Calderone not only became a model and an inspiration for many men and women over the course of three decades, but she also embodied the dilemma of the modern professional woman in an era of sexual "revolution" and intimate display: Calderone *promulgated* sexual liberation while performing professional competence and feminine composure. This applied to her private performance as a sexual being as much as to her public performance as a professional expert. Thus, in 1982, the year she retired from SIECUS, Calderone declared, "I have grown up myself along with the human sexuality movement which is now all over the world." This is fair. But the full quotation yields the truer picture: "When I helped to form SIECUS I was sixty, and without being aware of it, I began to recover the sexuality I had experienced as a child because by then I was giving myself permission. I began then to be myself."[17] This is the core of Calderone's carefully nourished confessional narrative of sexual repression and recovery. It would behoove us to take her words with the utmost seriousness and respect, but equally, to engage in the necessary practice of deconstruction and revision.

To Calderone, public confessions were tools of personal reconstruction as much as public persuasion. Over time, with increasing explicitness, she linked her arguments for sexual health to her own experience of sexual oppression. As she told an interviewer in 1974 about her work with SIECUS, "We have . . . found ourselves freed in our personal sexual lives in terms of getting rid of some inhibitions. Our lives have been changed by many of these experiences."[18] Calderone's private story became the core of her public story.

Calderone's "sexual stories" never drew on neo-Freudian commentators on feminine sexuality of the postwar era such as Helene Deutsch or Karen Horney, and she clearly sided with William Masters and Virginia E. Johnson and Alfred Kinsey against traditional Freudians in the debates over female sexual responsiveness. But in crafting her sexual story over the course of her long career, she always began, in classic Freudian fashion, with her parents and the "enchanted" house in Voulangis.[19] The following "story" returns to her vivid watercolor: Imagine a sunlit afternoon at the house in Voulangis, imbued with the added glory of her father's garden boasting clusters of blue delphiniums. Amid all this, imagine the young Mary Steichen, about eight years old, tall for her age with a lush beauty, large, deep-set eyes, and an insatiable curiosity. Picture her following the family gardener, a seventeen-year-old from the village, to the back of the tool shed behind their house. Fascinated and puzzled, as she recalled many years later, she observed him expose himself "in full erection." Twice more the scene was repeated until her horrified parents discovered the two. In Calderone's retelling during an interview for *Playboy* in 1970, her father was completely appalled. Steichen "dragged" her into the house. Tearfully, he lamented, "'Now you have lost your innocence!'" He summoned the boy's father, made a fuss, and summarily fired the fellow. Although she insisted that the boy never touched her ("he wasn't even masturbating, as I remember"), Mary felt keenly that she was "bad" and "dirty," feelings that persisted for decades. Nevertheless, she kept her watercolor of the man in the garden at Voulangis—the gardener? her father?—all her life.[20]

According to her recollection, at that same age, her mother was still forcing her to wear aluminum mittens (with air holes) at bedtime to prevent masturbation, a measure begun, she believed, when she was three or four years old. There is considerable literature on the Victorian cultural and technological apparatus for control of masturbation, as well as medical rationalizations for such interventions. But Calderone needed no commentators to tell her that she had been traumatized.[21] She took every opportunity to remind her audiences that "my own 'liberation' from the effects of this trauma, and consequent sexual flowering, did not begin until I was well into my sixties."[22] At the age of seventy-five, she wrote:

> Until I was 9, I lived with my mother and father in a sort of magic, enchanted garden in France, a walled garden which my father has made famous in some paintings and photographs. It was a very free existence. He believed that nude bodies were beautiful for children. Nudity wasn't practiced so much by adults in those days. But I have a lovely photograph of him holding my sister's hand, my hand, one on each side of him, and our little nude bodies are prancing along with him in the garden. It's a lovely photograph.
>
> My mother had a bad influence on me. She was a sad, very unhappy woman. She was compulsive. She was angry. She was hostile. I've analyzed it looking backwards, and from my own knowledge, because she was very jealous of my father, who was a womanizer in a very sensitive and wonderful way. She took it out on me by being destructive and hostile and jealous and cutting me down all the time, with truly destructive phrases which I prefer not to repeat . . . I was fortunate to be separated from my mother by the time I was 10.[23]

Consider the girl's confusion. Calderone recalled no time in her life when, as she often put it, "I was not sexual," when she didn't derive pleasure from masturbation. She and her sister were encouraged by their father to appreciate the beauty and scientific wonder of nature—including the human body.[24] (Both girls also modeled in the nude for their uncle, the sculptor Willard Paddock, when they were about ten or twelve.)[25] But their mother heeded the darkest suspicions of Victorian doctors that unregulated sexuality was a sign of perversion, demanding constant vigilance, and iron control. Calderone's mother communicated the belief that sexuality, however pleasurable, was laden with embarrassment, shame, and risk. Her daughter's striking watercolor, however,

Photograph (possibly by Edward Steichen) of Mary Steichen, 1912, age 8. Reproduced with permission from the Schlesinger Library, Radcliffe Institute, Harvard University.

reminds us that childhood fascinations are not so easily governed. When Calderone recalled these childhood sexual traumas, she aimed her accusations at her mother, not her father or the gardener. As her sister later observed, "Mary was always Daddy's girl."[26]

Practitioners of oral history often remark on the way subjects "weave and reweave" their stories in the telling. Calderone, though, always associated her own sexual difficulties with her mother's oppressive intervention against her masturbating, never with her father's embarrassment and shock over her voyeuristic rendezvous (and abuse) in their Voulangis garden. Calderone was convinced that "you carry your parents around with you to the day you die; you internalize a parent figure who may not be exactly like your parents but who is probably based on them, who says you should or you shouldn't." The father she "carried around" was clearly an idealized figure; Calderone's papers and public utterances are replete with references to Edward Steichen as a "tremendous man, passionate, marvelous, gifted." Her mother was "a terrible incubus I'd been carrying around all those . . . years."[27]

It seems no less than the truth to say that she did carry her parents around with her throughout her life. These formative events eventually energized her career, as she campaigned for sexual pleasure but also for sexual responsibility. In 1977, for example, in preparation for a talk at the Brearley School in New York City (her alma mater) titled, "Look Where I Am—How Did I Ever Get Here?" Calderone began her outline with "Heredity," followed by "Early Environment—neg. att. Re masturbation." Two years later, at the age seventy-five, at the Schlesinger Library during a lunch in her honor, she spoke about the genesis of her work with SIECUS, beginning with her "remarkable background . . . my father, Edward Steichen." But, she continued, "Sexually speaking [it was] a very damaging environment." Turn-of-the-century physicians were "up in arms over the very dangerous act of sexual self-pleasuring in children." As for her oppressive relationship with her mother, she concluded, "This had a very severe effect later on in my sexual adjustment in marriage."[28] Three months before her eightieth birthday, she responded sharply to a written request for marital advice that harkened back to these childhood experiences: "Your letter," she wrote, "indicates lack of thought and also total lack of awareness of the terrible difficulties that many people have in experiencing sexual pleasure in marriage. It has been identified as having two major causes—interference with the normal process of experiencing one's own genital pleasure as a child, and being brought up in a very authoritarian way with no understanding of why certain things are demanded."[29]

Edward and Clara Steichen separated within months of the family's return to New York in 1914. Mary stayed with her father, and Kate returned to France with her mother. Fortunately for Mary, her father managed to board her with Alfred Stieglitz's brother and wife, Dr. Leopold (Lee) and Elizabeth (Lizzie) Stieglitz, while he sent her to the rigorous as well as progressive Brearley School. Mary attended Brearley from her first year back in New York until she left school a year early in 1921 to begin college at Vassar. She majored in chemistry, intending to go into medicine, but as an expression of a growing need for self-expression and experimentation, she changed her major to theater arts. When she graduated in 1925, she became an intern at the American Laboratory Theatre in New York. There she soon met the handsome W. Lon Martin, a fellow intern with whom she soon entered into a passionate relationship. In 1926, after she became pregnant with their daughter Nel, they married. Their second daughter, Linda, was born two years later, by which time Mary's theatrical ambitions were put aside for good. But the strain of having two young children, little money, and no career prospects added to what Calderone later described as her own inhibitions, quickly cooled their ardor. They separated within a few years of Linda's birth and divorced around 1933. With two young girls to help support and no particular job skills, she entered a period of directionless drift.

Mary Steichen's headlong rush into sexual intimacy and unplanned pregnancy never sat easily beside her desired self-image as someone in control of her destiny. Despite her proximity to her father's bohemian circle and her own colleagues at the Lab theater, she seemed to have had no inner compass, no real sense of her own potential abilities. Sociologist Joseph Kirk Folsom, who was on the faculty of Vassar during the same years that Steichen was a student there, wrote in 1928 that a "woman may conscientiously allow herself to *feel* passion to the same extent as the man, if she controls its expression." An unpublished study by Kinsey, cited in Modell, documented a steep rise in the percentage of white male and female adolescent foreplay, or "erotic petting," something Modell describes as the "sexualization of the whole path to marriage."[30] Quoting from an article on the sexual revolution by sociologist and SIECUS board member Ira Reiss during a presentation in 1967, Calderone underlined his words: "What was done by a female in 1925 acting as a rebel and a deviant can be done by a female in 1965 as a conformist." But Steichen's rebellion looks a lot like desperation, not the self-conscious act of a social revolutionary. Indeed, she readily admitted feeling she "had trespassed for having had premarital sex with my first marriage."[31]

During this rocky period, she underwent psychoanalysis, apparently taking advantage of the low-cost treatment provided by analytic candidates at the New York Psychoanalytic Institute. She later remembered, "I underwent an '*echt*' [pure, unadulterated] analysis, so I talked for three years and never knew my diagnosis or prognosis. I rarely talked about my sexual feelings except the ones of anger at my mother for making me wear those outrageous mitts."[32] As she reflected, half a century later, "Nothing seemed to go right. So my analyst suggested that I go to . . . the Stevens Institute to take the Johnson-O'Connor Aptitude Tests . . . They showed me to have too many skills and my tendency was to become restless if I didn't use all of them. So they showed me how to avoid that . . . That was [going into] medicine."[33]

Her season of underachievement and self-doubt, what she called a "messy period," ended with medical school.[34] Despite the sudden blow of her daughter Nel's death from pneumonia in 1935, Calderone attended the University of Rochester, graduating thirteenth in her class in 1939. Her identity as a public health crusader began to crystallize.[35] Becoming a physician represented more than a way to make a living or even a code of conduct by which to live. It was a script for conducting the rest of her life, a white coat, literally and figuratively, with which to contain her unruly life. "As to professionalism," she wrote in 1984, "let me say that professionalism was my most important tool for advance, and refuge from attacks, in dealing with two highly charged and undeveloped areas like birth control (1953–1964) and sexuality (1964 on). I never go beyond the data I have, and I always disclaim competency in any area . . . that I know I do not have or have only marginally or intuitively."[36] Yet the problem of roles persisted. For a beautiful and highly sensual woman to become a physician, and then to identify herself with sexuality, took an astonishing blend of self-confidence and self-denial. Calderone occasionally repeated what Dean George Whipple once said of her, that she never made use of being a woman but never forgot it either. Addressing popular audiences she tried to appear warm and grandmotherly, though with a touch of glamour. To physicians, she emphasized her commitment to the science of sexuality medicine. But in person, some associates remembered her much as she was portrayed in a damaging passage in the book *Oh! Sex Education!* by the young reporter Mary Breasted: as cold, somewhat arrogant, the "personification of *noblesse oblige*"—the same mixture of intelligence, condescension, and insecurity her mother struggled to subdue in Voulangis many years earlier. Sociologists might call this "role strain." Postmodernists might see her as the avatar of the multilayered performance. For Calderone, her jostling identities sometimes left her exhausted.[37]

Following graduation, she and daughter Linda returned to New York where she completed an internship in pediatrics at Bellevue and then began a master's degree in public health. During her fellowship, she met her second husband, Dr. Frank Calderone, who was in charge of the Health Department's Lower East Side Health Center where Mary completed her public health training. The two were married in 1941. After her marriage and her graduation in 1942, Mary Calderone was employed briefly by the American Public Health Association. But, at age thirty-nine, she became pregnant with their first daughter, Francesca, who was born in 1943. Frank had become deputy commissioner of Public Health for New York, but left the health department in 1946 to direct the Interim Commission for the World Health Organization and, later, the Health Service for the United Nations Secretariat in New York. With another daughter, Maria, born in 1946, and Linda just finishing high school, Mary was content—for a few years—to live on an estate on Long Island Sound and play the part of "glamorous housewife."

Most accounts of Calderone's life during the decade between her second marriage and her decision to work for Planned Parenthood convey a gracious, upper-middle-class opulence. As *Playboy* editor Nat Lehrman, who lived nearby and knew her during this period, noted, "Dr. Calderone is a moderately well-to-do woman who could be spending her days at home or in the serenity of a 'safe' job." It took some ingenuity to flout postwar expectations of women, even professional women, to stay home and be "good mothers," especially if one cared about one's reputation.[38] But as soon as the younger girls were in school, Calderone began work as a school physician in Great Neck ("to keep her hand in," she said), while also volunteering for the Mental Health Association of Nassau County to give PTA lectures on parents' roles in sex education for school-age children.

"That, Too, Is Sex Education"

A typescript manuscript in her papers from 1948 through 1950 records one of these informal lectures to parents. Remarkably enough, this early speech, delivered more than fifteen years before the creation of SIECUS, contains the essence of most of Calderone's ideas on sexual education as well as strong hints of why she determined to make the subject her own.[39]

Calderone's lecture occurred during a deceptive moment in the long cultural debate over how to talk and think about human sexuality. Although few overt changes in sexual mores were visible to the culture at large, below the surface, seismic shifts had been occurring since at least the end of World War I. Popular marriage manuals published in the 1920s and early 1930s reflected the public's new acknowledgment that marital "satisfaction" encompasses *sexual* satisfaction. Marital guides by Margaret Sanger, Theodoor Van de Velde, Robert Latou Dickinson and Lura Beam, and Hannah and Abraham Stone argued for equality of sexual satisfaction between husband and wife, an egalitarianism based on a frank understanding of sexual anatomy.[40]

Historians like Beth Bailey have also demonstrated that, with quiet persistence, sexual behavior among college students and other youth in the 1940s had become more experimental, more daring, more defiant of longstanding verities that forbade erotic foreplay—much less, outright coitus—between unmarried young adults. Even before publication of the Kinsey reports, postwar geographic dislocation, the rush to make up for years of sexual privation in the military, and the greater sexual maturity of the GIs now visible on most campuses, together produced a low-decibel sexual explosion—heterosexual and homosexual alike—on college campuses and in cities nationwide. Yet, as sociologist John Modell concluded, "Most data on adolescent sexuality between 1945 and the mid-1960s point to no major enlargement during this period of the role of coitus or even heterosexual genital play more generally—even though attitudes apparently shifted enough in an accepting direction to alarm many adults into inferring a real increase in adolescent sexual behavior."[41]

Publication of the first of the so-called Kinsey Reports, a multivolume study sponsored by the National Research Council's Committee on Research in Problems of Sex and funded by the Rockefeller Foundation, unleashed a surge of public self-searching and debate.[42] *Sexual Behavior in the Human Male*, published the same year as Calderone's PTA lectures, forced the public to engage openly with the changes in sexual attitudes it claimed to document.[43] Among the most disconcerted by Kinsey's findings may have been those who ought to have been the least surprised, professional sex educators, many affiliated with the American Social Hygiene Association. The ASHA had been working to reform the public schools' approach to marriage and family life education, including sex, since the founding of its parent organization, the Society for Sanitary and Moral Prophylaxis, in 1905. To many ASHA members, the report of widespread sexual activity was plain evidence of the moral decline of the current generation of youth, not, as Kinsey intended, data documenting slowly evolving sexual mores.[44] Teachers, too, were unhappy with the report's implications. Few were comfortable including sex education in public school curricula even in the staid form of family life education propounded by the ASHA. Despite two decades of curriculum development, including uplifting essays in popular magazines and the like, most teachers disdained discussing sexuality in the classroom.[45]

Calderone's lectures thus were one response to the simmering unease in the years after World War II over just what adolescents were thinking and doing and what role their elders could play in keeping them under reasonable control. But PTA lectures like hers, which were sponsored by the county Mental Health Association, represented a different agenda from the curricular efforts of the ASHA, although both had as their immediate object the improvement of sex education for pre- and postpubescent youth. Rather, it was part of an initiative by "mental hygienists" to advance what came to be known as "positive mental health." As one of its chief apostles, Frankwood E. Williams of the National Committee for Mental Hygiene, wrote, "mental hygiene activities . . . center less on the matter of [psychiatric] hospitals and more on the schools, the colleges, and the home." The mental hygiene movement reached out to professionals other than psychiatrists since they conceived of their task as essentially preventive, much in the spirit of public health.[46] Thus, Calderone's PTA sex education lectures were invested with the effort to update—but not revolutionize—sexual attitudes in light of the widespread social changes and perceived stressfulness of postwar life.

The personal and professional meaning of such a role for her should be examined in light of the symbolically ambiguous standing of the "woman physician" in the postwar cultural landscape. As previous works on twentieth-century women professionals have suggested, women physicians, like scientists or lawyers, were not immune to conflicting gender expectations for professional women who wanted to do more than remain at home and raise good and loyal Cold War American children. Calderone, comfortably married with school-age children and a prominent and successful husband, already possessed of modest professional success, agreed to lecture on one of the riskiest subjects on which women could face a public audience.[47]

Yet, if her surviving manuscript is any guide, Calderone deviated significantly from the typical script. Her own life experiences had diverged dramatically from the abstinence-till-marriage message of most family life educators, and she was not loath to say so: "I knew about birth control because by [the 1930s] I had been to see the great [Dr.] Hannah Stone as a patient. She and her husband [Dr.] Abraham Stone were pioneers in the family planning and marriage and sex counseling movements. She fitted me with a diaphragm, of course."[48] Calderone's own turbulent past plus the bohemian world she grew up in may have prepared her to take on sexuality in a fresh and relatively unabashed way. But more than that, Calderone's conviction that sexual honesty, not hypocrisy, would lead to sexual health was the impetus for a strikingly different approach to sexuality as a public health issue. Echoes of the "sexual enthusiasts," as Paul Robinson termed them, can be heard in her insistence on a positive attitude toward sexuality as a precondition for positive sex education.[49]

Yet Calderone's views are in fact more radical, not with respect to adult sexuality but with respect to the sexual experiences and sensibilities of children. Nothing exemplified this more than her assertion of the child's right to masturbate—in private—without suffering the oppressions of adult ridicule, shame, or punishment. Indeed, Calderone heartily recommended the practice as an emotional outlet both for children and adults. Even a passing glance at the literature cited in her own works reveals the individuality of her attitudes.[50] Margaret Sanger, for example, could not be outdone in her fervor for the joys of marital passion. As she exhorted her readers, "Never be ashamed of passion. If you are strongly sexed you are richly endowed." Still, she wrote that, to build up one's creative and sexual resources, "the young must learn to refrain from lesser sex experiences and temptations." Hannah and Abraham Stone's marriage manual, published over a decade later than Sanger's and the product of careful study and unusually wide reading, flatly stated that no physical harm could come from masturbating. Yet, they still felt it prudent to add that, "when carried out at frequent intervals and over a long period of time . . . masturbation readily lends itself to excesses." They concluded that such consequences were "rare" and not applicable to "the average individual who resorts to the moderate practice of masturbation during youth and adolescence," but they strongly implied that masturbation in adults, at least as an end in itself, was nothing to celebrate.[51]

While Calderone cannot be interpreted as a radical feminist avant la lettre, propounding masturbation as a highway to sexual autonomy for women—she never conceptualized her advocacy of masturbation in politicized language nor was she ever comfortable being labeled a feminist; nevertheless, she called for accepting masturbation as a suitable form of sexual release. Masturbation was healthy, she believed, both for the individual and, ultimately, for society in that it generated the sense of bodily ease necessary for a positive sexual adjustment in interpersonal sexual relations.[52] Facing an audience of largely middle-class, suburban parents with, presumably, conventional ideas about sex education, Calderone jettisoned the typical, cautious approach to masturbation. Her primary concern was that children be given a positive message about the rightness of bodily pleasure, a stance that would characterize her whole career. For Calderone, indeed, masturbatory pleasure provided the foundation for all subsequent sexual pleasure.

She began on a deceptively conventional note, a well-dressed, attractive, well-to-do matron addressing others much like herself. Parents were ideal sex educators, she began: "Who else should give the child these facts, if not the parent? Who else does the child have the same kind of confidence in, and close feeling about?" She also advised that children not talk about these matters with their friends because their friends' parents "like to give this information to their own children [themselves] . . . They are the child's parents and they have the right to say how the child shall be told."[53] But soon she unveiled an unconventional frankness and enthusiasm for her subject, observing that parents often feel "they don't have any idea of how to talk to their children about [sex]—how to let their children in on the mystery, so to speak, the wonderful mystery. So, being scared, or embarrassed, or ashamed, they do nothing. And this is tragic." Then she proceeded to prod her audience gently into admitting that "sex education" might mean more than "how the baby grows in the mother" or "how the baby gets out of the mother," or even "how the baby got into the mother in the first place." As she candidly told her audience, "My job, I think, is to help you achieve good feelings about sex. If necessary—to change your feelings. Once you feel that sex is right, and warm, and a good part of life, you will have no difficulty in letting your child in on this right and warm and good thing. At the right time, the right words will come to you."[54]

Unlike most sex education speakers, then and now, Calderone emphasized that reproduction and sexuality were distinct, if related, subjects. Beginning with the facts of reproduction, she suggested that a young child's questions about *pregnancy* might be a good opportunity to begin the dialogue about *sexuality*. But her lecture had nothing to say about the process of sexual arousal nor gave advice to parents on how to talk about it with their children. This is not surprising. Discussions of sexual technique were to be found nowhere in mainstream literature for children and youth. More than half a century later, parents still are uncomfortable discussing such matters with their children.

Calderone's stated goal was to break the "vicious chain" of negative sexual attitudes that, she believed, were unconsciously passed down from parents to children, ultimately inhibiting their capacity for healthy, loving sexual relationships as adults and warping the attitudes they pass on to the next generation. Here, she drew on memories of her experiences with her mother. Although she expressed the hope that fathers would participate in educating the child in an equable acceptance of sexuality, Calderone mostly addressed the mothers of the audience. For example, she observed that boys often begin their "sex" education with the discovery of their own bodies at bath time. She emphasized that "how the mother reacts to that—what she does, the tone in her voice, the look in her face—that's all part of the baby's sex education. If the mother slaps his hand and shouts at him . . . that, too, is sex education," although surely not a desirable one.[55]

Mary Steichen Calderone, 1967. Photograph by Jack Marshall and Co. Reproduced with permission from the Schlesinger Library, Radcliffe Institute, Harvard University.

A few minutes later, she returned to the theme of parental attitudes, this time tackling the subject of masturbation directly. Calderone was at pains to convince her audience that, according to psychiatrists, young children have sexual feelings, which they experience unwittingly at first and then deliberately through masturbation. "Psychologists tell us that [very young children's sexual feelings] are apt to be quite strong, and they are also quite normal.

> Most little children play with their sexual organs. Parents are apt to react rather violently to this, so most children learn to do it in private . . . Psychiatry assures us that in the young child this is normal and harmless—that is, masturbation is normal and harmless. Some psychiatrists even say that the only harmful thing about it is . . . especially, the sense of guilt that can fill a child with shame and fear.

> Here, again, I have had letters from young people about this. Shame, fear and guilt are terrible things for a young child to struggle with all by himself. And they are entirely unnecessary.[56]

How Early Experiences Can Affect You

Twelve years later, when Calderone published her own book of sexual advice aimed at married couples, she had been the medical director of Planned Parenthood Federation of America for six years. Her book, *Release from Sexual Tensions*, drew together many of her responses to the letters requesting marital advice she had received over the years. The book was well received, with the exception of Goodrich Schauffler, a Portland gynecologist, a columnist for the *Ladies' Home Journal*, and a friend. Schauffler praised the author's dexterity in "skating on thin ice between common horse sense and 'ultradynamic psychiatry.'" But, infuriating Calderone, he complained that she dealt "too warmly with the subject of masturbation."[57] The following vignette from Calderone's book may illustrate what alarmed Dr. Schauffler as well as the psychological sleight of hand by which Calderone's childhood ghosts, her mother, her father, the young gardener, became the rhetorical mainstays of her professional life. It also reminds us of the many destructive pressures on women during the 1950s, pressures to leave the workplace, marry, have children, and be a spectacular sex partner (defined as having vaginal orgasms) with your martini-sipping husband, home after a long day out in the lonely crowd. It occurs in a chapter titled "Doctor, My Husband Says I'm Frigid." Calderone's approach certainly does nothing to counter the systemic or structural forces arrayed against the American woman at the time, or the "myth of the vaginal orgasm."[58] But neither did the book accept the idea that inorgasmic women are either inherently "frigid" or are suffering the ill effects of deviant, unwomanly behavior:

> First, you must return in your mind to when you were a little girl. What do you remember about the sex instruction that you received? . . . For instance, can you remember once when your mother "caught" you playing with yourself in the genital region? This must have happened at least once—it is such a common occurrence. Do you remember how you were feeling before she saw you? Didn't it feel warm and comfortable? Didn't you have a lovely sensation while playing with yourself? Try to recapture the actual feeling, try to feel that way now . . . Now try to remember back to that moment when your mother caught you. Did you feel a

> stunned surprise at her onslaught? Did she scold you? Did she slap your hands or perhaps threaten to cut them off? (Hard to believe, isn't it? But some mothers have done this.) Did she tell you that you were a bad, dirty girl and God would punish you? Did she say or imply that if you persisted . . . you might become insane?[59]

Here I will give Calderone the last word, taken from her reply to Schauffler: "As far as masturbation goes," she wrote to him, "I think you are all wrong. I don't mind saying so to you. I don't think you can deal too warmly with the subject. It is the first sexual activity in the lives of most human beings and, therefore, the one most likely to be associated with guilt and trauma in their memories—conscious or unconscious. If left alone, it is a self-regulated activity—a real safety valve . . . and I will not concede that it is not normal."[60]

Conclusion

For another twenty-five years, Calderone continued her work for sex education to promote "sexuality as a health entity," in the words of the SIECUS mission statement. Crisscrossing the United States to meet with educational and lay organizations, appearing at conferences around the world, on television, and in newspaper stories, she effectively became a symbol for changing sexual attitudes among Americans of all ages and political persuasions and especially her fellow physicians. Her success can be measured against the virulence of her radical right-wing opponents, beginning with the John Birch Society and continuing with the Moral Majority. They vilified her work, but more than that, they vilified her person—her character and beliefs.[61] Calderone's successful evocation of her private trauma to create a complex performance of feminine, but professional, activism evoked the kind of rage that all iconoclasts bring on themselves. Her portrayal of the woman physician as feminine, sexual, yet professionally expert, drew applause and attacks; it drew blood. Calderone successfully transformed her private trauma into a professional advocacy for sexual health through sex education, and she reached a worldwide public.

NOTES

Epigraphs. David Mace, "A Quaker Portrait: Mary Steichen Calderone," *Friends Journal*, March, 16, 1971, pp. 166–168, quotation p. 166; Dorothy L. Sayers, *Clouds of Witness* (1926; reprint New York: Harper & Row, 1964), 316.

1. I wish to express my gratitude to the Radcliffe Institute for Advanced Study for the

award of a Fellowship in 2000–2001, to the National Endowment for the Humanities for a Senior Research Fellowship in 2004–2005, and to the University of Texas Medical Branch at Galveston for a President's Faculty Development Leave in 2005, without which my work on Calderone would not have been possible. For the quotation, see Heather Love's review of Eve Kosovsky Sedgwick, *Touching Feeling: Affect, Pedagogy, Performativity* (2003): "The Performative and the Peri-," *Women's Review of Books*, 2004, *20*: 12–13.

2. Regina Morantz-Sanchez, *Conduct Unbecoming a Woman: Medicine on Trial in Turn-of-the-Century Brooklyn* (New York: Oxford University Press, 1999); and Michael Sappol, *A Traffic in Dead Bodies: Anatomy and Embodied Social Identities* (Princeton, NJ: Princeton University Press, 2002).

3. Ellen S. More, *Restoring the Balance: Women Physicians and the Profession of Medicine, 1850–1995* (Cambridge, MA: Harvard University Press, 1999).

4. See the literature for women in law, such as Mona Harrington, *Women Lawyers: Rewriting the Rules* (New York: Alfred A. Knopf, 1994), 97–150, 231–255.

5. Subhead quotation from Mary Steichen Calderone, in Mary Brannum, *When I Was Sixteen* (New York: Platt and Munk, 1967), 147–166, quotation on 162. Although she stoutly resisted the "feminist" label, she would have been pleased to know she was inducted posthumously into the National Women's Hall of Fame at Seneca Falls where she was fulsomely memorialized. See SIECUS Vertical Files, "Mary Steichen Calderone, National Women's Hall of Fame, 1998," SIECUS Library, New York City.

6. Jean Otto, "Sex Education Is Happening, Dr. Calderone Says," *Milwaukee Journal*, January 20, 1971, p. 7, Carton 23, unprocessed, fol. "January 20, 1971," in Mary Steichen Calderone Collection, Schlesinger Library, Radcliffe Institute for Advanced Study, Harvard University, Cambridge, Massachusetts [hereafter, MSC/SL]. In 1971, Calderone spoke on sex education at Marquette University. She was picketed. During the question-and-answer period following her talk, a woman approached the microphone and said, "Dr. Calderone, I accuse you of rape of the mind." Also see Harold I. Lief, "In Memoriam: Mary Calderone, MD, MPH," *Journal of Sex Education and Therapy*, 1998, *23*: 113–114. Lief also hailed her as "our field's sexiest grandmother" at the award ceremony for the American Association of Sex Educators and Counselors, April 1973, in Carton 24, unprocessed, fol. "Annual Award, AASEC," p. 3, MSC/SL; Jane E. Brody, "The Grande Dame of Sexual Education Is Dead at 94," *New York Times*, October 10, 1998, p. 47.

7. For an overview of Calderone's career, see More, *Restoring the Balance*, 205–212, and idem, "Mary Steichen Calderone," *Notable American Women: Completing the Twentieth Century*, ed. Susan Ware (Cambridge, MA: Belknap Press, 2004), 99–101. As examples, see James R. Newman, "A Conference on Abortion as a Disease of Societies," *Scientific American*, 1959, *200*: 149–254; Fowler V. Harper, "Book Review," *Yale Law Journal*, 1958, *68*: 395–398; David J. Garrow, *Liberty and Sexuality: The Right to Privacy and the Making of* Roe v. Wade, 2nd ed. (1994; Berkeley: University of California Press, 1998), 275, 280; Jeffrey P. Moran, *Teaching Sex: The Shaping of Adolescence in the 20th Century* (Cambridge, MA: Harvard University Press, 2000), 161–165. Donald T. Critchlow treats Calderone as a minor figure, although he acknowledges her influence over John D. Rockefeller III. Critchlow, *Intended Consequences: Birth Control, Abortion, and the Federal Government in Modern America* (New York: Oxford University Press, 1999), 194.

8. Author's interview with Mary S. Calderone, Rochester, New York, February 2, 1984; Penelope Niven, *Steichen: A Biography* (New York: Clarkson Potter, 1997), 302–304, 636; interview of Mary Steichen Calderone, August 7, 1974, Family Planning Oral

History Project, OH-1/Calderone, transcript, James W. Reed, interviewer, pp. 1–47, esp. p. 1. Cf. Mary Steichen Calderone, *Particular Passions: Talks with Women Who Have Shaped Our Times*, ed. Lynn Gilbert and Gaylen Moore (New York: Clarkson N. Potter, 1982), 255–263.

9. [Mary S. Calderone], "My Name Is Mary Calderone," typescript, pp. 1–21, quotation on p. 19, in Box 13, fol. 222, MSC/SL. Calderone annotated this as probably from 1948 to 1950, when she was giving sex education talks to the local PTA on behalf of the Nassau County Mental Hygiene Association. Also cf. Mary S. Calderone, "Draft Autobiography," ca. February 22, 1984, p. 9, typescript in author's collection. This was the first of two short drafts Calderone sent to me over four months. Although the directory for her computer lists thirty-five file names, none of them was found in her collected papers at the Schlesinger Library as of 2001, Box 25, loose papers, unprocessed, MSC/SL.

10. On Margaret Sanger, see Ellen Chesler, *Woman of Valor: Margaret Sanger and the Birth Control Movement in America* (New York: Simon & Schuster, 1992), 95, 96.

11. Calderone attributed the phrase "re-sexualization of women" to the sociologist Jessie Bernard. Cf. Nat Lehrman, "The Playboy Interview," *Playboy*, 1970, *17*: 63–78, 154, 236–240, quotation p. 154. For the intellectual context, see Mari Jo Buhle, *Feminism and Its Discontents: A Century of Struggle with Psychoanalysis* (Cambridge, MA: Harvard University Press, 1998).

12. Description of the Steichen house from photographs, ca. 1925, in fol. "Antiquities"; MSC, "A Walk thro' the 20th Century: I'm Bill Moyers," handwritten notes," fol. "Reference Papers: 'my own voice' / Autobiography and Handwritten Jottings," 1984; both in Carton 25, unprocessed, MSC/SL. The "Antiquities" and "Autobiography" folders in these unprocessed papers include the following items: Mary Steichen's poems, ca. 1913, two photographs of the house, one showing Mary standing in the front, holding a cat; royalty statements and market research for her first books; photography books for pre-readers for which she wrote the text and her father took the black-and-white photos.

13. Dolores Alexander, "A Look at Mary Calderone: The Grandmother of Modern Sex Education," *Newsday*, February 22, 1966, pp. 27–29, Box 17, fol. 286, "The Grandmother of Modern Sex Education," in MSC/SL, quotation on p. 27. Letter from Clara Steichen to Alfred Stieglitz, September 18, 1916, as quoted in Niven, p. 440. Caps in original.

14. Mary Steichen Calderone to Monsieur et Madame Michel Brousse, Bourdeaux-Cauderan, France, April 12, 1984, Carton 1, fol. "Organization A-M, 1984–5–6," unprocessed, MSC/SL.

15. Franz Alexander, "Psychoanalysis and Medicine," *JAMA*, 1931, *96*:17 (April 25): 1352–1358; Edmund S. Bergler, "Some Recurrent Misconceptions Concerning Impotence," *Psychoanalytic Review*, 1940, *27*: 450–466. Bergler's article was signed, "With the writer's compliments." Both are in Carton 25, fol. "Historical," unprocessed, MSC/SL. Bergler, a strict Freudian, published a book titled *Neurotic Counterfeit Sex* in 1951, which described the feminine inability to have vaginal orgasms—that is, *genuine* orgasms—as "a mass problem." Cf. *Our Bodies, Our Selves*, 1st ed., 3rd printing (Boston: Boston Women's Health Course Collective, 1971), 9–24, quotation on p. 16. For his homophobia, see Jennifer Terry, *An American Obsession: Science, Medicine, and Homosexuality in American Society* (Chicago: University of Chicago Press, 1999), 308–313.

16. *Steichen's Garden*, a painting by Arthur B. Carles, memorializes the dense cultivation.

17. Calderone, *Particular Passions*, 255–263, quotation on p. 261.

18. Transcript, Interview of Mary Steichen Calderone by James W. Reed, August 7, 1974, Family Planning Oral History Project, OH-1/Calderone, pp. 1–47, quotation on pp. 39–40, SL.

19. Buhle, *Feminism and Its Discontents*. Calderone undertook Freudian psychoanalysis in the 1930s and in the 1940s. See note 32. The phrase "sexual story" comes from Ken Plummer, *Telling Sexual Stories* (London: Routledge, 1995).

20. Nat Lehrman, "The Playboy Interview," 154.

21. In her "Draft Autobiography," Calderone wrote that during her first encounter with psychoanalysis she dwelt on her anger over "those outrageous mitts," p. 9. For late-nineteenth-century attitudes toward masturbation, see Thomas A. Laqueur, *Solitary Sex: A Cultural History of Masturbation* (New York: Zone Books, 2003), 46, 431 n. 42; Helen Lefkowitz Horowitz, *Rereading Sex: Battles over Sexual Knowledge and Suppression in Nineteenth-Century America* (New York: Alfred A. Knopf, 2002), 106–107; John S. Haller and Robin M. Haller, *The Physician and Sexuality in Victorian America* (New York: W. W. Norton, 1974), pp. 105, 202–211; C. D. W. Colby, "Mechanical Restraint in a Young Girl," *Medical Record in New York*, 1897, *52*: 206; Vern L. Bullough, "Technology for the Prevention of '*Les Maladies Produites Par La Masturbation*,'" *Technology and Culture*, 1987, *28*: 828–832; Moran, *Teaching Sex*, 9–10.

22. Mary S. Calderone, "Draft Autobiography," 12; Lehrman, "Playboy Interview," 154; Calderone, *Particular Passions*, 259; Gold, *Until the Singing Stops*, 318–319; Audio Tape, Dr. Mary Calderone, "Lecture/Luncheon, October 22, 1979," Acc. no. 79-m261, T-42 (2), SL.

23. "Dr. Mary Calderone," in *Until the Singing Stops: A Celebration of Life and Old Age in America*, ed. Don Gold (New York: Holt, Rinehart and Winston, 1979), 311–328, esp. p. 315. About the imagery of the garden, Thomas Laqueur points out in *Solitary Sex* that feminist literature on masturbation from the 1970s often employed the image of a "secret garden," possibly in tribute to Frances Hodgson Burnett's children's book, *The Secret Garden*, published in 1909, when Mary was five years old. Cf. Laqueur, pp. 79, 438 n. 99.

24. Mary Steichen Calderone, " . . . My Father," *Infinity*, 1955, December 1954–January 1955, pp. 4, 5, 16, quotation on p. 4.

25. Mary S. Calderone to Donelson F. Hoopes, April 20, 1967, Carton 1, unprocessed, fol. "B," MSC/SL.

26. Quoted in Niven, *Steichen*, 390.

27. The description of her father concludes with the words, "and a great womanizer, as [my mother] must have known by then"; Calderone, *Particular Passions*, 260.

28. Mary Steichen Calderone, handwritten outline, "Look where I am—how did I ever get here?" September 22, 1979, Carton 1, unprocessed, fol. "Brearley 1977," MSC/SL. Cf. Audio Tape, Dr. Mary Calderone, "Lecture/Luncheon, October 22, 1979," SL.

29. Mary Steichen Calderone to "JK" [pseud.], April 5, 1984, Carton 1, unprocessed, fol. "Individual I-Z, Pinks, 1984–5–6," MSC/SL.

30. John Modell, *Into One's Own: From Youth to Adulthood in the United States, 1920–1975* (Berkeley: University of California Press, 1989), 75–77, 82–93, quotation on p. 97; Joseph Kirk Folsom, "Observations on the Sex Problem in America, *American Journal of Psychiatry*, 1928, *8*: 529, quoted in Modell, *Into One's Own*, 93. Also see Beth Bailey, *Sex in the Heartland* (Cambridge, MA: Harvard University Press, 1999), 1–80;

Elaine Tyler May, *Homeward Bound: American Families in the Cold War Era* (New York: Basic Books, 1988), 101, 116; John D'Emilio and Estelle B. Freedman, *Intimate Matters: A History of Sexuality in America* (New York: Harper & Row, 1988), 239–274; Paul Robinson, *The Modernization of Sex* (New York: Harper & Row, 1976).

31. Dr. Mary Calderone, "Human Sexuality—Attitudes and Education," Keynote Address delivered at a symposium sponsored by Ortho Pharmaceuticals, September 23, 1967, Toronto, Canada, Carton 1, unprocessed, fol. "Vita, contd.," MSC/SL. Italics in original. Mary S. Calderone, "To Live or to Die," *Theory into Practice*, 1969, *8*: 302–303, Box 17, fol. 275, MSC/SL.

32. Draft autobiography, Typescript, p. 9; "Eunice B. Armstrong Is Dead; A Writer and Psychoanalyst," *New York Times*, June 30, 1971, p. 50, in Box 1, fol. 1, "Biographical," MSC/SL. A handwritten annotation in Calderone's hand reads, "Mrs. Armstrong was MSC's analyst from around 1930–1932." Cf. Brannum, *When I Was 16*, 153. Calderone also undertook analysis in the 1940s, and a signed copy of a 1940 article on impotence by the ultra-Freudian analyst Edmund Bergler in her files might indicate that he was her physician. If so, and there is no direct evidence for it, he may have been the subject of her comment regarding her "*echt*" analysis. On Bergler, see Terry, *An American Obsession*, 308–314.

33. Interview with Ellen More, February 2, 1984.

34. Calderone to Dr. Hans Clarke, June 2, 1958, Box 1, Folder 5, "Correspondence, personal, 1942, 1958, 1969," MSC/SL; *Thirteenth Announcement, 1936–1937*, University of Rochester School of Medicine and Dentistry, p. 12. Cf. *Particular Passions* (1982), 255–256. Calderone told me (and others), "So naively, when I got to Rochester just two days before school opened, I went to see the dean." Cf. Carton 1, unprocessed material, folder, "W," Mary S. Calderone, "About George Hoyt Whipple," March 16, 1976, MSC/SL. On Whipple, see George W. Corner, *George Hoyt Whipple and His Friends* (Philadelphia: J. B. Lippincott, 1963), 160–162. Also see letter of Mary S. Calderone to George W. Corner, Rockefeller Institute of Medical Research, January 25, 1956, Box 9, folder 149, MSC/SL; George W. Corner, "Science and Sex Ethics," *Saturday Evening Post*, October 10, 1959. On women physicians in the 1930s, see More, *Restoring the Balance*, chaps. 6, 7. My appreciation to Christopher Hoolihan, historian and medical librarian, Edward G. Miner Library, University of Rochester School of Medicine and Dentistry, for supplying me with the relevant University of Rochester medical school announcements and card catalogue entries for Mary Steichen and George Corner in the 1930s.

35. During her third year, she reverted to using her maiden name, Steichen. Cf. her listing in the University of Rochester School of Medicine and Dentistry *Twelfth Announcement, 1935–1936* and the listing for the 1938 edition.

36. Mary S. Calderone to Ellen S. More, September 1, 1984, author's collection.

37. Mary Breasted, *Oh! Sex Education!* (New York: Praeger, 1972), 211–212. For gender performance, see, for example, Sedgwick, *Touching Feeling*; Joan Cassell, *The Woman in the Surgeon's Body* (Cambridge, MA: Harvard University Press, 1998); and the discussion of the concept of *habitus* in Robert A. Nye, "The Legacy of Masculine Codes of Honor and the Admission of Women to the Medical Profession in the Nineteenth Century," Chapter 6 in this volume.

38. On the "feminine mystique" and women physicians, see Ellen S. More, *Restoring the Balance*, 182–193.

39. Subhead quotation from "My name is Mary Calderone," typescript manuscript,

n.d., pp. 1–21, Box 13, fol. 222, MSC/SL. Quotation from handwritten insert, verso p. 10, emphasis in original. Nat Lehrman, "Playboy Interview," 64. "My name is Mary Calderone," with this handwritten annotation by author: "probably 1948–1950 when I was speaking to PTA's as a school physician—through the Mental Health Ass. of Nassau County."

40. Margaret Sanger, *Happiness in Marriage* (1926; reprint Old Saybrook, CT: Applewood Books, 1993); Theodoor H. Van de Velde, *Ideal Marriage: Its Physiology and Technique*, trans. Stella Browne (New York: Random House, 1926); idem, *Sexual Tensions in Marriage: Their Origins, Prevention and Treatment*, trans. Hamilton Marr (New York: Random House, 1928); Robert L. Dickinson and Lura Beam, *A Thousand Marriages* (Baltimore: Williams and Wilkins, 1931).

41. Linda Gordon, *Woman's Body, Woman's Right: Birth Control in America* (1976; New York: Penguin Books, 1977), 356; Modell, *Into One's Own*, 305. Bailey, *Sex in the Heartland*.

42. James H. Jones, *Alfred C. Kinsey: A Public/Private Life* (New York: W. W. Norton, 1997), 418–436.

43. Lionel Trilling, "The Kinsey Report," 1948, reprinted in *An Analysis of the Kinsey Reports on Sexual Behavior in the Human Male and Female*, ed. Donald Porter Geddes (New York: New American Library, 1954), 212–229; Bailey, *Sex in the Heartland*, 1–80.

44. Moran, *Teaching Sex*, 23–31, 135–137.

45. Moran, *Teaching Sex*, 91–101, 110–112.

46. Hans Pols, "Divergences in Psychiatry during the Depression: Somatic Psychiatry, Community Mental Hygiene, and Social Reconstruction," *Journal of the History of the Behavioral Sciences*, 2001, *37*: 369–388, accessed at www.usyd.edu.au/hps/staff/hans/jhbsart.htm, December 29, 2004, pp. 1–19; quotation on p. 5.

47. Margaret W. Rossiter, *Women Scientists in America: Before Affirmative Action, 1940–1972* (Baltimore: Johns Hopkins University Press, 1995); More, *Restoring the Balance*, 182–215. On women in the law, see Harrington, *Women Lawyers*.

48. Quotation from "Mary Steichen Calderone," in *Particular Passions*, 257. Also see "Mary Steichen Calderone," *Encyclopedia of the American Woman*, typescript, p. 2, Box 1, fol. 1, MSC/SL.

49. Hannah M. Stone and Abraham Stone, *A Marriage Manual*, rev. ed. (1937; New York: Simon and Schuster, 1939); Chesler, *Woman of Valor*, 288–289, 303–308. Cf. Linda Gordon, *Woman's Body, Woman's Right*, 369–374; Paul Robinson, *The Modernization of Sex*, 2.

50. Her title, *Release from Sexual Tensions*, is a clear response to Van de Velde's *Sexual Tensions in Marriage*, published in English translation in 1928. Moran, *Teaching Sex*, 164–165.

51. Sanger, *Happiness in Marriage*, 21, 33; Stone and Stone, *A Marriage Manual*, 273–276. Cf. Gordon, *Woman's Body, Woman's Right*, 373.

52. Mary S. Calderone, "Above and Beyond Politics: The Sexual Socialization of Children," in *Pleasure and Danger: Exploring Female Sexuality*, ed. Carol S. Vance (Boston: Routledge and Kegan Paul, 1984), 131–137. But see Sharon Thompson, "Search for Tomorrow: On Feminism and the Reconstruction of Teen Romance," also in Vance, p. 376. See also Jane Gerhard, *Desiring Revolution: Second-Wave Feminism and the Rewriting of American Sexual Thought, 1920–1982* (New York: Columbia University Press, 2001).

53. "My name is Mary Calderone," 10, 12–13.

54. Ibid., 8–9.

55. Ibid., 3–4, 20–21.

56. Ibid., 7–8.

57. Mary Steichen Calderone and Phyllis and Robert Goldman, *Release from Sexual Tensions: Toward an Understanding of their Causes and Effects in Marriage* (New York: Random House, 1960), 57–62. The Goldmans are described as "science writers" on the flyleaf. GCS [Goodrich C. Schauffler], "Review of *Release from Sexual Tensions*," in *Western Journal of Surgery, Obstetrics and Gynecology*, 1960, *68*: 6, 8; Mary S. Calderone to Goodrich C. Schauffler (Dear Gig), June 12, 1960; Mary S. Calderone to Goodrich C. Schauffler, September 9, 1959. All in Box 9, fo. 155, MSC/SL. Also see Heather Munro Prescott, "'Guides to Womanhood': Gynaecology and Adolescent Sexuality in the Post-Second World War Era," in *Women, Health, and Nation: Canada and the United States since 1945*, ed. Georgina Feldberg, Molly Ladd-Taylor, Alison Li, and Kathryn McPherson (Montreal: McGill-Queens University Press, 2003), 199–222.

58. Anne Koedt, "The Myth of the Vaginal Orgasm," in *Notes (From the Second Year): Radical Feminism*, ed. Shulamith Firestone and Anne Koedt (New York: New York Radical Feminists, April, 1970), 37–41. My thanks to Elizabeth Singer More for a copy of the Koedt essay. Cf. Hannah Stone and Abraham Stone, *A Marriage Manual*, 258–262.

59. Calderone, *Release from Sexual Tensions*, 170–172.

60. Mary S. Calderone to Goodrich C. Schauffler (Dear Gig), June 12, 1960, in Box 9, fol. 155, MSC/SL.

61. For an extended consideration of Calderone's career, see Ellen S. More, "Mary Calderone, Sex Education, and American Society" (under contract with Beacon Press).

Part II / Challenging the Culture of Professionalism

CHAPTER SIX

The Legacy of Masculine Codes of Honor and the Admission of Women to the Medical Profession in the Nineteenth Century

Robert A. Nye

I will look at the history of women in medicine from the perspective of the historical gendering of the professions. Until recently, the history of the professions has been dominated by analyses that define professions as specialized monopolies of knowledge supported by licensing and educational standards defended by regulation and law. Scholarly studies of the strategy of "occupational closure" have focused on the way that these formal strategies marked women or others as part of a class of "ineligibles," "through excluding them from routes of access to resources such as skills, knowledge, entry credentials, or technical competence."[1] These legal strategies have generally been taken as sufficient explanations of how male professionals managed for years to lock out women from full professional status.[2] Closure strategies are not inherently strategies of gender discrimination. They have been used by white women against women of color and by Christian men against Jewish men, and they have proven to be ineffective against male minorities who have gone on to positions of leadership in historically female professions such as nursing.[3] More recently, gender scholars have expanded the range of explanations for the slow integration of women into the professions by studying the ways that education, the social division of labor, professional practices, and various enduring cultural beliefs helped ensure a masculine identity for the modern professions that has prevailed until recent times. We have learned not only how a gendered society shapes professional and academic life but also how the work and the content of medicine, law, and science came to be gendered male.[4] As formal and legal barriers to the professions have begun to disappear in much of the world, attention has shifted to other explanations for the paucity of women surgeons, senior partners, engineering project bosses, and full professors, and for the "glass ceiling" that keeps most women in lower rank and salary echelons and sequestered in particular specialties.[5]

The related "pipeline" argument holds that, at every stage of education, training, and practice, women have had to cope with obstacles that resulted in a cumulative career disadvantage. Ellen S. More, who has studied the history of American women physicians, acknowledges that men can also accumulate disadvantages from the outset of their educations, but such disadvantages are experienced much more uniformly among women.[6] There is a tendency among scholars to see the "pipeline" analysis as a combination of formal barriers and the "differences in women's preferences and values" that orient them toward certain specialties and lower tracks of career success—the so-called Matilda effect.[7]

These approaches make assumptions about women's motives and desires that risk reintroducing a friendly version of essential gender difference, but, more problematically, they do not directly consider the culture of work practices and sociability that have historically constituted the professional work experience. Though work and sociability are governed by rules, such cultural regulations are typically informal or tacit, an indispensable feature of their authority. Historians should take cues from recent work done by sociologists, ethnographers, and organizational theorists who study this phenomenon.[8] Some scholars have noted the ad hoc defenses thrown up by male-dominated groups, stressing that the issue now turns "not so much on the *exclusion* of women, but on a particular form of their *inclusion*, and on the way that this inclusion is masked in a discourse of gender that lies at the heart of professional practice itself."[9] As Joan Acker has argued, scholars in the 1970s and 1980s often portrayed organizations as neutral when, in fact, "gender is not an addition to ongoing processes, conceived as gender neutral. Rather, it is an integral part of those processes, which cannot be properly understood without an analysis of gender."[10] We cannot overlook, she writes, that workers' and professionals' bodies are already gendered male before women arrive on the scene. Indeed, as R. W. Connell has been saying for years, these "neutral" bodies are gendered according to a prevailing discourse of hegemonic masculinity that operates as a device for selecting, regulating, and discriminating between individuals.[11]

In this perspective, it is the historically masculine gendering of professions and the skills and work within them that constituted the chief obstacle for women seeking full professional status. Far from weakening the prevailing gender orthodoxy, the admission of women into previously all-male organizations has sometimes sharpened gender and gendered discriminations and provoked creative and subtle tactics for segregating, isolating, or undervaluing women and their work. Even where women have become a majority presence in a profession, their exclusion from key positions, or the clustering of men in high status subfields, has kept them from enjoying equal salary and power.[12] The historical and sociological literature on this phenomenon is steadily growing.[13] I want to argue here that the pervasive "masculinization" of professional culture that took place within the medical profession in the nineteenth century was a consequence of social practices that were not aimed originally at women but at admitting, controlling, and retaining a certain kind of man.[14] The historic *ethos* that has anchored masculine professional culture operated informally to limit access to "outsiders" and reinforce solidarity in the ranks.[15] It was these tacit rules that women confronted when they entered the classrooms, workplaces, and voluntary organizations of professional life.

The honor culture that regulated male conflict and sociability in nineteenth-century Anglo-American and European professional culture had ancient European roots.[16] From the late Middle Ages through the early modern era, the status of a gentleman reposed on three closely related kinds of personal independence. (1) The military and service origins of the nobility stressed the personal courage and virtue of the nobleman and valued unambiguous demonstrations of loyalty.[17] (2) His patrimony and property, including his wife and progeny, allowed him to maneuver in the world.[18] (3) Intellectual autonomy forged links between personal honor, civility, and reliable testimony about natural facts.[19]

Despite the growing dominion of modern forms of masculinity, violence remained a crucial element of modern life. Gentlemen continued to fight duels nearly everywhere in the West, and the elaborate rituals of the *point d'honneur*, remained a last resort in the resolution of conflicts until the early twentieth century, except in Britain and in the United States.[20] In the course of the modern era, this etiquette was deeply embedded in the parliamentary rules of order adopted by private clubs and public assemblies, as has been pointed out by Wilbert

Van Vree. It appears that procedural rules evolved as much to thwart violence as to assure that all voices could be democratically heard.[21] Nonetheless, even where it was formally forsworn, the duel remained the last resort in the hierarchy of social practices that mediated competition and conflict in continental Europe. At stake was not so much the truth of an allegation against a man as his willingness to back up words with deeds, which testified, at the minimum, to the sincerity, if not the accuracy, of his views.[22] Since these codes depended on cues and behavior that circulated within relatively narrow social networks, they posed translation problems for outsiders. Men of nonelite class, religious, educational, or linguistic backgrounds or exotic racial or national origins were unlikely to know the fine points of these codes, which placed them outside the charmed circle. In brief, violence, or the threat of it, undergirded much of the structure of public discourse in the nineteenth century, haunting the forms and proprieties of masculine sociability like an unruly visitor who has left the table but might return at any time.[23]

The first private clubs were modeled on Masonic Lodges, which sprang up everywhere in Europe and America in the post-Revolutionary era.[24] Club sociability in the nineteenth century was regulated by statutes that laid out the criteria and obligations of membership and the conditions for its abrogation, recalling to members their duty to respect the dignity of others, limiting discussions to issues not likely to provoke controversy or conflict, and defending the interests and solidarity of the club on pain of exclusion. What we know about the statutes by which professionals governed themselves suggests that they were informed in their decisions by the tacit knowledge their members possessed as men of a certain class; in other words, the "ethical" standard that guided them was really an adaptation of the *ethos* of the upper-class male honor culture.

Notaries, lawyers, doctors, the clergy, and even engineers already operated under the authority of such corporate principles in the eighteenth century, and they continued to do so in their modern incarnations as professions. They took oaths to uphold the solidarity of the group, collected honoraria instead of fees for their services, and regularly proclaimed their disinterest, probity, discretion, and "delicacy" to their clients and the general public; the essential quality of an honorable man was his independence, which guaranteed his autonomy and disinterestedness and protected his prospective clients against the venalities of greed or collusion.[25] Honor thus became more democratic in the course of the nineteenth century, in the sense that more men could claim to possess it and demonstrate a grasp of its procedures.[26] But it also seems likely that an unspoken threat of violence and some requirement of personal reparation acted to discourage some men and all but the boldest women from attempting to join male clubs, professional societies, academies, or societies of emulation, at least until near the end of the century.[27] Newcomers in such a setting of tense male rivalries may not have felt personally threatened, but they almost certainly felt excluded.

All-male organizations everywhere in the West followed tacit rules for admitting, expelling, and regulating their members; professional organizations were even more exigent in this respect since their professional integrity was additionally at stake. We have some pertinent evidence for how this process worked for British medical professionals, who engaged in self-regulation in a way that was directly inspired by the culture of honor. As in North America, medicine in Britain was a "free" profession without statutory protection or protected monopoly. Because they competed in the same market for patients as quacks, wise women, charlatans, and faith healers, practitioners sought to distinguish themselves by class, training, and decorum from their lowborn rivals and by a form of ethical practice that assured professional accountability and solidarity. From the late eighteenth century, British medical professionals had improvised ways of enforcing ethical guidelines that would allow the profession to discipline or expel doctors who violated the etiquette of professional courtesy and competition or who dishonored the profession by infamous behavior. Because physicians possessed no centralized police powers over members, they engaged in a form of highly localized self-regulation through what were called the "medical ethical committees" of their local medical societies. These societies were as much social as professional. Members read medical papers and socialized and dined together, but time was invariably set aside for a kind of "medical court" that heard, discussed, and sometimes arbitrated differences between practitioners.[28]

At the national level an independent, physician-staffed General Medical Council was created in 1858 as part of a wave of reforms in medical licensing and practice. The council's responsibility was to hear charges brought against physicians (by other physicians) for violations of professional ethics, including such things as practicing without a license, advertising, or "unethical" behavior, serenely undefined. There was no written code that might serve as a statutory benchmark, hearings were closed, and no transcript was published. Procedures of this kind illustrate the extent to which medical professionals until recent times engaged in such intraprofessional regulation in a relatively informal way, according to a code that physicians, as gentlemen, were supposed to naturally understand. As Margaret Stacey has written, "until very recently, . . . the Council had much of the appearance of a London Gentleman's Club, concerned only that its members behave like gentlemen and that 'bounders' should be evicted."[29] Donald MacAlister, the GMC's turn-of-the-century president proclaimed that the GMC was not a "high court of medical conduct" but a clearinghouse of "warning notices" to "quicken the conscience" of men who already know what is right and proper.[30]

Toward the end of the nineteenth century, the number of registered physicians was increasing faster than the population, medical incomes stagnated, and medical clubs and friendly societies were able to hire medical services for fees below the going rates.[31] In response to this crisis, local medical associations sprang up throughout the country to defend their "medicopolitical" interests and to promote "the social intermingling in professional life, chiefly through the medium of the medical society [which] begets a better knowledge of confreres and evokes fraternal relationships in a true understanding of the needs of the time."[32] There was much debate about the danger that organizational activity would dilute the "honor and interests" of the profession, but there was general agreement that there was need of "more gentlemen" or, as one medical professor said, "it is incumbent on all to see that anyone desiring to enter its ranks should, in addition to acquiring the necessary technical instruction . . . , possess those qualities that are summed up in the word 'gentleman.' Mere knowledge or purely professional attainments are not sufficient; other qualities of the heart are necessary—he must be a gentleman in manners, but also a man of feeling and sympathetic withal."[33]

To meet this crisis, in 1902 the British Medical Association (BMA), to which the majority of British practitioners belonged in the early twentieth century through a network of branch societies, formed a Central Ethical Committee (CEC) as part of a major constitutional reform. Initiative was to be wholly local and good ethical practice judged by local customs. As an internal memo of the association's solicitor put it, "it would be impossible to expect a similar high standard of ethics in a mining district in South Wales as one might expect in the West End of London, and it is always an elementary point of justice that a man should be judged by his peers."[34] However, to bring divisions and branches up to speed, the CEC drafted a set of "model ethical rules" in 1907, occasionally revised before 1950. The rules, however, were entirely procedural and offered no substantive advice on ethical principles. Branches or divisions could censure a miscreant after a proper hearing and recommend "ostracism," which required all members to refuse collaboration or consultation with him unless a patient's life was at stake. For more serious offenses, units forwarded a recommendation of expulsion—grandly referred to as the "supreme penalty" or "vocational death"—to the CEC, which had sole power to apply it.[35]

Individual accusations and their resolution assumed the form and the discourse of affairs of honor. As late as 1950, the committee was still ruling that complaints must first be addressed by the complainant to the offender demanding "*his* explanation"; only when this proved "unsatisfactory" must the charge be forwarded to the committee. Legal counsel was forbidden to a man against whom a complaint had been brought, though he could be supported at a hearing by his "friends." The committee hesitated to revoke membership in most cases. Instead, it mobilized a host of informal mechanisms, ranging from inviting disputants to seek an "amicable resolution," where "amends" or "regrets" were expressed, to urging "ostracism," and extending censure to members who maintained professional or friendly relations with offenders. When a physician from Pembrokeshire wrote to the committee complaining of another BMA member poaching his patients, the committee's secretary advised him to have their differences arbitrated at the branch office in Wales "or in clubs you both belong to."[36] As late as 1946, there were no written ethical rules, but one member still defended this informal procedure "rather much as the committee of a social club which has the power to get rid of a member who offends against the *recognized* code."[37]

Astonishingly, this language and body of procedures remained in force even though the number of registered female practitioners in Britain in 1921, many of whom belonged to local branches of the BMA, was 1,500, or 3.3 percent of the total, rising to 7,520 by 1951, or 9.1 percent. However, women were brought before the disciplinary panel of the General Medical Council proportionately less often than men, and, in particular, men trained abroad.[38] As I have argued elsewhere, a similar crisis in medical overpopulation was occurring in France at the same time, with a remarkably similar outcome.[39] In the United States, medical societies and medical specialty organizations excluded women from their ranks without legally barring them, even after the late nineteenth century when women gained admission to formerly all-male medical schools. They could obtain degrees of equal worth to men but were discouraged from participating in any of the noneducational aspects of professional life and were thus isolated from the often lucrative aspects of social networking.[40]

When we isolate and examine this masculine *ethos* of professional sociability and informal self-regulation, we can better understand how difficult it was for women to make a place for themselves in medicine. By the twentieth century, the most determined women in Europe and America could follow a legal pathway to medical practice almost everywhere, but no dramatic increases in the number of female physicians occurred over the next sixty years. Women physicians were shunted into acceptably "feminine" areas of practice: public health, hygiene, gynecology, pediatrics, and obstetrics. In the United States, when women's medical colleges began to close in the late nineteenth century, mainstream schools applied informal quotas for admission and, in some cases, denied women admission outright. However, women were not impeded solely by these formal barriers to medical education and licensing but also by the masculine culture of medicine.

I have attempted to establish the ubiquity and power of male honor codes in all-male settings and the latent violence that often simmered just beneath the surface of fraternal solidarity. When women began to enter the professions in significant numbers, a hierarchy of disincentives, ranging from brutal to subtle, was firmly rooted in the culture of masculine sociability. In this atmosphere, educated middle-class women likely felt discomfort at minimum, and at most, felt a thoroughgoing disenfranchisement and personal humiliation. Where the "point of honor" either did not exist or had died out, its imprint on the forms of sociability remained. Thus, as we have seen, "satisfaction," "making amends," the mediation of "differences," and the premium put on deft but "frank" interchange remained important aspects of medical sociability. It seems likely that such frankly masculine environments might have silenced the voices of women, whose bodies could not unambiguously represent masculine notions of resolve or who did not know or care to adopt the gestural repertory of assertiveness.[41] As Carla Bittel (Chapter 1) and Regina Morantz-Sanchez (Chapter 3) have shown in their respective studies in this volume of Mary Putnam Jacobi and Mary Dixon Jones, when women took firm stands against the grain of traditional gender scripts, they could suffer considerable public and professional resistance.

Further down the hierarchy but no less redolent of violence was the holy (or unholy) trinity of smoking, drinking, and profanity, all salient expressions of male exclusivity if not aggression. Such homosocial behavior flourished in the dissecting theater, the medical school classrooms, and in male circles and societies as an indispensable aspect of the dynamics of male solidarity. There is excellent evidence that men fully grasped the gendered meanings it conveyed. Cigars,[42] heavy drinking, and sexually blunt talk, a normal part of anatomy classes in nineteenth-century America, were used to discourage women when they first appeared at the doors of the dissection theaters. Michael Sappol has written about how men displayed their masculinity in the anatomy classes as a test of their "poise and courage," joking about death and posing their cadavers in grotesque ways.[43] One famous incident involving the dissection of corpses provoked much public comment for and against women in 1869 when women students at the Woman's Medical College of Pennsylvania attended a dissection at the neighboring all-male University of Pennsylvania. The men shouted, jumped on their chairs, insulted the women, and raucously escorted them from the building when the class concluded. As Steven Peitzman explains it, this decidedly ungentlemanly behavior was acceptable to otherwise polite fellows when addressed to women who had already "desexed" themselves by viewing a naked male cadaver.[44]

In Europe there were similar experiences. Sophia Jex-Blake, a pioneering medical graduate of the University of Edinburgh, described the experience of being locked out of the final examination hall by a group of male medical students "who stood within, smoking and passing about bottles of whiskey, while they abused us in the foulest possible language."[45] Another medical pioneer, the American Mary Putnam Jacobi, who attended the École de Médecine in the late 1860s, was excluded de facto from a hospital internship because the interns' common room was notorious for erotic drawings and a vulgar atmosphere.[46] On various occasions, efforts were made to bring greater levels of respectability to professional life by disciplining violence and vulgarity and channeling youthful high spirits into sport, but educational venues and prime hospital residencies remained male preserves well into the twentieth century and retained a male ethos long after that.[47]

In addition, the system that controlled medical school admissions, residency, and consultancy appointments was an informal procedure that heavily favored the old-boy network, rather, as Rosemary Pringle observes, "like joining a club."[48] Such networks originated in professor/student relations, in which young men were often included in their mentor's family life, and in the bonding activities of student associations and sports groups.[49] Carole Dyhouse has written about the system of patronage that was cultivated at British medical schools in the 1920s and 1930s: "The authoritative stance was often one of a stern or benevolent paternalism, shading, as the students grew towards professional competence, into a shared 'gentlemanly' code of honour. Where male students appealed against female admissions to their deans and the boards of management in the 20s, the tone was often governed by intimations that gentleman's agreements or codes of honour had been threatened by their esteemed patrons, and in some cases the tone verged towards that of filial complaint."[50] The picture in the United States, where accreditation, medical school admissions, and licensing were local matters, was more complex. In some cases, state legislatures, such as Michigan's, forced the state university's unwilling medical faculty to accept women, but many private universities bowed to male professors' assertions that large numbers of women students would cause disorder.[51]

Local medical societies proliferated throughout the West in the nineteenth century. They performed a host of functions, including the reading of medical papers, medical-ethical discussions, dining, travel, sport, and other forms of sociability. At first, pioneer women practitioners thought it worthwhile for women to join these groups, despite stubborn opposition within some of them, but as their numbers grew, women physicians found the atmosphere of these male-dominated societies discouraging, and parallel all-female societies sprang up wherever there were enough women medical professionals to fill them. As a founder of one of these groups (in Rochester, New York) said of male physicians, "The medical societies are under their control; we have been admitted to these after the persistent knocking of the pioneer women of the profession, but we are not at home there as in our own circles. We need the general societies to broaden our minds and give us lines of thought but our work and growth should be free where we are without embarrassment or restraint."[52] Characteristically, men's and women's groups recruited differently. Women carefully screened and consulted others about prospects to avoid wounding feelings or vanity, while men summarily blackballed nominees to their groups.[53] In the twentieth century, when women began to rejoin the male societies, one woman doctor observed, "A generation earlier, women doctors were on the outside standing together. Now they were on the inside sitting alone."[54]

As Rosemary Pringle has argued, the strong social reformist and feminist influences that previously brought women into the field were being replaced by a painful accommodation to the masculine ethos of the profession.[55] The women who "made" it in this era were remarkable in every way.[56]

At least since World War II, the qualities expected of physicians have had less to do with personal character and more to do with mastery of the scientific and technical aspects of modern medical practice.[57] This great shift ought to have deemphasized gender and provided a more neutral terrain on which women could compete. But, as several observers have noted, mathematics, various scientific disciplines, and many technological fields have their own highly gendered identities. Field sciences like geology have long had a manly image, as have the laboratory sciences of physics and chemistry.[58] And we know that scientific and technological knowledge is partly transmitted in tacit forms related to gendered practices rather than explicitly through text and demonstrations.[59]

The most convincing work on the transmission of knowledge in this tacit dimension has been done by the French social anthropologist Pierre Bourdieu and his disciples. Bourdieu has explained how we use a form of tacit or practical knowledge when we are obliged to function in a particular sociocultural situation, which he calls a habitus. This tacit dimension of knowledge may take a profoundly gendered form, which, in Bourdieu's analysis, operates more or less unconsciously to transmit knowledge about appropriate comportment. Gender, in other words, is treated as culture, and operates linguistically as a verb, not a noun.[60] In his 1998 book *La domination masculine*, Bourdieu showed how masculinity reproduces historically as a mode of practical and symbolic domination. He did not think that the postmodern notion of "performativity" could appreciate the extent to which the gender order is incorporated into bodies as "schemata of perception, of thought, and of action," as physical attitudes and dispositions, and even as "corporeal emotions."[61] Such is the power of what he calls the "immense symbolic machine" that performs "symbolic violence" on women—and on men who do not embody acceptable masculine qualities—that hegemonic masculinity seems to all observers to be both neutral and natural.

Analyzing gender dominance this way allows us to appreciate both the largely unconscious nature of these processes of incarnation and how they operate to maintain masculine dominance in social life. A woman who enters a public space dominated by a masculine culture will feel uneasy because the way she walks, talks, assumes postures, carries her head, gestures, looks (or does not look) at her interlocutor, and displays (or does not display) confidence may mark her as "other." She may also be at a disadvantage, as I have argued, in easily mastering a new laboratory technique or clinical procedure, not because she is incapable of it, but because she simply takes for granted, as do the men in this "habitus," that men have a natural sense of how things work that everyone takes for granted. Bourdieu explicitly credits the "sense of honor" possessed and incorporated into men's bodies and the "games of honor" that men continue to play with one another for producing a climate in which, as Leslie McCall has written, women feel themselves reflexively to be in an alien social situation that most men find quite natural.[62]

In *Sex and Medicine*, Rosemary Pringle makes excellent use of the Bourdieuian categories of habitus and practical sense in explaining how women are obliged to negotiate their way through the masculine territory of medicine in contemporary British and Australian society. She shows how masculine rites of passage and institutionalized forms of verbal abuse work obvious hardships on women and identifies medical specialties based on teamwork modeled on sports or that require familiarity with particular techniques, especially surgery. Few women were believed to "naturally" possess the qualities of decisiveness, endurance, and courage that the surgeon has been held historically to possess.[63] Women who have succeeded in male specialties find they must adapt to a culture in which to be "one of the boys" actually means to be treated decidedly as "one of the girls," as Frances K. Conley, a neurological surgeon at Stanford, has described in her best-selling *Walking Out on the Boys*.[64] Another Bourdieuian scholar, the anthropologist Joan Cassell, summarizes the dilemma of the woman surgeon in *The Woman in the Surgeon's Body*:

> A Woman surgeon has little choice. She is a woman as well as a surgeon . . . She has a woman's body, a woman's movements, a lifetime of experiences as a woman. She can perceive herself as one of the guys; she can epitomize the mystique of the iron surgeon; she can be more macho than the men, defining caring and compassion as the subjective, unscientific, touchy-feely province of nurses and social workers. Nevertheless, she will always be a macho *woman*, with a different habitus from that of a macho man—just as the fatherly habitus of a compassionate man differs from the motherly habitus of a compassionate woman. Seniors, colleagues, patients, and nurses will react to her as a woman (not as a neutral category—not as one of the guys—but as a woman).[65]

There are plenty of success stories in contemporary medicine, though many of the informal barriers remain remarkably intact in medical education and in medical honorary societies and social organizations. It is still difficult for women to be elected to higher office in medical societies, and some of the oldest British medical societies have only admitted a woman in the past few years, despite their growing numbers in the profession.[66]

The "pipeline" is no longer the problem; rather, "the pipeline empties into territory that women and faculty of color too often experience as uninviting, unaccommodating, and unappealing."[67] Cumulative career disadvantage still haunts women whose pregnancies and child-rearing activities are not compensated in custom or law; things are never quite the same when they return from maternity leave. When a formerly "feminine" profession like nursing began to admit larger numbers of men, they often gravitated into nursing specialties that pay higher wages and have more prestige than ordinary staff positions.[68] The masculine cultures that have dominated the professions are not just pockets of male privilege scattered here and there, but the persistent expressions of developments that have historically favored a certain kind of masculinity in all Western societies. The modes and nature of this domination have changed over time, but slowly and in a way that in many fields preserves masculine advantage.

NOTES

1. Anne Witz, *Professions and Patriarchy* (London: Routledge, 1992), 46. See also Margaret W. Rossiter, *Women Scientists in America: Struggles and Strategies to 1940* (Baltimore: Johns Hopkins University Press, 1982), 73–99; Samuel Haber, *The Quest for Authority and Honor in the American Professions* (Chicago: University of Chicago Press, 1991); for criticisms of the closure argument, see Wai-Fong Chua and Stewart Clegg, "Professional Closure: The Case of British Nursing," *Theory and Society*, 1990, *19*: 135–172; Paul Atkinson and Sara Delamont, "Professions and Powerlessness: Female Marginality in the Learned Occupations," *Sociological Review*, 1990, *38* (1): 90–110.

2. A major assumption of these analyses has been the central role given to the claims that professionals make for the scientific or technical superiority of their work. In this genre of writing, the most influential have been Andrew Abbot, *The System of Professions* (Chicago: University of Chicago Press, 1988); Magali Sarfatti Larson, *The Rise of Professionalism: A Sociological Analysis* (Berkeley: University of California Press, 1977); Thomas L. Haskell, ed., *The Authority of Experts: Studies in History and Theory* (Bloomington: Indiana University Press, 1984); and for medicine, Eliot Freidson, *The Profession of Medicine* (Chicago: University of Chicago Press, 1970).

3. See Liz Walker, "'Conservative Pioneers': The Formation of the South African Society of Medical Women," *Social History of Medicine*, 2002, *14* (3): 483–505; Christine L. Williams, *Still a Man's World: Men Who Do Women's Work* (Berkeley: University of California Press, 1995).

4. For an overview, see Londa Schiebinger, *Has Feminism Changed Science?* (Cambridge, MA: Harvard University Press, 1999); Lani Guinier, Michelle Fine, and Jane Balin, *Becoming Gentlemen: Women, Law School, and Institutional Change* (Boston: Beacon Press, 1997).

5. Ann Crittenden, *The Price of Motherhood: Why the Most Important Job in the World Is Still the Least Valued* (New York: Metropolitan Books, 2001).

6. Ellen S. More, *Restoring the Balance: Women Physicians and the Profession of Medicine, 1850–1995* (Cambridge, MA: Harvard University Press, 1999), 231.

7. Ibid., for a summary of this argument. Among the studies of women in the professions that pose the issue as one of "deficit" or "choice" is Gerald Holton and Gerhard

Sonnert, *Gender Differences in Scientific Careers* (New Brunswick, NJ: Rutgers University Press, 1995).

8. See the pioneering work of Rosabeth Moss Kanter, "Some Effects of Proportions on Group Life: Skewed Sex Ratios and Responses to Token Women," *American Journal of Sociology*, 1977, *82*: 965–990. Also, Kanter, *Men and Women of the Corporation* (New York: Basic Books, 1977).

9. Celia Davis, "The Sociology of Professions and the Profession of Gender," *Sociology*, 1996, *30* (4): 663.

10. Joan Acker, "Hierarchies, Jobs, Bodies: A Theory of Gendered Organizations," *Gender and Society*, 1990, *4* (2): 146.

11. R. W. Connell, *Gender and Power* (Stanford, CA: Stanford University Press, 1987); and especially *Masculinities* (Cambridge: Polity Press, 1995); and *The Men and the Boys* (Berkeley: University of California Press, 2000). See also Jeff Hearn, *Sex at Work* (Brighton: Wheatsheaf, 1987).

12. On this situation, see Henry Etkowitz and Carol Kemelgor, "Gender Inequality in Science: A Universal Condition?" *Minerva*, 2001, *39*: 153–174; Donald A. Barr and Elizabeth Heger Boyle, "Gender and Professional Purity: Explaining Formal and Informal Work Rewards for Physicians in Estonia," *Gender and Society*, 2001, *15* (1): 29–54.

13. See, for example, Ellen Kuhlmann, "The Rise of German Dental Professionalism as a Gendered Project: How Scientific Progress and Health Policy Evoked Changes in Gender Relations," *Medical History*, 2001, *45*: 441–460; Amy Sue Bix, "Feminism Where Men Predominate: The History of Women's Science and Engineering Education at MIT," *Women's Studies Quarterly*, 2000, nos. 1 and 2: 24–45; Elaine Thomson, "Physiology, Hygiene and the Entry of Women to the Medical Profession in Edinburgh c. 1869–c. 1900," *Studies in the History and Philosophy of Biological and Biomedical Sciences*, 2001, *32* (1): 105–126; Wendy Bottero, "The Changing Face of the Professions: Gender and Explanations of Women's Entry to Pharmacy," *Work, Employment and Society*, 1992, *6* (3): 329–346; Annie Canel, Ruth Oldenziel, and Karin Zachmann, eds., *Crossing Boundaries, Building Bridges: Comparing the History of Women Engineers, 1870s–1990s* (Amsterdam: Harwood, 2000); Sally Gregory Kohlstedt and Helen Longino, eds., *Women, Gender, and Science: New Directions, Osiris*, 1997, *12*.

14. Ted Porter has discussed these operations of community "boundary-drawing" in science and other "technologies of trust," in *Trust in Numbers: The Pursuit of Objectivity in Science and Public Life* (Princeton, NJ: Princeton University Press, 1995), 220–231.

15. For a brief historiographical review and example of this kind of analysis, see Thomas Broman, "Rethinking Professionalization: Theory, Practice, and Professional Ideology in Eighteenth-Century German Medicine," *Journal of Modern History*, 1995, *67*: 835–872.

16. For France, see Robert A. Nye, *Masculinity and Male Codes of Honor in Modern France* (New York: Oxford University Press, 1993); Kevin McAleer has treated Germany in *Dueling: The Cult of Honor in Fin-de-Siècle Germany* (Princeton, NJ: Princeton University Press, 1995); for the United States, see Bertram Wyatt-Brown, *Southern Honor: Ethics and Behavior in the Old South* (New York: Oxford University Press, 1982).

17. On the instability of relations in a noble honor culture, see Kristen Neuschel, *Word of Honor: Interpreting Noble Culture in Sixteenth-Century France* (Ithaca, NY: Cornell University Press, 1989); Jonathan Dewald, *Aristocratic Experience and the Origins of Modern Culture* (Berkeley: University of California Press, 1993); Mervyn James, *Society,*

Politics and Culture: Studies on Early Modern England (Cambridge: Cambridge University Press, 1986).

18. Nye, *Masculinity and Male Codes of Honor in Modern France*, 6–7, 20–21; Pierre Bourdieu, *The Logic of Practice*, trans. Richard Nice (Stanford, CA: Stanford University Press, 1990), 68–78. See also Arlette Jouanna, *Ordre social: mythes et hiérarchies dans la France du XVIe siècle* (Paris: Hachette, 1977).

19. Mario Biagioli, *Galileo, Courtier: The Practice of Science in the Culture of Absolutism* (Chicago: University of Chicago Press, 1993); Steve Shapin, *The Social History of Truth: Civility and Science in Seventeenth-Century England* (Chicago: University of Chicago Press, 1994).

20. Nye, *Masculinity and Male Codes of Honor*, 23–30, 127–147. See also, Nye, "The End of the Modern French Duel," in *Men and Violence: Gender, Honor, and Rituals in Modern Europe and America*, ed. Pieter Spierenburg (Columbus: Ohio State University Press, 1998), 82–102.

21. Wilbert van Vree, *Meetings, Manners, and Civilization: The Development of Modern Meeting Behavior*, trans. Kathleen Bell (Leicester: Leicester University Press, 1999); Michael Curtin, "A Question of Manners: Status and Gender in Etiquette and Courtesy," *Journal of Modern History*, 1985, *57*: 395–423; Nye, *Masculinity and Male Codes of Honor*, 127–130.

22. For a particularly egregious example that occurred at the *Société de biologie* in Paris in 1858, see the Letter of April 1858 of Edouard Claparède to his mother in *Lettres de René-Edouard Claparède*, annotées par Georges de Morsier, fasc. XXIX, in Basler Veröffentlichungen zur Geschichte der Medizen und der Biologie (Basel: Schwabe & Co., n.d.), 56–58.

23. William M. Reddy, "Condottieri of the Pen: Journalists and the Public Sphere in Postrevolutionary France (1815–1850)," *American Historical Review*, 1994, *99* (5): 1546–1570. See also Nye, *Masculinity and Male Codes of Honor*, 187–199.

24. For the adoptive lodges and their role in the spread of Enlightenment ideals, see Janet M. Burke and Margaret C. Jacob, "French Freemasonry, Women, and Feminist Scholarship," *Journal of Modern History*, 1996, *68*: 513–549. On the origins of clubs in France, see Maurice Agulhon, *Le cercle dans la France bourgeoise, 1810–1848* (Paris: Colin, 1977).

25. Nye, *Masculinity and Male Codes of Honor*, 128–129.

26. McAleer, *Dueling*, 119–158; also Peter Gay, *The Cultivation of Hatred*, Vol. 3, *The Bourgeois Experience: Victoria to Freud* (New York: Norton, 1993), 9–34.

27. Carol E. Harrison has discussed the masculine club ethos of early nineteenth-century France in *The Bourgeois Citizen in Nineteenth-Century France: Gender, Sociability, and the Uses of Emulation* (Oxford: Clarendon, 1999).

28. See, in particular, the fine study of Scottish Medical Societies by Jacqueline Jenkinson, "The Role of Medical Societies in the Rise of the Scottish Medical Profession," *Social History of Medicine*, 1991, *4* (2): 259–261. See also Witz, *Professions and Patriarchy*, 73–103. For both Europe and America, see Thomas Neville Bonner, *To the Ends of the Earth: Women's Search for Education in Medicine* (Cambridge, MA: Harvard University Press, 1992); for America, see Samuel Haber, *The Quest for Authority and Honor in the American Professions*.

29. Margaret Stacey, "The British General Medical Council and Medical Ethics," in *Social Science Perspectives in Medical Ethics*, ed. George Weisz (Dordrecht: Kluwer, 1990),

175. See also Margaret Stacey, *Regulating British Medicine: The General Medical Council* (London, 1992); and Russell G. Smith, *Medical Discipline: The Professional Conduct Jurisdiction of the General Medical Council, 1858–1990* (Oxford: Clarendon, 1994).

30. Donald MacAlister, "The General Medical Council: Its Powers and Its Work," *Lancet*, 1906 (Oct. 6): 12.

31. Anne Digby, *Making a Medical Living: Doctors and Patients in the English Market for Medicine, 1720–1911* (Cambridge: Cambridge University Press, 1994), 11–16, 50–51.

32. William Muir Smith, "The Eastbourne Provident Medical Association," *British Medical Journal*, 1902 (Oct. 4): 1037–1039.

33. See *BMJ*, May 4, 1901, p. 1110; and Professor Stirling, "The Noble and the Ideal in Medicine," *BMJ*, Oct. 11, 1902, 1187. At this time, middle-class professionals still identified with the essence of the traditional gentleman: personal and occupational independence and personal virtue. See Harold Perkin, *The Rise of Professional Society: England since 1880* (London: Routledge, 1989), 16–17, 83–84, 119.

34. Wellcome Library SA/BMA D191 E319, letter of J. H. Stamp, Jan. 4, 1937. The reluctance to engage in centralized oversight and to defer to local authority is where most of the mischief has always been made in matters of discrimination against women and other "outsiders." See Mary Frank Fox, "Women, Science and Academia: Graduate Education and Careers," *Gender and Society*, 2001, *15* (5): 654–666.

35. The papers of the C.E.C. contain the debates over revision of the rules. For the revisions of 1912, see Wellcome Library SA/BMA D183 E307; for those of 1938, see Wellcome Library SA/BMA C99 E10.

36. My emphasis. Minutes of the Central Ethical Committee, April 3, 1950, SA/BMA D199 E10 Wellcome Library, London. See also a copy of procedural rules written in 1919, discussed at a meeting of the Ethical Committee Feb. 8, 1944, SA/BMA D99 E10, Wellcome Library, London; Copy of letter of Secretary of Central Ethical Committee to Dr. J. H. Elliot, May 21, 1938 SA/BMA D177 E213, Wellcome Library, London.

37. Letter of C.O. Hawthorne, Feb 5, 1946, ibid.

38. Smith, *Medical Discipline*, 116–124.

39. See Robert A. Nye, "Honor Codes and Medical Ethics in Modern France," *Bulletin of the History of Medicine*, 1996, *69*: 91–111. Most of the material from the subsequent section is taken from this article.

40. See More, *Restoring the Balance*, 96–112.

41. In some professional settings, such as engineering, there were historic links to the military that encouraged the adoption of military dress and comportment. At the École Polytechnique in France, engineering students wore swords as part of their regalia and engaged in duels. When engineers died in the performance of their duties, their obituaries indicated them as "morts au champs d'honneur." This was one of the filters employed to attract and retain men who could lead other men in projects according to a military conception of leadership. See John H Weiss, "'Bridges and Barriers': Narrowing Access and Changing Structure in the French Engineering Profession, 1800–1850," in *Professions and the French State, 1700–1900*, ed. Gerald L. Geison (Philadelphia: University of Pennsylvania Press, 1984), 15–65. Annie Canel has noted this and formal obstacles to women in France attempting to enter the profession that obliged them to follow other alternatives. Annie Canel, "Maintaining the Walls: Women Engineers at the École Poly-

technique Féminine and the Grandes Écoles in France," in *Crossing Boundaries, Building Bridges*, 127–158.

42. Nye, *Masculinity*, 131. On smoking see also Rossiter, *Women Scientists in America*, 92–94, and Amy Sue Bix, "Feminism Where Men Predominate: The History of Women's Science and Engineering Education at MIT," *Women's Studies Quarterly*, 2000, *28*: 25.

43. See the summary of male bonding activities in nineteenth-century anatomy classes in Michael Sappol, *A Traffic of Dead Bodies: Anatomy and Embodied Social Identity in Nineteenth-Century America* (Princeton, NJ: Princeton University Press, 2002), 72–89.

44. Steven J. Peitzman, *A New and Untried Course: Woman's Medical College and Medical College of Pennsylvania, 1850–1998* (New Brunswick, NJ: Rutgers University Press, 2000), 34–38.

45. As quoted in Witz, *Professions and Patriarchy*, 90. See, for a Canadian example, Bonner, *To the Ends of the Earth*, 145.

46. Joy Harvey, "La Visite: Mary Putnam Jacobi and the Paris Medical Clinic," in *French Medical Culture in the Nineteenth Century*, ed. Ann La Berge and Mordechai Feingold (Amsterdam: Rodopi, 1994), 355, 369 n. 21. For a revealing look at the sexually and anatomically graphic nature of the "art" of the *salle de garde* in the twentieth century, see the collection in Patrick Balloul, *La salle de garde ou le plaisir des dieux* (Paris: Editions de Loya, 1993).

47. See Keir Waddington, "Mayhem and Medical Students: Image, Conduct, and Control in the Victorian and Edwardian London Teaching Hospital," *Social History of Medicine*, 2002, *15* (1): 45–64.

48. Rosemary Pringle, *Sex and Medicine: Gender, Power and Authority in the Medical Profession* (Cambridge: Cambridge University Press, 1998), 57.

49. Sociologists and historians have noted the processes in which these homosocial paternal and filial networks are constructed in various professions and scientific disciplines. See Mary Frank Fox, "Women, Science, and Academia: Graduate Education and Careers," *Gender and Society*, 2001, *15* (5): 654–666; Paula Mählck, "Mapping Gender Differences in Scientific Careers in Social and Bibliometric Space," *Science, Technology, and Human Values*, 2001, *26* (2): 167–190; Boel Berner, "Educating Men: Women and the Royal Swedish Institute of Technology," in *Crossing Boundaries*, 75–102.

50. Carol Dyhouse, "Women Students and the London Medical Schools, 1914–39: The Anatomy of a Masculine Culture," *Gender and History*, 1998, *10* (1): 124.

51. Bonner, *To the Ends of the Earth*, 140–149.

52. Quoted in Cora Bagley Marrett, "On the Evolution of Women's Medical Societies," *Bulletin of the History of Medicine*, 1979, *53* (3): 446–447.

53. Ellen S. More, *Restoring the Balance*, 47.

54. Quoted in Virginia G. Drachman, "The Limits of Progress: The Professional Lives of Women Doctors, 1881–1926," *Bulletin of the History of Medicine*, 1986, *60* (1): 71.

55. Pringle, *Sex and Medicine*, 29.

56. Carol Dyhouse, "Driving Ambitions: Women in Pursuit of a Medical Education, 1890–1939," *Women's History Review*, 1998, 7 (3): 321–341.

57. See, on this matter, Marc Berg, "Turning a Practice into a Science: Reconceptualizing Postwar Medical Practice," *Social Studies of Science*, 1995, *25*: 437–476.

58. On geology, see Naomi Oreskes, "Objectivity or Heroism? On the Invisibility

of Women in Science," *Osiris*, 1996, *11*: 87–113; Bruce Hevly, "The Heroic Science of Glacier Motion," *Osiris*, 1996, *11*: 66–86; on physics, see Sharon Traweek, *Beamtimes and Lifetimes: The World of High-Energy Physicists* (Cambridge, MA: Harvard University Press, 1988). On oceanography, see Naomi Oreskes, "Laissez-tomber: Military Patronage and Women's Work in Mid-20th Century Oceanography," *Historical Studies in the Physical and Biological Sciences*, 2000, *30* (3): 373–392; for Antarctic science, see Robin Burns, "Women in Antarctic Science: Forging New Practices and Meanings," *Women's Studies Quarterly*, 2000, *28* (1–2): 165–180.

59. See, in general, on these issues, Londa Schiebinger, *Has Feminism Changed Science?* 65–104. On the "culture" of the sciences from the point of view of particular women's experiences, see Margaret A. M. Murray, *Women Becoming Mathematicians: Creating a Professional Identity in Post–World War II America* (Cambridge, MA: MIT Press, 2000); Evelyn Fox Keller, "The Anatomy of a Woman in Physics," and Aimee Sands, "Never Meant to Survive, a Black Woman's Journey: An Interview with Evelyn Hammonds," in *Women, Science, and Technology*, ed. Mary Wyer et al. (New York: Routledge, 2001), 9–16, 17–25.

60. The distinction of Sally Johnson, "Theorizing Language and Masculinity: A Feminist Perspective," in *Language and Masculinity*, ed. Ulrike Hanna Meinhof and Sally Johnson (Oxford: Blackwell, 1997), 22.

61. Pierre Bourdieu, *La domination masculine* (Paris: Editions du Seuil, 1998), 14–15, 29, 44.

62. Leslie McCall, "Does Gender Fit? Bourdieu, Feminism, and Conceptions of Social Order," *Theory and Society*, 1992, *21*: 837–867. As I have argued, masculine gestural repertories were originally constructed to provide a kind of gendered ideal for male professionals and exclude the "wrong kind of man." As one mid-nineteenth-century doctor observed of medical students, "If a student does not carry himself just so, he is hissed and cheered (by stomping) in a manner that is not at all pleasant." Quoted in Michael Sappol, *A Traffic of Dead Bodies*, 81. See also Christopher Lawrence's excellent analysis of the bodily incorporation of masculinity into physicians and surgeons in Great Britain and America in "Medical Minds, Surgical Bodies: Corporeality and the Doctors," in *Science Incarnate: Historical Embodiments of Natural Knowledge*, ed. Christopher Lawrence and Steven Shapin (Chicago: University of Chicago Press, 1998), 156–201.

63. Pringle, *Sex and Medicine*, 69, 85–90.

64. See Frances K. Conley, *Walking Out on the Boys* (New York: Farrar, Straus and Giroux, 1998), esp. 20–21, 34, 48–49, 55.

65. Joan Cassell, *The Woman in the Surgeon's Body* (Cambridge, MA: Harvard University Press, 1998), 210–211.

66. Two important collections of oral testimony from medically trained women testify to the continued presence of gendered discrimination. See, in this connection, Regina Markell Morantz, Cynthia Stodola Pomerleau, and Carol Hansen Fenichel, eds., *In Her Own Words: Oral Histories of Women Physicians* (New Haven, CT: Yale University Press, 1982), 70, 149, 171, 180–181, 191, 221, 238. French women practitioners have many of the same complaints, arguing that medical education and practice is still "a world of man with all the clubby values of men's gossip and corridor decision-making." See Thérèse Planiol, *Herbes folles hier: Femmes médecins aujourd'hui* (Paris: Cheminements, 2000), 190; also 89, 98, 111, 114, 123, 127, 135, 150–154, 167–168, 193–198, 200, 269, 309. On society exclusion, see also David Hay, *Honest Talk and Wholesome Wine: A His-*

tory of the St. Albans Medical Club: 1789 to the Present (London: St. Albans Medical Club, 2004), 215, 241–243.

67. Cathy A. Trower and Richard P. Chait, "Faculty Diversity: Too Little for Too Long," *Harvard Magazine*, 2002: 2–3. See also Kate Zernike, "The Reluctant Feminist," *New York Times Education Life*, April 8, 2001, pp. 34–35.

68. For this point, see Christine Williams, *Gender Differences at Work: Women and Men in Nontraditional Occupations* (Berkeley: University of California Press, 1989), 129–130.

CHAPTER SEVEN

Women Physicians and the Twentieth-Century Women's Health Movement in the United States

Sandra Morgen

The dramatic growth in the number and proportion of women physicians in the United States in the past several decades cannot be understood apart from the larger political, economic, and cultural changes shaped significantly by second-wave feminism. Feminists critiqued gender-based occupational segregation, challenging discriminatory practices that had long restricted women's access to well-paid, high-status careers, including medicine. But beyond challenging employment discrimination, feminist activists sought broader changes in the health care system. They built a powerful grassroots movement that has had a dramatic influence on women's health care.

In the late 1960s, tens of thousands of grassroots activists across the country declared that ordinary women, not physicians or lawmakers, deserved greater control of fundamental decisions about their bodies and reproductive lives. They formed local and national organizations that constituted a vibrant, growing social movement. The women's health movement flourished in the 1970s, and, despite significant challenges and changes in the movement, it remains a significant political force.[1] In its early years, movement activists focused on decriminalizing abortion and expanding women's reproductive rights. But their political horizons were broader. They wanted more knowledge about and control over their own bodies and health care. They demanded changes in the inequitable power relationships between physicians and patients. They sought to change the marginalization of women in the practice and administration of medicine and in health policy. They envisioned and fought for a health care system in which the health needs of people took precedence over the profits of the health care industry and the wealth of physicians.

Of the many gains scored by the movement, one of the most visible has been opening the doors of clinical and academic medicine to more women.[2] In 1970, there were 25,401 women physicians in the United States, making up a mere 7.6 percent of the profession. By 1980, women accounted for 11.6 percent of physicians, and by 1990, women were 16.9 percent; more recent data show that more than 25 percent of physicians and half of the medical students in the United States are women.[3] Despite such impressive gains, racial discrimination has limited equitable progress for African American, Hispanic, and Native American women. According to recent data, 70 percent of women physicians are white and 18 percent are Asian American, but only 7 percent are African American, 5 percent are Hispanic, and less than 1 percent are American Indian/Alaskan Native.[4] Progress has been slower in academic medicine. The proportion of women from underrepresented minorities on medical school faculties grew from 4 percent to only 6 percent over the past two decades, while the proportion of all women on these faculties increased from 15.2 to 26.6 percent.[5] And despite unequivocal progress, discrimination, sexual harassment, and a professional culture built by and for white men still confront women physicians.[6]

Most women physicians in practice today entered the profession during or following the advent of concerted feminist health activism.[7] Although most women physicians were not personally involved in the women's health movement, this does not mean that their career opportunities or the ways that they practiced medicine were unaffected by the movement. A 1998 Council on Graduate Medical Education (COGME) report on women in medicine concludes that "women in medicine bring a unique perspective to women's health concerns . . . [and] therefore, these women can substantially influence the shape and direction of changes in the health care system."[8] In this chapter, I suggest that such changes are not simply a result of demographic changes in *who* practices medicine. I argue that these changes have been catalyzed and shaped by the concerted and continuing activism of the women's health movement and the institutional, cultural, and political changes the movement helped provoke and shape.

This article focuses on the complex relationship between women physicians and the women's health movement,[9] a relationship addressed by other chapters in this book, especially those by Naomi Rogers (Chapter 9) and Susan Wells (Chapter 8). I will discuss evolving attitudes about women physicians within the movement, the involvement of women physicians in the movement, and the critical role the movement played in articulating and promoting values and practices that are now taken for granted in what has come to be known as the field of "women's health." Beyond illuminating the often unacknowledged role of grassroots feminist health activism in expanding opportunities for women in medicine, I conclude that recognizing the role of this social movement challenges and enables us to move beyond essentialist[10] explanations of gender differences in the way (many) men and women practice medicine, that is, to understand these differences as socially and politically constructed.

The Women's Health Movement in the United States: A Brief Historical Overview

The women's health movement emerged in the late 1960s in the context of a larger political environment of social activism peopled by women, African Americans, Latinos, Native Americans, gays and lesbians, welfare recipients, labor, college students, and critics of the Vietnam War and U.S. imperialism. This broad political left encompassed revolutionary, radical, and liberal organizations and visions of change. Although the political goals, constituencies, and strategies of these various movements differed, as a whole they offered a radical critique of sexism, racism, homophobia, and class inequality in U.S. society. No major social institution escaped this radical critique, including the health care system.

The main focus of feminist health activism in the late 1960s was on women's reproductive rights, especially decriminalizing abortion and increasing women's access to birth control and safe abortion. In the aftermath of the pivotal *Roe v. Wade* Supreme Court decision in 1973, movement activists broadened their critique of the health care system. Physicians, hospital administrators, pharmaceutical companies, and health regulatory agencies came under fire for exposing women to unnecessary medical procedures (including hysterectomies, Cesarean sections, coercive sterilization, and one-step mastectomies), inadequately tested estrogenic products (high estrogen oral contraceptives, DES [diethylstilbestrol], hormone replacement therapy), and intrauterine devices. Activists challenged American medicine for putting "profits above people" and for excluding the voices of health care consumers, especially women, in deliberations about how to provide health care services and health policies that served the best interests of women.

The activities, political strategies, and organizations of the emerging women's health movement were diverse. Some groups produced information, seeking to raise women's consciousness and empower women. The best known of these is the Boston Women's Health Book Collective. This collective wrote the influential book *Our Bodies, Ourselves*, the first edition of which circulated in the movement in 1970. By 2005, eight English-language and eighteen foreign-language editions of the book had been published and circulated worldwide. *Our Bodies, Ourselves* articulated a radical political perspective on women's health. Along with presenting information from medical books and journals in an accessible manner, as Susan Wells discusses in this collection in "Narrative Forms in *Our Bodies, Ourselves*," the book validated women's knowledge and experiences about their bodies and health care.

Another key movement strategy was the provision of alternative health services. These services were designed to empower the women they served. Feminist health activists, most of whom were not health professionals, founded and staffed women-controlled health clinics or worked within community health centers to create feminist services for women. An authoritative source documented at least fifty feminist clinics in U.S. cities and towns in 1976.[11] The clinics constituted alternative models of health care delivery and were an important organizational base that activists worked from to change mainstream medical care.[12] They promoted preventive and other health education, self-help, and the message that the gender inequalities at the heart of the U.S. health care system could be changed.

Armed with a belief that routine passages of women's reproductive lives were overmedicalized, alternative health care providers sought to shift the balance of authority and control away from physicians and toward women in areas such as childbirth, family planning, routine gynecological care, and menopause. Clinic staff often organized their services and organizational structures to minimize hierarchy and power differences among staff, and between staff, volunteers, and clients. In response to the ubiquitous movement concern that many women lacked the financial resources necessary to access quality health services, clinics usually offered free, low-cost, or sliding-scale services for abortions, family planning, and routine gynecological health care. Services were designed to provide high-quality health care and to empower women as health care consumers and in their daily lives.

A third major movement strategy was political advocacy. Advocacy groups sought to change local, state, and federal health policy. Some, such as the National Women's Health Network and the public policy office of the National Black Women's Health Project, located their offices in Washington, D.C., symbolizing and facilitating their intent to "watchdog" the health watchdog agencies, such as the Food and Drug Administration, and to serve as a women's health-policy lobby group. Many others addressed state and local health policy.

Drawing on a radical political analysis of how gender, race, and class inequalities translated into women's individual and collective disenfranchisement as health care consumers, health care providers, and health care policymakers, these advocacy groups took on a variety of issues. Some argued that women's health was overmedicalized and that physicians and lawmakers had disproportionate say over women's bodily integrity and health decisions. At the same time, they recognized that many low-income women and families lacked access to high-quality primary health services, especially when they had no health insurance or were forced to rely on overcrowded public clinics. Advocates decried the abuses of medical experimentation and training that subjected women, especially women of color (who relied on care at public or teaching hospitals), to procedures without providing them with the information they needed to give informed consent. Collectively, the critiques by feminist health advocates exposed the ways sexism, racism, and a market-based health care system resulted in inequitable, and often inadequate or inaccessible, health care.

From the movement's earliest days, some women activists of color argued—in print and in organizational contexts—that the health issues affecting women of color and poor women had to do with more than gender. Racism and economic inequality, they argued, also created significant obstacles to reproductive rights and access to high-quality health care.[13] Their critique of racism focused not just on the health care system but also on predominantly white feminist reproductive rights and health movement groups that failed to address the intersection of racism, class inequalities, and gender in women's health.

The failure of many women's health movement groups to address racism led women of color to organize a new generation of women's health organizations in the 1980s. These included the National Black Women's Health Project, the National Latina Health Organization, the Native American Women's Health Education Resource Center, and the National Asian Women's Health Organization. Some of the women who founded these groups had been (and some remained) active in predominantly white, middle-class health organizations; others were not. Their visions of a health agenda for poor women and women of color had multiple sources, including feminist health and reproductive rights, civil rights, community health, welfare rights, racial identity movements, and their own experiences as health consumers and providers.[14] These groups organized within their racial or ethnic communities and worked collaboratively with predominantly white health movement organizations. Their impact on and leadership in the larger movement has been substantial.

The women's health movement flourished during the 1970s; at least one thousand organizations were directly involved in diverse forms of women's health activism during the decade.[15] However, these health activists faced new challenges in the 1980s, a period of significant political backlash and economic retrenchment. The movement endured despite the powerful political assault on feminism, and, especially, on reproductive rights, with massive cuts in health and human services enacted during the Reagan administration. Many feminist groups, including women-controlled health clinics, however, did not survive this political backlash or the fiscal effects of the federal budget cuts that eroded funding for many community-based health clinics and advocacy groups.

The movement and its allies managed to maintain the legality of abortion, although access to abortion was increasingly restricted. While many of the feminist clinics that provided abortions closed their doors in the 1980s, some of these clinics as well as other outpatient abortion facilities continued to provide services. Moreover, even abortion facilities that were not explicitly feminist often provided services that were influenced, to some degree, by the service model pioneered in the early feminist clinics.

Today many more women, including women of color, are in medical schools and medical practices, in health care management and administration, and in medical research labs. Informed consent has become institutionalized, providing a degree of protection against sterilization abuse, medical experimentation without the patient's knowledge, and unnecessary medical procedures. Childbirth practices today generally permit both greater involvement in and control over labor and delivery by women and their partners. Many women, especially those whose health care plans or financial resources give them a choice of providers, can expect to be treated with more respect and dignity by health care professionals and are encouraged to play a more active role in decision making about their health care. These are some of the changes in women's health care that can be attributed to the activism of this grassroots movement and its allies.

By the early 1990s, the politics and landscape of the women's health movement looked somewhat different than in the movement's younger days. What began as a radical grassroots movement morphed into a diverse constellation of organizations and activities with different goals, strategies, and constituencies, each shaped in different ways by its origins in women's health activism. The advocacy groups and the few original women-controlled clinics that still exist have had to adapt to a difficult and evolving political, fiscal, and regulatory environment. Moreover, they compete now with a new breed of "women's clinics" owned and operated by hospitals or physicians, often women physicians. These clinics have adapted aspects of the feminist health care model most amenable to mainstream medicine, and while they owe their origins and some of their consumer base to changes the movement initiated, these clinics do not share or articulate the powerful critique of the health care system that was integral to the feminist clinics that came before them.[16]

Indeed, the women's health movement has left its mark on the institutions of medicine and health care policy. In the 1990s, some movement groups moved toward what Ruzek and Becker call a "professional" orientation,[17] with a focus on policymakers, the biomedical research community, and the growing number of women in positions of some authority and power vis-à-vis health policy. They formed alliances with women in Congress, at the National Institutes of Health (NIH), and elsewhere in the federal health care bureaucracy to advocate for changes in the allocation of federal funding for medical research and in biomedical research policies, such as the greater inclusion of women and minorities in clinic trials. These strategic alliances have won increased allocation of dollars to women's health research, especially with the founding of the Office of Research on Women's Health in the NIH. In turn, women health professionals have revived or created new professional organizations and journals to represent their professional interests and their concerns about women's health more broadly.

This brief overview can only gesture to the diversity and achievements of a movement that, for all its accomplishments, rarely gets credit for the changes it helped catalyze. Many women physicians know little about the women's health movement, including the roles played by women physicians in the movement.

They are often unaware that what is now called the specialty of women's health owes its roots to feminist grassroots health activism. This history is an important facet of the history of women physicians in the last half of the twentieth century.

The Women's Health Movement and Physicians: Critique and Collaboration

Physicians, especially obstetrician-gynecologists, were the target of a searing critique by feminist activists, and as Rogers's essay (Chapter 9) in this volume explores, by women medical students in particular. The most radical critique came from outside the profession, from women patients on the receiving end of sexism and racism. An early edition of *Our Bodies, Ourselves* rebuked the overwhelmingly white and male medical profession for failing to "take responsibility for the health of the people" and defined doctors' attitudes toward their women patients as "terribly condescending."[18] Source after source decried the excessive wealth and power of doctors, their failure to emphasize preventive health care or to ensure that underserved communities received medical care, the cultural deification of physicians, and doctors' gender, race, and class biases.

Feminist scholars and activists unearthed a history of American medicine that revealed the virtual exclusion of women of all races and men of color from the ranks of medicine.[19] For example, historians documented the role played by the fledgling American Medical Association in marginalizing women health providers and healers in the late nineteenth century, analyzing their actions as a means of consolidating the power of "regular" medicine.[20] They identified numerous examples of physician action, inaction, or complicity in denying reproductive rights to women, engaging in unethical medical experimentation (especially on poor women of color), and monopolizing health care decision making and health policy.[21] Physicians' practices were analyzed as routinely stripping women patients of their dignity, their right to bodily integrity, and their freedom to choose appropriate treatment.

As these analyses proliferated in books and in scholarly and popular articles, as the movement began to score policy victories, and as self-help groups and women-controlled clinics opened for business, physicians, especially the obstetrician-gynecologists who took the brunt of the feminist critique, could no longer ignore the movement completely. In 1974, Barbara Kaiser and her physician husband Irwin Kaiser presented a paper entitled "The Challenge of the Women's Movement to American Gynecology" at the annual meeting of the American Gynecological Association. The Kaisers were generally sympathetic to the movement's goal of "redistribut[ing] power between doctors and patients"; they also supported women's right "to define their femininity in their own terms . . . and to assume some control and decision functions with respect to the social and personality effects of gynecologic care."[22] While they were concerned about grassroots activists' lack of medical training and questioned some of the self-help practices promoted in the movement, they urged their colleagues to listen to the concerns articulated by feminist critics. They also suggested that physicians provide more medical information to patients and that they learn not to expect their patients' "unquestioned trust." The Kaisers advised them to emerge from "behind the drapes [the standard practice of draping for the gynecological exam] to teach and demonstrate self-examination of patients," to share records and lab reports, allow "loved ones in labor and delivery rooms," and express "candor about malignant disease." These changes, they believed, would produce a "vast improvement in medical care."[23]

In response to their paper, six physicians, including Georgeanna Jones, a female physician, gave prepared responses. Several were defensive, including Dennis Cavanagh who dismissed the women's health movement as the "lunatic fringe." Jones deflected the criticisms of physicians, claiming that "good doctors," male and female, already practiced medicine in the manner the feminist activists advocated.[24] Others were more open to the critique. Although he bristled at what he saw as "overgeneralizations" about physicians, Clay Burchell agreed that changing times and cultural mores meant that medicine also needed to change.[25] Whether this exchange is an accurate barometer of the attitudes of the profession at large, the attention given to the Kaisers' paper is evidence that the women's health movement was increasingly viewed as a powerful force.

Of the growing number of women medical students and physicians in the 1970s, only a small minority became active in the women's health movement. As Rogers's essay shows, some did individually and collectively resist sexist practices in medical school. But others went further, working with feminist health organizations and, sometimes, taking on leadership roles in the movement. Most physicians who participated did so as service providers in feminist clinics. Although health movement activists believed in modulating physician control of women's health, feminist clinics had to comply, at least on paper, with strict health regulatory policies, including licensing regulations and the reimbursement policies of health insurance companies and government health programs.

These regulations ensured that physicians retained considerable control over health care delivery. Although nonprofessional health workers in feminist clinics performed more of the health care service delivery than was standard in mainstream medical facilities, physicians still oversaw and often provided those services, sometimes functioning as titular or actual medical directors of the clinics.

Feminist clinics were keen to employ women physicians, but they were in short supply, especially in the movement's early years, so sympathetic male doctors were also recruited. Some doctors served as medical directors of clinics or participated in the collective administration of these organizations, sharing significant decision-making power with the laywomen who founded and staffed the clinics. The involvement of others was restricted to working shifts to provide routine gynecological care, family planning, or abortions, following protocols developed by health collectives or feminist directors of these clinics. Relationships between physicians and lay staff of feminist clinics varied, depending on the values and beliefs of the individual physician on the role of nonprofessionals in health care delivery, self-help, and other issues, and the attitudes and values of clinic staff about how they wanted doctors to function within their organizations.[26]

Not all women's health movement organizations shared the same views about physicians or about the priority of equitable power sharing between physicians and the laywomen staff. For example, while predominantly white health movement organizations tended to articulate an explicit critique of male/professional dominance of health care, leading to a priority on using alternative health care practices and practitioners, this was less true of health movement organizations developed by women of color. These groups were often more concerned about increasing the access of low-income women and their families to primary health care services, including physician-provided services. Sometimes women health activists of color could also count on the strong support of physicians of color whose own experiences of racism or sexism led them to share the movement's social justice goals.

Physician Leadership and Involvement in the Women's Health Movement

A handful of women physicians were not only involved in the women's health movement, they were also movement leaders.[27] In this section, I briefly discuss some of these physicians, not with the intent of producing an adequate biography of these important leaders but to elucidate some of the ways women physicians provided leadership to the movement. One of the earliest and best-known leaders was Mary Howell, a physician who sought to reform medical education and who collaborated with other feminist health activists to organize one of the leading national feminist health advocacy organizations, the National Women's Health Network.[28] Howell, who earned her M.D. from the University of Minnesota in 1962, was one of the tiny cohort of women medical students during the 1950s–1960s. She joined the faculty of Harvard Medical School in 1969 as an assistant professor of pediatrics. In 1972, she became the first woman to serve as an associate dean at Harvard Medical School when she was appointed associate dean of student affairs.

From this vantage point, she saw plainly and painfully that little had changed for women medical students since her own years of education, despite the upsurge of feminist activism. She undertook a survey of 146 women students from forty-one medical schools in the United States to document their experiences, and in 1973, under the pseudonym Mary A. Campbell, she published *Why Would a Girl Go into Medicine? Medical Education in the United States: A Guide for Women.* The book was a scorching analysis of the ways in which medical education perpetuated sexist beliefs about women patients and imposed overt and covert forms of discrimination on female students.[29] Howell was dedicated to challenging sexism both in medical education and in the profession at large. But her political commitments extended beyond improving the position of women physicians. Unlike many sister women physicians or medical students, Howell had a broader and more radical vision of change. She worked closely with feminist health activists outside the medical profession, speaking at conferences, writing articles, and, in the early 1970s, joining four women in founding the National Women's Health Lobby, which was soon renamed the National Women's Health Network.

Another physician who left a lasting mark on the women's health movement was Helen Rodriguez-Trias, a Puerto Rican physician, born in New York City and raised in Puerto Rico. Rodriguez-Trias was one of the most visible physician-leaders in the women's health movement. She was also one of the boldest in articulating an analysis of women's health that spoke to the intersection of sexism, racism, and class inequalities as they affected women's health and experiences with the health care system. Unlike most women physicians of her cohort, she was a woman of color and, when she graduated from medical school in 1960, she was also a wife and mother of three young children. Her own experiences undoubtedly influenced her radical vision, as did her relationship to the Puerto Rican nationalist movement (and its critique of U.S. imperialism). She was involved in health activism through her work at Lincoln Hospital in the Bronx, where she practiced pediatrics and, eventually, served as head of the Pediatrics Unit. There she saw firsthand the failure of the health care system to address the most basic health needs of poor women and their families. Rodriguez-Trias understood gender, race, ethnicity, and class as inextricably connected in shaping a two-tiered system of health care.[30]

One of her many important contributions to the movement was the leadership she provided in organizing against the widespread coercive sterilization of poor women of color, a too common occurrence, especially in public teaching hospitals. She helped found and lead the New York–based Committee to End Sterilization Abuse (CESA). CESA successfully fought, first locally, and then in a national coalition with other groups, to require hospitals to implement a series of practices to protect women from sterilization abuse. Rodriguez-Trias also worked to expand the political vision and activities of predominantly white reproductive rights and women's health organizations beyond the issue of abortion. She advocated for the incorporation of childbearing rights, including protection from coercive sterilization. A charter member of the National Women's Health Network, Rodriguez-Trias served on its board of directors in the 1970s. From her multiple positions of leadership in the movement, she helped shape movement priorities, especially a growing awareness of and commitment to addressing racism and the particular needs of poor women.

The leadership of some physicians was directed more locally. For an example, consider Mary Jane Gray, a founder of one of the first feminist abortion clinics in the United States. Gray, a faculty member at the University of Vermont Medical School, had seen how women's lack of access to safe, legal, affordable abortions led to suffering and ill health. After spending years chairing a group working to liberalize abortion legislation in Vermont, and when some of her colleagues blocked her proposal to start an abortion clinic at the hospital, she worked with others in the community to found the freestanding Vermont Women's Health Clinic. The clinic, which involved both health professionals and laywomen, was a pioneer facility, providing legal abortions before 1973 (Vermont was one of only a handful of states that legalized abortion before the *Roe v. Wade* decision) and pioneering a feminist approach to abortion services.

Gray located herself on the border between mainstream medicine and the women's health movement. In an interview she said, "I was as radical as you could be as a doctor and as conservative as you could be in the movement." Her own experiences as a physician and a woman supported her belief that there was "a lot of truth" in the basic feminist critique of doctors. She often thought of herself as a "translator between physicians and women's health activists."[31] For example, in addition to her leadership at the clinic, Gray published a paper about the clinic's policies and practices in the *American Journal of Obstetrics and Gynecology*. The article described the clinic's philosophy, its services, and the internal and external challenges its organizers and staff faced over several years. The piece concluded that "physicians working in the center were threatened initially by the lack of a rigid medical hierarchy coupled with the existence of an atmosphere in which their judgment could be freely questioned." But eventually, "recognition of the genuine concern for the patient felt by all persons working in the clinic, respect for the increasing level of medical information held by all staff members, and the fact that the physician maintained the right to make a final decision in medical matters led to acceptance of the working conditions and a sense of sharing."[32] In other words, beyond her own involvement in a feminist health clinic she worked as an advocate among her medical peers to legitimize the alternative health care model.

Some physicians who participated in the movement had previous social movement experience. And for some of them, the decision to pursue a medical degree was simultaneously a political and a career decision. One such physician I interviewed linked her decision to go to medical school with the political consciousness she developed through her activism in the antiwar, civil rights, and early feminist movement while she was in college. She later worked at several feminist and community clinics in the San Francisco Bay Area as a way of extending her political commitments into her professional life. For her, working at feminist clinics was an "antidote to establishment medicine."[33] At the Berkeley Women's Health Collective, she oversaw a gynecology practice staffed largely by lay medics. She believed these nonprofessional health providers "got pretty good at doing just regular, general gynecology checkups and doing Pap smears and solving routine gynecology problems."[34] Her involvement in the women's health movement also shaped her own practice of medicine. And it was in the movement, not medical school, that she learned "certain ways of dealing with patients that were . . . valuable in terms of treating them as equals, . . . sharing information with them, the kind of thing that is not particularly the way we are taught in medical school . . . [And] I learned a lot more about contraception and minor GYN problems than I learned in medical school."[35] Clearly, this young physician not only contributed to the movement, but also gained a great deal of additional training from her participation.

While most of the physicians involved in the movement practiced reproductive health care, especially in its first decade, some did not. For example, psychiatrist Judith Herman, a respected medical authority on domestic violence, incest, and sexual abuse and an associate professor at Harvard Medical School, worked at a feminist health clinic in Somerville, Massachusetts, in the mid-to-late 1970s. Herman was a founding member of a feminist mental health collective that shared space and worked closely with the Somerville Women's Health Project. In addition to her leadership in the feminist mental health movement, she was also a member of the collective that ran the clinic, which provided a wide range of community-based primary health services, including gynecological care. Herman also served as the clinic's medical director. This clinic, like many others, was a site of intense political struggle during its short organizational life. Class differences within the staff produced tensions and conflicts in a group that believed in the importance of equality among staff and between staff and clients. In some instances, as in Herman's case, physician-leaders became lightning rods for such conflicts, as the relative power they accrued from their professional credentials, experiences, and training were both a resource and a basis for inequality within organizations that eschewed hierarchy.[36]

Another well-known physician-activist who both contributed to and drew on the resources of the women's health movement is Susan Love, a towering figure in the field of breast cancer surgery and advocacy. Love merged innovative medicine and advocacy in her work at Beth Israel Hospital and the Dana Farber Cancer Clinic in Boston. She later founded the Faulkner Breast Center, which was staffed entirely by women surgeons, radiotherapists, and plastic surgeons. She also helped to found the National Breast Cancer Coalition. The Santa Barbara Breast Cancer Institute was renamed the Susan Love MD Breast Cancer Research Foundation in honor of her work as director. As the author of widely acclaimed books on breast health, breast cancer, and hormone replacement therapies. Love's work has had a significant effect on breast cancer treatment and has spurred the organization of breast cancer patients and survivors into a powerful lobby.[37]

Some feminist physicians who shared some of the goals and political analysis of the women's health movement chose to work in family planning or community health clinics rather than the new feminist health clinics formed during the 1970s. Some of them, especially women of color, chose to lend their energies to community-based clinics serving poor communities of color. One example is Helen Barnes, an African American obstetrician-gynecologist who joined the staff of the community-based Delta Health Center in rural Mississippi in the late 1960s. Her work shows that women physicians of color sometimes forged alternative approaches to women's health outside feminist clinics. Like feminist clinics, community-based clinics, such as the Delta Health Center, often envisioned health care as a tool for "community development and individual empowerment."[38] Historian Jennifer Nelson suggests that the physician-activists and other health care providers at this clinic worked energetically to involve community residents both in their own health care and in the management and staffing of the clinic. Services were based on the idea that health care is a right, not a race and class privilege. Clinic practitioners, many of whom were African American women in and from the community, understood nutrition, human services, and environmental issues to be an integral part of health care delivery.[39] In this context, Barnes provided different obstetrical, family planning, and gynecological services than were otherwise available to the African American residents of the region, including being part of a team that included "poor women [who] provided and managed much of their own care in Mound Bayou."[40]

With the founding of national women of color health movement organizations in the 1980s and 1990s, more women physicians of color found homes in the movement, lending some of their energies to these new organizations. In some cases, they provided clinical services through movement-related groups. They also served on boards or participated in conferences and other programs sponsored by these organizations. For example, two of six members of a recent slate of board members of the Black Women's Health Imperative (the new name of the National Black Women's Health Project) were physicians: Marilyn Gaston, former director of the Bureau of Primary Health Care and a former assistant surgeon general, and Janet Taylor, a clinical instructor in the department of psychiatry at Harlem Hospital. Similarly, at the time of this chapter's publication, almost one-third of the board of directors of the National Asian Women's Health Organization are physicians: Yi-Shin Kuo, a gynecological oncologist in New York City, Descartes Li from the University of California–San Francisco Department of Psychiatry; and Pamela Wang Anderson, a physician affiliated with the Lilly Research Laboratory. In fact, the women's health movement became a less contested home for women physicians by the 1990s as movement attitudes toward physicians were changing and as more women physicians laid claim to the label "feminist."

"Trickle Up"? The Professionalization of the Women's Health Movement in the 1990s

Health politics in the 1990s, including women's health movement politics, were changing with the times. By the 1990s, many more women were in medicine and in other health professions, and more women held positions of authority in health policy formation and implementation. When President Clinton announced his intent to reform U.S. health care in his first term, many of these women and veteran women's health movement activists poured their energy into what was ultimately an unsuccessful political campaign. Nevertheless, by the 1990s, even the mainstream medical establishment was more prone to recognize the existence and injustice of gender and race inequalities in health status and health care.

New and continuing collaborations between movement activists, women physicians, and other women health professionals were forged in this changed political climate. At the same time, the lines between grassroots feminist health activists, feminist physicians, and the new breed of feminist health professional organizations were blurred and the balance of power shifted. As feminist health advocates turned their attention to the growing epidemics of breast cancer and AIDS, physician/professional dominance of health care became a less salient issue. Increasingly, feminist health advocates turned a critical eye to biomedical research priorities and protocols. Moreover, decades of activism by feminists, civil rights activists, and consumer health advocates finally forced the federal government to define as politically important what began to be called gender and race "disparities"[41] in health. Finally, the equity concerns of feminists and other progressive health care activists were "trickling up" to the federal agencies responsible for public health and medical research.

Reports by federal health agencies, often instigated by feminists or advocates of civil rights within federal agencies, documented serious and continuing inequalities in health status and access to quality health care between men and women, whites and people of color, and the insured and uninsured. A 1985 report of the U.S. Public Health Service criticized the U.S. Department of Health and Human Services for failing to respond adequately to women's health needs.[42] Another 1985 report, from the Department of Health and Human Services Task Force on Black and Minority Health, documented racial and ethnic "health disparities" for many health conditions across the United States.[43] In response, the National Institutes of Health recommended new policies to encourage the inclusion of women and racial minorities in clinical research. But several years later, when a General Accounting Office (GAO) report found these recommendations resulted in little change, Congress passed the Women's Health Equity Act of 1990, establishing the NIH Office of Research on Women's Health. In 1991, Bernadine Healy, the first woman physician to head the NIH, announced a new Women's Health Initiative that funneled hundreds of millions of dollars to research on breast cancer, cardiovascular disease, and osteoporosis. In 1998, the surgeon general announced the Initiative to Eliminate Racial and Ethnic Disparities in Health, and in 2001, the prestigious Institute of Medicine documented what many scholarly and movement publications had been arguing for decades—the existence and consequences of racial disparities in health and health care.[44]

Because some concerns of health movement activists were "mainstreamed"—as more physicians and other health care professionals took on and, sometimes, took over leadership and responsibility for these issues—the grassroots organizations that had sustained the movement were sometimes sidelined and their radical critique was watered down. While movement activists had envisioned a fundamental transformation of American medicine, the professionalized women's health organizations that developed in the late 1980s and 1990s often followed a "consumerist" model aimed at providing "more and different services for women and women's access to positions of power within medicine."[45] An example is the proliferation of hospital and physician-owned women's health centers that numbered in the thousands by the mid-1990s.[46] These women's health centers, often for-profit operations, adapted some of the ideas and practices developed in the feminist, women-controlled clinics that were founded in the 1970s and 1980s. They marketed themselves as staffed by women and as offering "caring" medical services designed to empower medical consumers.[47]

Despite this rhetoric, many of these clinics are different from the feminist health clinics of their forebears. These new physician- and hospital-managed clinics are certainly responses to the "raised" consciousness of women in the wake of feminism, but they are also the result of "diversification strategies" in the highly competitive health care marketplace today.[48] They orient their services to patients with insurance and their client base tends to be middle and upper middle class and white, unlike the feminist clinics that aimed to provide affordable health care for all women.[49] In addition, while these clinics talk about the priorities of prevention and primary care, their practices are often skewed toward high-tech detection and treatment strategies, especially for cancers of the breast, cervix, uterus, and ovaries; pregnancy, labor, and delivery; and treatments of osteoporosis and menopause.[50] The capital resources these clinics command far exceed those most feminist clinics have at their disposal. Community-based clinics often struggle to compete (and many have had to close) in this context, as the more economically advantaged patients are attracted to these new clinics, eroding even further the marginal financial health of clinics committed to serving those with the least financial resources.

The growth of the "women's health" field has created new opportunities for women physicians who can capitalize on what women consumers now expect or want from physicians, including the growing preference by women for female physicians. Some doctors who have been active in helping to create this field, such as Lila Wallis, acknowledge feminism as the catalyst for this new clinical and academic specialization,[51] but many do not. Some feminists, including Elianne Riska, are concerned that this "professional project of . . . women physicians who present themselves as the professional advocates of women's health and defenders of women's interests in medicine"[52] has overshadowed the catalyzing and more radical perspective of the feminist health movement.

Like the physician-managed or hospital women's clinics discussed earlier, the approach of increasing professional opportunities for women still falls short in terms of the key concerns of movement activists. Changing the physician-patient relationship and expanding the knowledge of physicians about women's health helps those women who have access to health care, but it does little for the growing numbers of women, many of them poor, women of color, or immigrants, who lack health insurance coverage and whose access to health care is thus limited. Nor is it clear that these reforms fundamentally address the critique of the medical hierarchy between doctors and other health care providers or between doctors and patients that was so important to early movement activists.

Women physicians sometimes express frustration and even anger in response to feminist criticism. For instance, Anne Colston Wentz, editor of the *Journal of Women's Health*, condemned what she calls an "anti-doctor" attitude in the women's health movement in a 1994 editorial in this journal. She expressed "discomfiture . . . alienation . . . sadness and . . . anger" in response to the common view that all physicians are regarded as "the same . . . and not good."[53] She proposes that feminist activists and doctors work as allies, not adversaries: "We need to get this women's health thing together and have people work together for change. It can't be a we-against-you situation when the bottom line is basically the same. We are not going to be able to reframe women's health without physicians, leaders in women's health, men and women all buying into the idea of change and working together to get the job done."[54] But for many movement activists the "bottom line" is *not* "basically the same." Wentz objects when physicians, or women physicians, are all seen as "the same." But within the field of women's health, women's diverse health needs and the differences related to race, ethnicity, class, and sexuality are often ignored and the growing success of women physicians sometimes obscures the limits of change for other women. These activists are aware that the class position of physicians differentiates them from many women whose economic insecurities and hardships translate into poor health, lack of access to good health care, and medical debt.

Conclusion

Wentz is right. It will take the energies and commitments of a variety of social actors to accomplish the goals of the women's health movement. Wentz may be disheartened by the continuing critique of women physicians by feminist health activists, but those activists are also disheartened and angry when women physicians dismiss or are ignorant of the women's health movement or its goals. Clearly, some women health professionals honor this history. Some advocate sustaining or recreating strong connections between women health professionals, including physicians and movement activists. As an example, planners of a University of Illinois institute on "reframing women's health" in the mid-1990s explicitly recognized the pioneering, visionary work of feminist activists and advocated the importance of "reconnecting to the women's health movement."[55]

But appreciating the role of grassroots women's activism is important not just in terms of the academic question of getting history "right" or of giving a social movement its due. When the Council on Graduate Medical Education suggests that "women in medicine bring a unique perspective to women's health concerns . . . [and] therefore, these women can substantially influence the shape and direction of changes in the health care system,"[56] the report leaves unanswered the question of the complex source of this "unique perspective." Essentialist explanations that presume women's so-called natural capacities for caring, empathy, collaboration, or relationships often lurk implicitly, and, sometimes, even explicitly, in discussions of what women bring to medicine.

I argue that the source of this "unique perspective" is not some essentially "feminine" essence. Nor is it an automatic reflex of women's gendered and racialized positions in society. Rather, this perspective has been developed, sustained, and disseminated from the trenches of a grassroots social movement and from the many individual and organizational branches that emerged from the movement's roots. Change was seeded, shaped, nurtured, and provoked by the vision, hard work, sacrifices, and struggles of activists in the women's health movement and in other social movements that shared the broad goal of challenging inequality and injustice.

Social change is a complex phenomenon. It never has one source. However, the women's health movement has played a pivotal role in transforming women's health and women's health care and in articulating and advocating radical, innovative ideas that now suffuse more of the culture, including the culture of medicine. The history of the women's health movement is an integral aspect of the history of women physicians in the last half of the twentieth century. If that history is better known and appreciated it not only illuminates the past, but it can be a source of future linkages between physicians and other women's health activists that can empower both groups in their quest for continuing health care reform on behalf of all women.

NOTES

1. Sheryl B. Ruzek, *The Women's Health Movement* (New York: Praeger, 1978); Judy Norsigian, "The Women's Health Movement in the U.S.," in *Man-Made Medicine: Women's Health, Public Policy, and Reform*, ed. Kary Moss (Durham, NC: Duke University Press, 1996); Carol Weisman, *Women's Health Care: Activist Traditions and Institutional Change* (Baltimore: Johns Hopkins University Press, 1998); Sandra Morgen, *Into Our Own Hands: The Women's Health Movement in the United States, 1969–1990* (New Brunswick, NJ: Rutgers University Press, 2002).

2. Ellen S. More, *Restoring the Balance: Women Physicians and the Profession of Medicine, 1850–1995* (Cambridge, MA: Harvard University Press, 1999). Clearly, the movement was not the only reason for this change. During the 1960s, changes in the economy meant that more families needed dual incomes; the broader civil rights and women's movements were transforming U.S. culture and politics; and women's own aspirations and needs had changed.

3. American Medical Association, *Physician Characteristics and Distribution in the U.S.*, 2005 edition. American Medical Association, www.ama-assn.org/ama/pub/category/12912.html.

4. I calculated these numbers by using the figures in the previous source. I excluded from the total number of women physicians those categorized as "other" or "unknown." Therefore, these percentages may be slightly off, but the basic pattern is accurate. African American, Latino, and American Indian / Alaskan Native men also continue to be underrepresented in medicine.

5. Emily Wong, Judyann Bigby, Myra Kleinpeter, Julie Mitchell, Delia Camacho,

Alice Dan, and Gloria Sarto, "Promoting the Advancement of Minority Women Faculty in Academic Medicine: The National Centers of Excellence in Women's Health," *Journal of Women's Health and Gender-Based Medicine*, 2001, *10* (6): 541–550.

6. See the scathing critique by Vanessa Northington Gamble, "On Becoming a Physician: A Dream Not Deferred," in *The Black Women's Health Book: Speaking for Ourselves*, ed. Evelyn White (Seattle, WA: Seal Press, 1990).

7. American Medical Association, Physician Characteristics and Distribution in the U.S.

8. Council on Graduate Medical Education (COGME), *Fifth Report: Women in Medicine*, www.cogme.gov/rpt5.htm.

9. This article draws on and extends the analysis of the women's health movement from my book *Into Our Own Hands: The Women's Health Movement in the U.S., 1969–1990*. For a much more in-depth understanding of the movement, the challenges it faced, and the changes it catalyzed, I refer readers to the longer study. Although the study of the relationship between women physicians and the women's health movement was not the main focus of my research, I learned a great deal about this topic. Nevertheless, there is much more to be learned, and it is my hope that this article will help spur new, in-depth, and systematic research about this topic by other scholars.

10. Essentialism is the belief that people have an unchanging essence, determined by their gender or race. Essentialist explanations often rely on an explicit or tacit biological determinism, in contrast to explanations that see gender or race as socially constructed. Much feminist theory is anti-essentialist. While recognizing biological differences between males and females, anti-essentialist explanations analyze gender and race as fundamentally sociocultural categories.

11. Ruzek, *The Women's Health Movement*.

12. See Morgen, *Into Our Own Hands*, or Ruzek, *The Women's Health Movement*, for descriptions of the alternative practices that constituted the alternative model of women's health care in these feminist clinics.

13. Toni Cade, *The Black Woman: An Anthology* (New York: Signet Books, New American Library, 1970); Combahee River Collective, "The Combahee River Collective Statement," in *Home Girls: A Black Feminist Anthology*, ed. Barbara Smith Latham (New York: Kitchen Table Women of Color Press, 1983); Angela Davis, "Sick and Tired of Being Sick and Tired: The Politics of Black Women's Health," in *The Black Women's Health Book*, ed. Evelyn C. White (Seattle, WA: Seal Press, 1990).

14. Loretta Ross, "African American Women and Abortion," in *Abortion Wars: A Half Century of Struggle, 1950–2000*, ed. Rickie Solinger (Berkeley: University of California Press, 1998); Jael Silliman, Marlene Fried, Loretta Ross, and Elana Gutierrez, *Undivided Rights: Women of Color Organize for Reproductive Justice* (Cambridge, MA: South End Press, 2004).

15. Ruzek, *The Women's Health Movement*.

16. Bonnie Kay, "The Commodification of Women's Health: The New Women's Health Centers," *Health/PAC Bulletin*, 1989 (Winter): 19–23; Nancy Worcester and Marianne Whately, "The Response of the Health Care System to the Women's Health Movement: The Selling of Women's Health Centers," in *Feminism within the Science and Health Care Professions*, ed. Sue Rosser (Oxford, UK: Pergamon Press, 1988).

17. Sheryl B. Ruzek and Julie Becker, "The Women's Health Movement in the U.S.:

From Grassroots Activism to Professional Agendas," *Journal of the American Medical Women's Association*, 1990, *54* (1): 4–8, 40.

18. Boston Women's Health Book Collective *Our Bodies, Ourselves: A Book by and for Women* (Boston: New England Free Press, 1971), 123, 125.

19. Regina Markell Morantz-Sanchez, *Sympathy and Science: Women Physicians in American Medicine* (New York: Oxford University Press, 1985); Judith Lorber, *Women Physicians: Careers, Status and Power* (New York and London: Tavistock, 1984).

20. Barbara Ehrenreich and Deidre English, *Witches, Midwives, and Nurses: A History of Women Healers* (Old Westbury, NY: Feminist Press, 1973).

21. Gina Corea, *The Hidden Malpractice: How American Medicine Treats Women as Patients and Professionals* (New York: Morrow, 1977); Claudia Dreifus, ed., *Seizing Our Bodies: The Politics of Women's Health* (New York, Vintage Books, 1977).

22. Barbara L. Kaiser and Irwin H. Kaiser, "The Challenge of the Women's Health Movement to American Gynecology," *American Journal of Obstetrics and Gynecology*, 1974, *120*: 653.

23. Kaiser and Kaiser, "The Challenge of the Women's Health Movement to American Gynecology," 661.

24. Dennis Cavanagh, "Discussion of 'The Challenge of the Women's Movement to American Gynecology,'" *American Journal of Obstetrics and Gynecology*, 1974, *120*: 664–665; Georgeanna Jones, "Discussion of 'The Challenge of the Women's Movement to American Gynecology,'" *American Journal of Obstetrics and Gynecology*, 1974, *120*: 665.

25. Dr. Clay Burchell, "Discussion of 'The Challenge of the Women's Movement to American Gynecology,'" *American Journal of Obstetrics and Gynecology*, 1974, *120*: 664.

26. In Morgen, *Into Our Own Hands*. I also examine the multiple ways opponents of the movement used various tactics to force clinics to close or to make it hard for them to recruit physicians to provide services.

27. My intent here is not to attempt to catalogue all the women physicians who played leadership roles in the movement or to attempt full biographies. I do not have all the data to accomplish that goal. Yet such biographies and historical research would contribute a valuable, missing perspective on both the history of this movement and of women physicians.

28. The National Women's Health Network is a leading feminist health advocacy organization. It was founded in 1974 and continues today in the role of serving as a powerful voice for women in federal health policy.

29. Mary Campbell [Mary Howell, pseud.], *Why Would a Girl Go into Medicine? Medical Education in the United States: A Guide for Women* (Old Westbury, NY: Feminist Press, 1973).

30. Interview with author, February 8, 1997. See also Laura Briggs, *Reproducing Empire: Race, Sex, Science, and U.S. Imperialism in Puerto Rico* (Berkeley: University of California Press, 2002).

31. Mary Jane Gray, 1999, Interview with author.

32. Mary Jane Gray and Judith Tyson, "Evolution of a Woman's Clinic: An Alternative System of Medical Care," *American Journal of Obstetrics and Gynecology*, 1976, *126*: 764.

33. Terry Brock [pseud.], 1991, Interview with author.

34. Ibid.

35. Ibid.

36. See Morgen, *Into Our Own Hands*, 207–217, for a detailed account of this case study.

37. Beginning in the 1980s, breast cancer activism politicized and worked to fundamentally change women's relationships with surgeons and oncologists. Whether breast cancer activism was seen as a branch of the women's health movement or a movement in its own right, breast cancer activists drew on the ideas, strategies, and resources of the larger women's health movement. See, for example, Ellen Leopold, *A Darker Ribbon: Breast Cancer, Women and Their Doctors in the Twentieth Century* (Boston: Beacon Press, 1999), and Barron H. Lerner, *Breast Cancer Wars: Hope, Fear, and the Pursuit of a Cure in Twentieth-Century America* (Oxford: Oxford University Press, 2001).

38. Jennifer Nelson, "'Hold Your Head Up High and Stick Out Your Chin': Community Health and Women's Health in Mound Bayou, Mississippi," *National Women's Studies Association Journal*, 2005, *17*: 99–116.

39. Ibid., 105.

40. Ibid., 114.

41. Elsewhere I analyze and critique the concept of health disparities as a gloss for racism and sexism in the health care system. See Sandra Morgen, "Movement Grounded Theory: The Politics of Gender, Race, and Class in Women's Health Activism in the U.S.," in *Race, Class, Gender and Health*, ed. Amy Schultz and Leith Mullings (San Francisco: Jossey-Bass, 2006).

42. U.S. Public Health Service, "Women's Health: Report of the Public Health Service Task Force on Women's Health Issues," *Public Health Reports*, 1985, *100*: 1–106.

43. Department of Health and Human Services Task Force on Black and Minority Health, *Black and Minority Health: Report of the Secretary's Task Force*. Vol. 1: *Executive Summary*, 1985 (Washington, D.C.: Department of Health and Human Services).

44. Brian D. Smedley, Adrienne Y. Stith, and Alan R. Nelson, eds., *Unequal Treatment: Confronting Racial and Ethnic Disparities in Healthcare* (Washington, D.C.: National Academies Press, 2003).

45. Elianne Riska, *Power, Politics, and Health: Forces Shaping American Medicine* (Helsinki, Finland: Finnish Society of Sciences and Letters Academic Bookstore, 1985), 121.

46. Carol Weisman, Barbara Curbow, and Amal J. Khoury, "The National Survey of Women's Health Centers: Current Models of Women-Centered Care," *Women's Health Issues*, 1995, *5*: 106.

47. Weisman, Curbow, and Khoury, "The National Survey of Women's Health Centers: Current Models of Women-Centered Care," 113.

48. Starting in the 1980s, except for occasional expansions, especially during the administration of President Bill Clinton, the federal government reversed those policies that had expanded access to health care, turning instead to cost containment as its top health care priority. Women's health centers attempted to capture a lucrative market of patients as cost-containment strategies resulted in reductions in inpatient days (largely a response to changes in insurance reimbursement policies) and other shortfalls attributable to changes in the demand for and use of obstetrical and gynecological care. See, for example, Patty Looker, "Women's Health Centers: History and Evolution," *Women's Health Issues*, 1993, *3*: 95–100; and Weisman, Curbow, and Khoury, "The National Survey of Women's Health Centers: Current Models of Women-Centered Care," 103–117.

49. Kay, "The Commodification of Women's Health: The New Women's Health Centers."

50. Worcester and Whatley, "The Response of the Health Care System to the Women's Health Movement: The Selling of Women's Health Centers."

51. Lila Wallis, "Why a Curriculum on Women's Health?" *Journal of Women's Health*, 1993, 2: 55–60.

52. Riska, *Power, Politics, and Health: Forces Shaping American Medicine*, 25.

53. Anne Colston Wentz, "Editorial," *Journal of Women's Health*, 1994, *3:* 249.

54. Wentz, "Editorial," 250.

55. Alice Dan, J. Jonikas, and Z. Ford, "Epilogue: An Invitation," in *Reframing Women's Health: Multidisciplinary Research and Practice*, ed. Alice Dan (Thousand Oaks, CA: Sage Publications, 1994).

56. Council on Graduate Medical Education, *Fifth Report: Women in Medicine*, 1998, www.cogme.gov/rpt5.htm.

CHAPTER EIGHT

Narrative Forms in *Our Bodies, Ourselves*

Susan Wells

Our Bodies, Ourselves is a central work in feminist health education. It tells many stories, beginning with the story of its writers, women who moved together from ignorance to knowledge and who brought their knowledge to others.[1] This "good story," in the words of the 1973 introduction, is a tale of self-cultivation, of transformative personal experience that becomes socially consequential.[2] The Boston Women's Health Book Collective also told stories about the female body: sexual explorations, the hormonal cycle, and childbirth. These stories inhabit the world of medical narratives: case histories (diagnostic narratives that punctuate and motivate medical decisions), cultural narratives of illness and health, and stories of medical education.[3] Finally, *Our Bodies, Ourselves* tells a story about the organization of medical care and about the hegemony of conventional medicine and the growth of alternative health movements. Science studies has examined the gendered consequences of stories like these, but we still have something to learn about their form, organization, and structure.[4] Where did the good stories of *Our Bodies, Ourselves* come from? What work did they do? This essay considers how these stories of bodily education, bodily processes, and medical institutions are told, and how their episodes are formed, ordered, and linked.

Our Bodies, Ourselves was written by women who saw films, read novels, and debated political questions. They encountered particular narrative forms, worked with the stories that they knew, and made changes as the book developed through various editions. The book was self-published in 1970 as *Women and Their Bodies: A Course*, by the Boston Women's Health Collective. *Women and Their Bodies* was designed so that readers could separate it into four bound booklets or insert its 198 pages into a ring binder. The opening and closing chapters about women, medicine, and capitalism framed material on anatomy, sexuality, myths about women, venereal disease, birth control, abortion, pregnancy, birth, and postpartum. From 1971 to 1973, the New England Free Press published and distributed *Our Bodies, Ourselves: A Course by and for Women* by the Boston Women's Health Course Collective (*Our Bodies, Ourselves* in the 1971 edition). Here, the social analysis of medicine was moved to a final chapter, and other changes were made in each of the many printings of the book. After 250,000 copies were sold, the New England Free Press could no longer keep pace with demand, and the collective moved publication to Simon and Schuster, negotiating a contract that provided for distribution of the book to women's groups at reduced prices.[5] The 1973 *Our Bodies, Ourselves: A Book by and for Women* included quickly written chapters on relationships, sexuality, nutrition, rape and self-defense, and menopause; it was revised and substantially expanded in 1976. By 1984, *The New Our Bodies, Ourselves* had become a comprehensive guide to women's health, including a broad range of information on alternative health care, environmental issues, and substance abuse, as well as an extensive chapter on "common and uncommon" health problems. Editions in 1998 (*Our Bodies, Ourselves for the New Century*) and 2005 (*Our Bodies, Ourselves: A New Edition for a New Era*) included even broader ranges of topics; the 2005 edition is augmented with a substantial website. Since its first publication, *Our Bodies, Ourselves* has been translated and adapted into nineteen languages. In this essay, I will focus on early editions of *Our Bodies, Ourselves*, since the text as it was produced from 1970 to 1976 offers particularly compelling examples of overlapping, richly structured narratives.[6]

Recent work in literacy studies has demonstrated the importance writing held for the women's movement of the 1960s and 1970s. Kathryn T. Flannery's compelling *Feminist Literacies, 1968–1975* has begun to account for the

Members of the Boston Women's Health Collective, 1970. Photograph by Phyllis Ewen.

fury of writing that marked second-wave feminism. She speaks of 560 feminist periodicals published from 1968 to 1973, including newspapers, newsletters, magazines, and journals; she recovers the importance of privately circulated polemics to the early women's movement and traces the outpouring of poetry and dramatic production during this period.[7] Like most feminist texts of the period, *Our Bodies, Ourselves* was written quickly, to meet a pressing need. Writers in the Boston Women's Health Book Collective drafted chapters, circulated them, and critiqued them in meetings that hybridized the consciousness-raising group and the graduate seminar. After it was published, *Our Bodies, Ourselves* was distributed through the active institutions of reading fostered by the women's movement; study groups, bookstores, literature tables, women's health centers, and women's schools offered political and cultural education and served as organizing centers for the new movement. It was reviewed, debated, and critiqued in feminist periodicals. *Our Bodies, Ourselves* was a transparent window through which women gazed, as if for the first time, on their own bodies; it was also, however, a book, produced and distributed through networks of writers and readers.

Consciousness-Raising and the Ramified Narrative

What was the structure of this text? Let me begin by considering one narrative form that the early women's movement sponsored: the ramified personal narrative as it was performed in consciousness-raising groups. From 1968 until the mid-1970s, consciousness-raising was recognized, for better or worse, as the characteristic mode of discussion among second-wave feminists. Consciousness-raising encouraged women to speak and to listen to one another; to build coherent, active, and small groups; and to construct knowledge about women's experience. It was signally important for the women of the Boston Women's Health Book Collective, whose primary face-to-face public activity from 1969 to 1971 was offering women's health courses. These courses looked a lot like consciousness-raising groups punctuated by short presentations of medical information. Forty years after the fact, collective member Ruth Bell Alexander characterized the collective as "essentially a consciousness-raising group."[8]

While group styles differed, consciousness-raising generally encouraged women to tell the stories of their lives. Women spoke with little or no interruption, and the group would reflect on their stories. Sometimes, the group would react to an initial question: one guide to forming a consciousness-raising group suggested that the group spend three months discussing its members' personal histories, taking up topics from family history, relationships with men, sex, age, ambition, and aspirations for the women's movement.[9] Other groups were clearly much less structured. A pamphlet from the Chicago Women's Liberation Union advised: "A different topic could be chosen each week, and everyone discusses it in terms of her own life. Go around in a circle, each woman talking in turn so that everyone speaks; this keeps anyone from dominating a discussion and helps keep on the topic. After everyone has talked (when you start your own group you will find it isn't hard to speak in a small, close group), you might want to discuss the information you gained as you went around the room."[10] The emphasis was not on answering a particular question but on telling a story that had been suppressed. A woman's consciousness was raised, specifically, when she understood that an experience she had seen as individual, private, and idiosyncratic was in fact widely shared, socially conditioned, and supported by the basic structures of gender oppression. Understanding released excitement; as another guide put it, "CONSCIOUSNESS-RAISING is not a confessional but intimate secrets may be spoken of when they are relevant. It is very consciousness-raising to discover that others' guilty secrets are the same as one's own."[11] Although many guides to consciousness-raising caution that no woman should be pressured to speak, all the common forms—"going around the room," offering each woman a chance to speak, supporting and responding positively to women's contributions, and accepting various styles of speech—encouraged participants to tell their life stories.

Nor was consciousness-raising limited to white feminist organizations. Some historians trace its beginnings to the civil rights movement.[12] Kimberly Springer observes in *Living for the Revolution: Black Feminist Organizations, 1968–1980*, "Common to all black feminist organizations' emergence was consciousness-raising."[13] In 1983, at a conference sponsored by the National Black Women's Health Project, participants were divided into groups by gender and race and asked to tell about times when they felt that they had control of their sexual and reproductive health and about times when they did not. Linda Villarosa's account of that meeting includes stories of systematically abusive health care that would have been unusual in consciousness-raising groups of white middle-class women, but the structure is the same: multiple narratives, encouragement to speak, and comments that connect individual stories to broad social structures.[14] While few multiracial political organizations survived into the late sixties without church sponsorship, discourses of the civil rights movement had broad political currency and unmatched authority within the whole left.

What kind of writing would this narrative structure have supported? How was it adapted to the written prose of *Our Bodies, Ourselves*? While consciousness-raising was characteristically an oral practice, it often prompted individuals or groups to write. Flannery points to a 1970 pamphlet by Pamela Allen, *Free Space: A Perspective on the Small Group in Women's Liberation*, which emphasizes the importance of writing, even in the face of fear or insecurity, as a way of "uniting the political and personal concerns of our group."[15] In the multiple narratives supported by consciousness-raising, stories accumulate, related to one another by a general topic or by associations among them. Personal accounts suggest generalizations or reflections; although individual stories were not subject to judgment or commentary, groups would often debate the significance of the stories they had told, advancing multiple explanations of them. Both stories and interpretations, therefore, were ramified in multiple, often independent, branching forms. A topic exfoliated and expanded, leading to reflections, commentaries, or generalizations. Because a ramified narrative contained so many experiences, and these experiences shared so many commonalities, it was easy to read the stories as a demonstration that individual experiences were socially connected or conditioned.

The ramified narrative was central to the political formation of many of the women in the Boston Women's Health Book Collective, and it was the hallmark of the political community in which they were formed. The nucleus of the group met at a conference workshop where women discussed their experiences with medical care, and many of its members participated in Bread and Roses, a Boston socialist feminist organization that promoted consciousness-raising.[16] It is not surprising, therefore, that the writers of *Our Bodies, Ourselves* considered the ramified personal narrative a productive form for political discourse and undertook to adapt it into a written text. Often, they moved from speech to writing seamlessly: During sessions of the courses on women's health that the collective taught, women would volunteer stories of their experience with medical care. Collective members would hand them index cards and ask them to write these accounts down; the index cards were often the sources for the inset personal narratives that characterized the book.[17]

Many ramified narratives occur in the book's chapters on pregnancy and childbirth. The 1976 edition includes an extended discussion of how women respond to finding out that they are pregnant, beginning, "At the beginning of a first pregnancy, of any pregnancy, there are so many variations in feeling, from delirious joy to deep depression."[18] The writers go on to list possible positive feelings, such as "an increased sexuality, a kind of sexual opening out toward the world, heightened perceptions like being in love," illustrated with quotations from readers.[19] Women's possible questions at this stage of pregnancy follow, also in a list: "What's going to happen to me?" Then, a list labeled, "Some negative feelings" ("Shock. I'm losing my individuality"), described as "all relevant and natural."[20] It is also fine not to think much about pregnancy at all or to be involved intermittently. The range of possibilities is large. Women might feel good about being pregnant, or bad, or both at once, or both in any sequence, or neither, or not much at all. And feelings about being pregnant can be considered separately from feelings about having a baby, which would also have their own exfoliating structure of possibilities. The traditional rhetorical figure of *copia* is relevant. The text works by opening itself out so that there is no necessary end to it at all. In this ramified structure, women are encouraged to write themselves into a rich and varied story of feminine experience that can be rewritten and transformed. The figure of choice was a central issue for members of the collective; when I asked collective members Ruth Bell Alexander and Wendy Sanford why the collective used the uncolloquial "Shall I Have a Baby?" as a chapter title instead of the more informal "Should I Have a Baby?" both of them responded independently that the collective wanted to write a book without "shoulds," one that would offer women alternatives rather than prescribe their experience.[21]

Like all narrative structures, however, the ramified narrative could turn against itself. A controlling or judgmental narrative voice could frame the multiplied stories as cautionary tales rather than a rich array of alternative choices. This issue vexed consciousness-raising groups, and there were complaints about stronger members dominating the group and judging others.[22] Although the Boston Women's Health Book Collective was supportive and nonjudgmental toward their readers, the correction of bad habits is a durable and tempting form of health advice. Sometimes, rarely, the ramified narratives in *Our Bodies, Ourselves* were transformed into lists of the many possible mistakes that a woman might make, as in this discussion of maternity clothes: "Some women try to hide their pregnancies from the world and even from themselves either by continuing for a time to wear the same clothes as before, though they no longer fit, or by wearing clothes so baggy that no one can see what is happening underneath. You've got to find yourself beyond and in spite of these myths. It's possible to feel comfortable and happy with the changes you are going through."[23]

Instead of branching out into multiple possibilities, the text's ramifications specify alternate expressions of the same false consciousness. Women hide pregnancy with either tight or loose clothes because they have bought into myths. If they reject the myths, they will reject the clothes and become comfortable and happy. Such narrative structures may well have left women feeling at odds with themselves and frustrated by the task they have been set: "You've got to find yourself." The reader should reject all the social conventions that define happiness and comfort, but she will only know that she has successfully moved past those conventions and "found herself" when she becomes happy and comfortable.

Narrative and Subjectivity: Realism and Modernism

Important as the ramified narrative was for the Boston Women's Health Book Collective, it was not the only narrative resource the group enjoyed. As a group, the collective enjoyed daunting literacy skills. Wendy Sanford, overall editor of *Our Bodies, Ourselves* during this period, won the English prize as a senior at Radcliffe, with the publication of her senior thesis, on *Hamlet*, by Harvard University Press. For the educated women of the Boston Women's Health Book Collective, the realist literature of the nineteenth and early twentieth century—the canonic school texts—would also have offered models of how stories can be told. And, like many activists of the sixties, members of the collective had been formed as readers and filmgoers by the modernist narrative of the subject.

Contemporary narrative theory has been concerned with defining modernist and realist narratives, and the literature on this subject is rich. Put simply, modernist narratives are concerned with subject formation. Stream of consciousness and other experimental ways of telling stories foreground the significance of individual perceptions and responses. Modernist narratives also enact the urgency of creating a new kind of subject; critic Fredric Jameson considers that a central idea of modernity is "the conviction that we ourselves are somehow new, that a new age is beginning, that everything is possible and nothing can ever be the same again; nor do we want anything to be the same again, we *want* to 'make it new,' . . . to be somehow transfigured."[24] Realist narratives locate the subject in broad social and historical networks, guiding readers' responses through a more or less omniscient narrator.[25] While realist narratives are generally associated with the nineteenth-century novel, and modernism with early twentieth-century novels and films, these critical terms can also refer to styles of reading: Older texts—*Hamlet*, for example—can be read as modernist, while contemporary fiction is rich in realist narratives.

In modernist narratives, the subject is dispersed and unstable, and consciousness is the object of investigation. The voice of the narrator, recounting experiences discontinuously and often unreliably, becomes the center of the text, an emphasis that resonated with early feminist preoccupations with women's speech and silence. Canonic modernist narratives—novels by Woolf and Fitzgerald, certain Hollywood films, plays by O'Neill and Williams—took up issues of sexuality and identity formation, important themes of the women's movement. These narratives were so well-known to second-wave feminists that one of the movement's first books, Kate Millett's *Sexual Politics*, was a critique of modernist and realist classics.

The structure of the modernist narrative of identity is episodic: In novels such as *Ulysses* or films such as *Citizen Kane*, events are connected by their significance to a central character or narrator. They are often presented discontinuously, and the sense of the narrative is constructed by attending to juxtapositions. Realist narratives of the nineteenth century aimed at comprehensively representing the whole social experience of a group through the eyes of an individual. Mary Steichen Calderone's story of child sexual formation, in which parents' attitudes and social mores determine a subject's whole future sexual life, is a paradigmatic realistic narrative. (See Ellen S. More's essay, Chapter 5 in this volume.) The writers of *Our Bodies, Ourselves* had a great deal of affection for such narratives. The overall structure of the book modulates between a realist story of development and a modernist account of identity formation.

The collective was interested in disseminating a story of development, of progress, of connections between the past and the future, of the formation of identity in the context of a group: These were the grand themes of the realist narrative. Such stories emerge in *Our Bodies, Ourselves* when the writers want to speak of social formation and socialization: "We are born loving our bodies . . . From the moment we are born, we are treated differently from little boys . . . When we become teenagers our developing bodies are usually a mystery to us."[26] When the issue is self-formation and autonomy, the writers often turned to discontinuous modernist narratives. The section "Feelings about Becoming Sexually Mature and Sexually Active" moves fluidly from childhood, menarche, and adolescent friendships to the collective's current thinking on sexual feelings toward women, the availability of birth control, and pressures to have sex. The grand narrative of sexual maturation loops, returns on itself, and is left incomplete. What is significant is the individual woman's struggle for autonomy and sexual expression, told from a variety of perspectives.

The Boston Women's Health Book Collective used these two forms to structure the book's overall presentation of medical information as a narrative of the life cycle. The overarching plan of the book, especially in the early editions, is the story of a female body's accession to and relinquishment of reproductive capacity. At the same time, the story of reproduction is discontinuous, interrupted, and necessarily incomplete. It represents the formation of women's identity as an incomplete and ongoing project.

From 1973 to 1985, the heart of *Our Bodies, Ourselves* was a series of chapters that present a segmented story of the reproductive life cycle. In 1976, this sequence began with sexually transmitted diseases and moved to birth control, abortion, parenthood, childbearing, pregnancy, childbirth, and postpartum. There was a chapter on exceptions to normal childbearing (stillbirth, miscarriage, certain birth defects) and then a quick look at menopause, clearly still a remote territory for these writers. And that was the end of the story. The final chapter is a critique of the health care system. A woman's life began with her initiation into sexuality, moved to her relation to reproduction, and ended with her reproductive capacity:

> In childhood our bodies are immature. Then during puberty we make the transition from childhood to maturity. In women, puberty is characterized by decreased bone growth; by growth of breasts, pubic and axillary (armpit) hair; starting of menstruation (menarche) and ovulation, and increase of sexual urges. The last stage of the cycle is when we are no longer able to reproduce. The climacteric is the transition between the reproductive and postreproductive stages. Menstruation stops (menopause) and ovulation stops. (Although "menopause" is commonly used to mean the whole transition period, technically this is incorrect.)[27]

It is not surprising that this book, written by very young women, presents a normal life cycle that does not end in death. It is as though proper prevention and a supportive collective might make us immortal. In 1976, the life cycle is a heteronormative, even monogamous, story. Every subsequent edition would become more inclusive and less conventional in its assumptions about the reader, including especially the assumption that the reader was white. With all its limits, this story powerfully communicates that women's relation to reproduction is not mysterious or ungovernable but knowable and open to investigation. This skeletal narrative also articulates physiological changes with stories of social relations among women, their partners, and their children. Hayden White, describing the cognate task of writing history, said, "Histories . . . are not only about events but also about the possible sets of relationships that those events can be demonstrated to figure. These sets of relationships are not, however, immanent in the events themselves; they exist only in the mind of the historian reflecting on them. Here they are present as the modes of relationships conceptualized in the myth, fable, and folklore, scientific knowledge, religion, and literary art, of the historian's own culture."[28] For the collective, physiological events were the armatures on which relations were constructed. Sometimes, that construction was realist. Take nursing, for example; lactation was not just a physiological process but also a practice of childcare deeply implicated with issues of class, caregiving, maternal dependence, and independence. Other narratives were modernist: The story of the reproductive life cycle is also the story of women's relationships and refusals of relationships, of their attempts to construct identity out of resistant materials.

Feminists had, from the first days of the women's movement, contested the common sense of the culture that women's sexuality, or the story of their reproductive lives, utterly determined their life choices and possibilities. *Our Bodies, Ourselves* sometimes argued against seeing anatomy as destiny. The book began with the lyrics to a feminist song that proclaimed, "Our faces belong to our bodies," but went on to say, "Our struggle is changing our faces, our bodies."[29] *Our Bodies, Ourselves* also recognized the importance of the feminist critique of sexuality that began with Anne Koedt's "The Myth of the Vaginal Orgasm."[30] The organization of *Our Bodies, Ourselves* as a narrative of the reproductive life cycle, therefore, could be seen both as a recognition of the received cultural narrative and as an early feminist critique of that story. The modernist narrative of self-formation and the collective's valorization of choice were mutually reinforcing. Both helped the collective to establish a clear alternative to a prescriptive undercurrent in the women's movement that surfaced in the celibacy line of Cell 16, an early Boston feminist collective that advocated a withdrawal from all sexual relations, or to the rigidity that led one woman to "apologize to [her] sisters" after telling a workshop on sexuality that she had experienced vaginal orgasms.[31] Within the pages of *Our Bodies, Ourselves* were many paths to personal autonomy and self-realization and many feminist sexualities recognized as authentic.

The book also tells a story much more humane than the "uterus as machine/woman as labor/doctor as supervisor" narrative that Emily Martin discerned in obstetric textbooks and medical accounts of women's physiology.[32] *Our Bodies, Ourselves* shows women negotiating reproduction in relationships, preserving a sense of their agency, working at reproduction in contexts that can be more or less supportive. Women take various paths through life, assuming and relinquishing roles at various moments and in various relationships. Although the order of the chapters assumes a "normal" life narrative, individual chapters are written to be freestanding, so that they could be read in any order that made sense, preserving the texture of choice. Compared with mechanistic stories Emily Martin analyzes, in which women march onto a conveyer belt to childbirth and are discarded as useless at menopause, this is a story rich in incident and possibility, a story in which many women can recognize themselves and their choices. In concert with the widespread, diverse women's health movement, *Our Bodies, Ourselves* sought to broaden and de-medicalize women's sense of life possibilities. (See Sandra Morgen's essay in Chapter 7 in this book.)

But the story of the reproductive life cycle is still only one story and in some ways a restricting one. It has no room for experiences of bodily effort, complication, or suffering that have nothing to do with reproduction but are experienced through the lens of femininity. It has little room for nonreproductive sexuality (although the sexuality chapter certainly affirmed the importance of women's pleasure). Within the collective's modernist versions of this story, social structures and causes are acknowledged, but the point of the reproductive life cycle is the formation of women's identity. A woman's choices to form sexual partnerships, to bear or not to bear children, to give birth or raise children in certain settings constitute her as a modernist center of consciousness, as self-integrated (however provisionally) and autonomous. In the collective's realist stories of the female life cycle, reproduction becomes the main vehicle for women's integration into society and history.

Chiasmic Narratives of Reversal and Transformation

Another narrative form used by the Boston Women's Health Book Collective was the chiasmic narrative. Chiasmus is a rhetorical figure, conventionally grouped with the "figures of words" rather than the "figures of thought" because it turns on a verbal reversal in the shape of the "X." Here is a simple chiasmus: She wanted to build a house; in the end, the house built her. Chiasmus is associated with energetic argument, even with satire. Consider Marx's opinion that "the weapons of criticism will give way to the criticism of weapons."[33] Often, as in this example, the two halves of the chiasmic pair are joined by a repeated word, but the repetition signals a reversal. One pair of terms (criticism/criticism) is an envelope for the other (weapons/weapons), forming a specular pattern that moves from the center of the phrase to the edges.[34]

Chiasmus seems an unlikely figure for thinking about narratives, in which one thing is told after another, carrying the reader to the end of the text, so that repetition is a pleasant retarding device, allowing us to savor the slow approach of a story's inevitable closure.[35] Chiasmic narratives, however, turn on reversal—what was up goes down, and what was down comes up. Consider the parable of the rich man and Lazarus (Luke 16). At the beginning of the story, the beggar Lazarus longs for the crumbs from the rich man's table; at the end, the beggar is in Abraham's bosom, and the rich man is in hell. When the rich man begs Lazarus to cool his tongue with water, he is reminded of his comfortable life and of Lazarus's misery. Abraham also tells the rich man that "between you and us there is fixed a great abyss, and those who might wish to cross from here to you cannot do so, nor can anyone cross from your side to us" (Luke 16). This story illustrates the narrative work of chiasmus. It establishes a tight connection between two things, and it places them in opposition to each other. The rich man is in hell *because* Lazarus lived a life of misery, but the rich man and Lazarus, in both halves of the story, are radically divided from each other: "there is fixed a great abyss."[36] Generations of Scripture readers have made good subversive use of the story of Lazarus, rejoicing in the fall of the rich man. But chiasmic stories can also turn on us. We expect to read, or tell, a progressive and cheerful "good story" that begins in one place and ends somewhere else, but we may find instead chiasmic reversal and opposition.

Chiasmus was a favorite figure of the New Left. Consider these examples, all included in Judith Clavir Albert and Steward Edward Albert's anthology, *The Sixties Papers: Documents of a Rebellious Decade*.[37] First, from the 1962 Port Huron Statement, the founding document of Students for a Democratic Society (SDS), the most influential New Left student organization:

> We are people of this generation, bred in at least modest comfort, housed now in universities, looking uncomfortably at the world we inherit.
>
> When we were kids the United States was the wealthiest and strongest country in the world . . . Many of us began maturing in complacency . . . As we grew, however, our comfort was penetrated by events too troubling to dismiss . . . We began to sense that what we had originally seen as the American Golden Age was actually the decline of an era.[38]

Second, from Herbert Marcuse's *One-Dimensional Man* (1964), often cited as the theoretical foundation of the New Left: "The advancing one-dimensional society alters the relation between the rational and the irrational. Contrasted with the fantastic and insane aspects of its rationality, the realm of the irrational becomes the home of the really rational—of the ideas which may 'promote the art of life.'"[39] Third, from the 1965 SDS leaflet calling for a national demonstration against the war in Vietnam in Washington, D.C.:

> In the name of freedom, America is mutilating Vietnam. In the name of peace, America turns that fertile country into a wasteland. And in the name of democracy, America is burying its own dreams and suffocating its own potential. . . .
>
> Our chance is the first in a generation to organize the powerless and the voiceless at home to confront America with its racial injustice, its apathy, and its poverty,

and with that same vision we dream for Vietnam: a vision of a society in which all can control their own destinies.[40]

Fourth, from a 1968 article in *New Left Notes*, the SDS paper, by Naomi Jaffe and Bernardine Dohrn: "Over the past few months, small groups have been coming together in various cities to meet around the realization that as women radicals we are not radical women."[41]

In these chiasmic narratives, an order that seemed benign is suddenly revealed as oppressive, but this exposure mobilizes a transformative movement. The women of the Boston Women's Health Book Collective heard and used this rhetoric. Many of them had been active in antiwar and civil rights movements; it was through those networks that they often made initial contact with the collective.[42]

Some of the most powerful political narratives in *Our Bodies, Ourselves* are chiasmic narratives of power and resistance, including the overarching narrative of conventional medicine, its hegemony consolidated by medicalizing reproduction, countered by women's desire to take reproduction back from doctors. Other chiasmic narratives echo archaic figures of the life story: old age recapitulates childhood or infancy. The 1976 edition includes a short discussion of the development of the breast in the chapter on anatomy and physiology. The text is illustrated with a series of six line drawings of the breast at various stages, showing how lactating glands grow at menarche and shrink at menopause. The series is a visual chiasmus. The adolescent breast mirrors the menopausal breast: both are small, with sparse milk-producing tissue. The pregnant breast and the lactating breast, both large and filled with glandular tissue, form the central crossing of the chiasmus. This chiasmic narrative resonates with familiar metaphors of the female body's relation to production and decay, although the writers of *Our Bodies, Ourselves* are careful not to describe the menopausal breast as "atrophied" or "shriveled." *Our Bodies, Ourselves* frames the breast as actively changing to meet the needs of new relationships and new life stages. Lactation is presented as an elaborate adaptation rather than something that happens to a woman. And the specular relation between adolescence and menopause offers some sense of possibility, of new beginnings, a sense that is reinforced in the text's treatment of menopause: "If a woman has spent a large part of her life to date raising a family, she now has some important decisions to make about what to do with her new freedom and the next thirty years of her life."[43]

These narrative structures—the ramified narrative, the episodic narratives of modernism and realism, the chiasmic narrative—did important work for the Boston Women's Health Book Collective. They organized both large sections of *Our Bodies, Ourselves* and smaller local narratives; these structures were combined, hybridized, rendered visually, and torqued in dozens of ways. They were used experimentally to sort out new ways of telling old stories of collective education, of identity formation, of reversal and transformation.

Narrative forms resist abstract investment with political or moral categories. Narrative structures do not exist outside of their many instantiations in language, and so no narrative form is inherently productive, static, or regressive. But narrative forms and structures also tell their own stories that are inscribed in cultural memory by the repeated experience of hearing them at our mothers' knees, in health classes, in doctors' offices, in presidential addresses, in our dreams. Forms can therefore work at cross-purposes to the intentions of writers, to the interests of readers, and to the possibilities of the narrated materials.

Hybrid Structures: The Birth Narrative

The multichapter unit on birth in *Our Bodies, Ourselves* (1976) is a series of nested narratives: the decision to become pregnant or to continue a pregnancy, the project of self-care in pregnancy, giving birth, caring for a baby, or dealing with the loss of a pregnancy. The collective moved among the narrative structures at its disposal in composing these narratives and modulating among them. Many are modernist stories of subject formation. Women are encouraged to take control of their lives by making choices that will shape who they are, as in this discussion of abortion: "With all these mixed feelings and not much time, deciding what to do with an unplanned pregnancy can feel like an impossible task. It may help somewhat to know that after you make a decision, the ambivalence and mixed feelings tend to subside. Whatever we decide, it is important for us as women to make an *active* decision, one which is ours, rather than passively slipping into one choice or another."[44] Modernist narratives in which a subject is born in crisis through an act of choice are juxtaposed with realist narratives in which the subject is formed in relation to others. Immediately after the quoted passage, the text reminds women who are unexpectedly pregnant that they are not alone and urges them to talk to their partners, to their friends, and to women who have had abortions. The seams between the story of identity-formation and the story of communal responsibility are sutured by figuring the autonomous woman as someone singularly able to contribute to family and community, "Only when we are in control of that choice [whether and when to have children] are we free to be all that we can be for ourselves, for children we already have or may have in the future, for our partners, for our communities."[45]

Other narratives of pregnancy and birth in *Our Bodies, Ourselves* are chiasmic stories of oppression and resistance: control of the birthing process passed from women to the medical establishment; birthing women will reclaim it. The reader who will manage her own child's birth has claimed control of the process for all women. The writers of *Our Bodies, Ourselves* were alive to the masculine culture of medicine and called on their readers to oppose it. (See Naomi Rogers's and Robert A. Nye's essays, Chapters 9 and 6, respectively, in this volume.) The chiasmic narrative is spliced into a realist narrative of progress, so that the reader's labor experience becomes part of a struggle for all women. The realist story offered women ways to experience their choices as socially consequential; the chiasmic story interpreted individual acts of resistance as reversing an imbalance of power.

Attempts to hybridize these two narrative forms were not always successful. For example, the section on anesthesia during childbirth includes this passage: "In actuality, the doctor or midwife should have very little part to play in a normal delivery. S/he checks the progress of labor and how the baby's heart is responding to it; looks to see if we have entered second stage (explained later) and can thus begin to push our baby out; and helps ease our baby out during the birth . . . Seldom is a baby born to an unanesthetized mother in as much need of medical assistance as a baby born to a mother who has had medication in labor."[46]

The physician's proffered pain relief is actually a means of assuming control of labor; when the physician's control is limited, there is less need for his or her aid. This bracing chiasmus modulates into a story of a natural past, which the reader will recover by dispensing with anesthesia. The transformative power of chiasmus is turned on the reader, who is individually assigned the responsibility for the reversal of decades of medical intervention. The awkwardness of the phrase "our baby" expresses the labor of representation that readers are asked to take on.

Chiasmus also structures larger narratives in *Our Bodies, Ourselves*. Supporting the book's modernist story of self-formation is its chiasmic account of the formation of a woman's health movement that would confront and transform conventional medicine. The final chapter of early editions of *Our Bodies, Our-*

selves (1971), "Women and Health Care," maintained that "the capitalist theory of disease" organized to treat symptoms and focused on the individual relationship of the woman to her physician. In a chiasmic reversal effected by the women's movement, a "social theory of disease" would reverse these vectors, focusing on women's self-help and their active participation in their health care, understanding health care as prevention, and illness as caused by environmental contamination as well as by germs, viruses, and bacteria.[47] For the writers of *Our Bodies, Ourselves*, the personal relationship between a woman and her doctor was a symptom of the shortcomings of the capitalist health care system. The book is an act of chiasmic reversal. Women were dominated by physicians because of the capitalist theory of disease; by dismantling that theory, *Our Bodies, Ourselves* opens the possibility of health care without domination.

Seeing this political project as a narrative does not diminish its significance. Francesca Polletta's recent study, *It Was Like a Fever: Storytelling in Protest and Politics*, demonstrates that narrative is a social and political resource, a way of constructing agreements and mobilizing action.[48] Like any other political resource, narrative can be studied in its own terms. Its forms, associations, and histories operate beneath the surface of political discourse. Narrative is particularly important in the history of medical women. Women physicians and their supporters have appropriated powerful stories of vocation and apprenticeship to motivate their entry into the profession. (See the essays in this volume by Judy Tzu-Chun Wu [Chapter 4] and Carla Bittel [Chapter 1] for examples of the canonic narrative, and the essay by Regina Morantz-Sanchez [Chapter 3] for an example of the costs of violating it.) Women's health organizations regularly invoke narratives of reversal and transformation to argue for a new distribution of resources. Political common sense is constructed from such stories. It is worth knowing everything we can about them.

Remarkably, the 1976 *Our Bodies, Ourselves* ended by counseling women to be skeptical of the book's authority: Chiasmus expanded to include the work of writers and readers. The reader is constructed as the equal and successor of the writer, responsible for extending the grand narrative of the book:

> It is perhaps a mistake to look to the doctor for confirmations of our feelings and identity as women in the first place. That is where only a group of women, being honest with themselves and with one another, can hope to sort it all out. But just sharing horror stories is not enough, though it is usually helpful. Learning about

> why things are the way they are and why things that have happened to us did happen, and getting the facts so that perhaps we can change our understanding of what actually happened, are equally important.[49]

The limits of the story of the reproductive life cycle are acknowledged:

> This book is a start in helping us to assume more responsibility for our health care, but we've only been able to touch briefly on the simplest aspects, the common medical events of a woman's life. The real toughies—the complicated diseases, the rare surgeries, death and dying—will have to be coped with and worked out individually over time. But everything we've said here applies: Don't let yourself be stampeded into any sudden decisions or forced to accept any medications or procedures you don't want. It's your body.[50]

Both these remarks are chiasmic invocations of the reader. Just as *Our Bodies, Ourselves* educated women, women will advance beyond the limits of the book, and educate the collective. While it has been conventional for writers since Chaucer to step out of the narrative frame at the end of a text, these remarks were not empty gestures. The women of the collective were actively organizing women's health classes, developing groups of writers and readers, donating their profits to women's clinics, and organizing networks of women's health activists.[51] At the end of *Our Bodies, Ourselves*, readers move, as it were, across the X, becoming producers of medical knowledge rather than consumers. The chiasmic narrative of confrontation and reversal seamlessly joins a modernist narrative of self-formation; the two stories augment and support each other, celebrating the reader's capacity to know herself and inciting her to learn more, perhaps by joining a consciousness-raising or self-help group. As they appeared in the 1976 edition, the final pages of the book organize ramified, modernist, and chiasmic narratives in a productive, stable ecology. Each structure supports and augments the others. It was, in fact, a very good story.

NOTES

I thank the participants in the seminar on "Women Physicians, Women's Health, and Women's Politics" for their useful comments, Ellen S. More for her warm support of this project, and Manon Parry for her tireless work on the collection.

1. For the fullest previous discussion of language in *Our Bodies, Ourselves*, see Robbie Pfeufer Kahn, *Bearing Meaning: The Language of Birth* (Urbana: University of Illi-

nois, 1995). Kathy Davis's *The Making of* Our Bodies, Ourselves: *How Feminist Knowledge Travels across Borders* (Durham, NC: Duke University Press, 2007) is a comprehensive discussion of the relation between the Boston Women's Health Book Collective and worldwide feminism.

2. Boston Women's Health Book Collective, *Our Bodies, Ourselves: A Book By and For Women* (New York: Simon and Schuster, 1973), 1.

3. See such early sources as Howard Brody, *Stories of Sickness* (New Haven, CT: Yale University Press, 1987), and Katherine M. Hunter, *Doctors' Stories: The Narrative Structure of Medical Knowledge* (Princeton, NJ: Princeton University Press, 1991). See also Katharine G. Young, *Presence in the Flesh: The Body in Medicine* (Cambridge, MA: Harvard University Press, 1997), and, more broadly, N. Katherine Hayles, *How We Became Posthuman: Virtual Bodies in Cybernetics, Literature, and Informatics* (Chicago: University of Chicago Press, 1999). The central source for gendered medical narratives of the body is Emily Martin, *The Woman in the Body: A Cultural Analysis of Reproduction* (Boston: Beacon, 1992).

4. For a survey of work in narrative theory, including feminist narrative theory, see Peter Brooks, *Body Work: Objects of Desire in Modern Narrative* (Cambridge, MA: Harvard University Press, 1993); Paul Cobley, *Narrative* (New York: Routledge, 2001); David Herman, *Narratologies: New Perspectives on Narrative Analysis* (Columbus: Ohio State University Press, 1999); Carol Jacobs and Henry Sussman, eds., *Acts of Narrative* (Stanford, CT: Stanford University Press, 2002); Carla Kaplan, *The Erotics of Talk: Women's Writing and Feminist Paradigms* (New York: Oxford University Press, 1996); Maureen Whitebrook, *Identity, Narrative, and Politics* (New York: Routledge, 2001).

5. *Our Bodies, Ourselves* Timeline, *Our Bodies, Ourselves* website, Kiki Zeldes, 2005, www.ourbodiesourselves.org/about/timeline.asp.

6. These editions include Boston Women's Health Collective, *Women and Their Bodies: A Course* (Boston: Boston Women's Health Collective, 1970), Boston Women's Health Book Collective, *Our Bodies, Our Selves: A Course by and for Women* (Boston: New England Free Press, 1971), *Our Bodies, Ourselves: A Book by and for Women* (New York, Simon and Schuster, 1971, 1973, 1976). Late printings of the New England Free Press edition include an exchange of letters about the decision to publish with Simon and Schuster. I have used the 1971 edition to represent the earliest state of *Our Bodies, Ourselves*, and the 1976 edition to represent the move into mainstream publishing. The 1973 edition was prepared for Simon and Schuster very quickly; many planned revisions were only implemented in 1976 (Boston Women's Health Book Collective Papers, Schlesinger Library, Harvard University, MC 503, Carton 1, Folder 13, Board Meeting Minutes). In this essay, I refer to all editions as *Our Bodies, Ourselves*, ignoring the changes in titles. Quotations are from the 1976 edition unless otherwise identified.

7. Kathryn T. Flannery, *Feminist Literacies, 1968–75* (Urbana: University of Illinois Press, 2005); discussion on periodicals, p. 50; polemics, pp. 149–150.

8. Ruth Bell Alexander, interview with author, Cherry Hill, NJ, October 2002.

9. "Consciousness Raising," anonymous document in Ann Koedt, Ellen Levine, and Anita Rapone, eds., *Radical Feminism* (New York: Quadrangle, 1973), 280–281.

10. Chicago Women's Liberation Union, "Consciousness Raising," by "a group of Connecticut women," CWLU Herstory Website, Historical Archive, Classic Feminist Writings, www.cwluherstory.com/CWLUArchive/crguidelines.html.

11. *How to Start Your Own Consciousness-Raising Group*, leaflet, 1971, in CWLU Herstory Website, Historical Archive, Consciousness, www.cwluherstory.com/CWLU Archive/crcwlu.html.

12. Sara M. Evans, *Personal Politics: The Roots of Women's Liberation in the Civil Rights Movement and the New Left* (New York: Knopf, 1979), 214.

13. Kimberly Springer, *Living for the Revolution: Black Feminist Organizations, 1968–1980* (Durham, NC: Duke University Press, 2005), 45.

14. Linda Villarosa, ed., *Body and Soul: The Black Women's Guide to Physical Health and Emotional Well-Being* (New York: HarperCollins, 1994), 5.

15. Flannery, *Feminist Literacies*, 89.

16. Bread and Roses, "Outreach Leaflet," reported in Rosalyn Baxandall and Linda Gordon, ed., *Dear Sisters: Dispatches from the Women's Liberation Movement* (New York: Basic Books, 2000), 35; "A Good Story," in *Our Bodies, Ourselves*, 1976, 11–14.

17. Vilunya Diskin and Paula B. Doress-Worters, interview with author, Boston, MA, 2002.

18. *Our Bodies, Ourselves*, 1976, 259.

19. *Our Bodies, Ourselves*, 1976, 259.

20. *Our Bodies, Ourselves*, 1976, 259.

21. Wendy Coppedge Sanford, interview with author, Cambridge, MA, October 2002; Ruth Bell Alexander interview.

22. See, for example, Jo Freeman, "The Tyranny of Structurelessness," in *Dear Sisters: Dispatches from the Women's Liberation Movement*, ed. Rosalyn Baxandall and Linda Gordon (New York: Basic Books, 2000), 73–75.

23. *Our Bodies, Ourselves*, 1976, 261.

24. Fredric Jameson, *Postmodernism, or, The Cultural Logic of Late Capitalism* (Durham, NC: Duke University Press, 1991), 310.

25. Paul Cobley, *Narrative*, chap. 4.

26. *Our Bodies, Ourselves*, 1976, 40.

27. Ibid., 32.

28. Hayden White, "The Historical Text as Literary Artifact," in *Narrative Dynamics: Essays on Time, Plot, Closure, and Frames*, ed. Brian Richardson (Columbus: Ohio State University Press, 2002), 191–210, esp. 204.

29. *Our Bodies, Ourselves*, 1976, 15.

30. For the reception of Koedt's essay on orgasm, see Alice Echols, *Daring to Be Bad: Radical Feminism in America, 1967–75* (Minneapolis: University of Minnesota Press, 1989), 111.

31. For Cell 16, see Sara M. Evans, *Tidal Wave: How Women Changed America at Century's End* (New York: Free Press, 2003); for the workshop story, see Echols, *Daring to Be Bad*, 111.

32. Martin, *The Woman in the Body*, chap. 1.

33. Karl Marx, "Critique of Hegel's Philosophy of Right," in *The Marx-Engels Reader*, ed. Robert Tucker (New York: Norton, 1972), 11–24, esp. 20.

34. For the surprisingly rich literature on chiasmus, see, for example, S. Budick, "Chiasmus and the Making of Literary Tradition: The Case of Wordsworth and 'The Days of Dryden and Pope,'" *ELH*, 1993, *60* (4): 961–87; Jonathan Xavier Inda, "Foreign Bodies: Migrants, Parasites, and the Pathological Nation," *Discourse*, 2000, 22 (3): 46–62;

Jeffrey Nealon, "Maxima Immoralia? Speed and Slowness in Adorno's *Minima Moralia*," *Theory and Event*, 2000, *4* (3); and Cecilia Sjoholm, "Crossing Lovers: Luce Irigaray's Elemental Passions," *Hypatia*, 2000, *15* (3): 92–112.

35. Peter Brooks, *Reading for the Plot: Design and Intention in Narrative* (New York: Vintage, 1985).

36. The literature on chiasmus in biblical studies is daunting. For the merest taste, see Elie Assis, "Chiasmus in Biblical Narrative: Rhetoric of Characterization," *Prooftexts*, 2002, 22 (3): 273–304.

37. Judith Clavir Albert and Steward Edward Albert, eds., *The Sixties Papers: Documents of a Rebellious Decade* (New York: Praeger, 1984).

38. "The Port Huron Statement," in *The Sixties Papers: Documents of a Rebellious Decade*, ed. Judith Clavir Albert and Steward Edward Albert (New York: Praeger, 1984), pp. 176–197, esp. 176–177.

39. Herbert Marcuse, *One-Dimensional Man*, in *The Sixties Papers: Documents of a Rebellious Decade*, ed. Judith Clavir Albert and Steward Edward Albert (New York: Praeger, 1984), 209–218, esp. p. 209.

40. "An SDS Antiwar Leaflet: November 27, 1965," in *The Sixties Papers*, 226–228, esp. 226–227.

41. Naomi Jaffe and Bernardine Dohrn, "The Look Is You," in *The Sixties Papers*, 228–232, esp. 228.

42. Judith Norsigian, interview with author, Philadelphia, PA, November 11, 2001; Diskin/Doress-Worters interview.

43. *Our Bodies, Ourselves*, 1976, 335.

44. Ibid., 222.

45. Ibid., 216.

46. Ibid., 268.

47. Boston Women's Health Collective, "Women and Their Bodies: A Course" (1970), "Women, Medicine, and Capitalism: an Introductory Essay," 6–9, esp. 8.

48. Francesca Polletta, *It Was Like a Fever: Storytelling in Protest and Politics* (Chicago: University of Chicago Press, 2006).

49. *Our Bodies, Ourselves*, 1976, 354.

50. Ibid., 355.

51. Sandra Morgen, *Into Our Own Hands: The Women's Health Movement in the United States, 1969–1990* (New Brunswick, NJ: Rutgers University Press, 2002).

CHAPTER NINE

Feminists Fight the Culture of Exclusion in Medical Education, 1970–1990

Naomi Rogers

> All of the professional trends of modern life are manifest in the medical profession and the medical school.
>
> BECKER ET AL., *BOYS IN WHITE*, 1961

> Her intelligence and aggressiveness are such that she stands out among, but does not antagonize, her male colleagues.
>
> LETTER OF REFERENCE FOR INTERNSHIP, EARLY 1970S

During the 1950s and 1960s, American social scientists lavished attention on the dynamics of professional socialization, a process they saw as epitomized by the American medical school. Medical educators, for their part, prided themselves on the thoroughgoing ways that students were transformed into fully fledged doctors. By the early 1970s, however, "socialization" had become a highly politicized term as feminists argued that, from preschool to professional school, boys and girls were taught discriminatory and stereotyped gender roles that shaped their work, family life, sexuality, and sense of self. Critics found abundant evidence that gender stereotyping was rampant throughout medical education and assailed its implications for the construction of a professional identity that left physicians unable to empathize with the world outside the privileged halls of medicine.

Women who entered American medical schools during the 1970s challenged this training and contested many of the fundamental conventions of American medicine. Their efforts to resist discrimination and harassment made visible a culture of exclusion. It was a shocking and at times nasty struggle, exposing ingrained elitism, condescension, and inflexibility. Women who pushed open the medical school doors sought not just equal opportunities in education but also equality in professional respect: not to be belittled as "ladies," ridiculed as mannish "hen medics," or scorned as "one of the boys." All these were public roles that women physicians of previous generations had chosen to carve out reputable identities as women professionals. Rejecting these roles and their ideological grounding, women students of the 1970s sought to transform not simply the system of medical training but also the profession—to alter accepted notions of the doctor-patient relationship, to make the culture and politics of the medical establishment more progressive, and to integrate into the medical citadel a feminist critique of the ways physicians sometimes deceived patients, infantilized them, and denied them knowledge about their illnesses and their bodies.

The ways humor was used to harass and to exclude women offers a promising but little explored platform for understanding twentieth-century American medical culture, especially at moments of instability to the professional status quo. Medical educators, sociologists, and other observers of medical student life have frequently called attention to the jokes told inside the gross anatomy lab, as students confront and dissect cadavers, and have tended to see them as providing an essential psychological function, part of "blowing off steam" after the uncomfortable encounter with death.[1] But joking outside the anatomy room has been largely ignored, and until the 1970s, victims of harassment rarely talked about it. When historians reexamine these jokes and other acts, it can illuminate in uncomfortable ways that what was considered funny depended on now unacceptable assumptions. As they spoke out against the ways their teachers and peers harassed and mocked them, women students began to retell the jokes that, outside the lecture halls and hospital wards, no longer sounded funny or harmless.[2]

Consider two jokes from the 1970s. In the first, from 1973, an associate dean told a group of men and women medical students at a cocktail party about the scale that faculty members on his medical school's admission committee used to rate female applicants:

1. I don't want her, you can have her.
2. I'll take her.
3. I'll take her, without reservation.
4. Save her for the dean.[3]

The second is a joke told at an internship interview in the 1976 novel *Woman Doctor*, based loosely on the experiences of Florence Haseltine, who graduated with a PhD in biophysics from Massachusetts Institute of Technology in 1970 and an MD from Albert Einstein College of Medicine in 1972, and then completed a residency in obstetrics and gynecology at the Boston Hospital for Women. The three men interviewing the main character spent most of the time discussing her ex-husband's specialty and the latest Elizabeth Taylor film. Finally, one of them asked her jovially, "Do you think we ought to pay women as much as men?" Phyllis Donnan "smiled broadly and said 'No, I think you ought to pay them much more!'" Later, Donnan felt proud that "at last I'd been able to speak up," but the men in this story did not laugh; in fact they were rendered speechless. For them, this joke was not funny.[4]

Gender, History, and the American Medical School

By the 1950s, a homogenized American medical profession had established a gendered medical culture that nurtured what were so aptly termed "boys in white." The few "ladies in white" were anomalies in every way; hen medics, as they were unkindly known, were "clearly unfeminine, that is unattractive and unmarried," two feminists later noted.[5] Those committed, smart, and frequently privileged women who did graduate from medical school before the 1970s understood that the "inner fraternity" of their profession, as a sociologist in 1948 named it, had "specific accepted mechanisms for both incorporating the newcomer and repelling the unwanted intruder" through hostility and "casual neglect."[6] This hostility and disrespect was replicated outside the medical school when women graduates sought to combine a professional with a personal life. In a society that saw a career as intrinsically opposed to femininity and motherhood, women physicians frequently adopted a fierce manner and style that reflected their professional isolation.

The young women who came to study medicine in the 1970s were a different breed from earlier generations. They were not all feminists, but they were all touched by the new feminist movement's "activism and affirmation of women's rights." And, feminists or not, they were mocked as "women's libbers" by male students and teachers.[7] With its deepening skepticism about the medical establishment, the burgeoning feminist health movement had not urged women to attend medical school. Nevertheless, the health centers, shelters, books, and workshops created by feminists trying to fashion new ways to provide health

Cover of *Woman Doctor* by Florence Haseltine and Yvonne Yaw (Boston: Houghton Mifflin, 1976).

care for women did embolden many women contemplating medical careers. Outside the halls of medicine, radical feminists dismissed the idea that medical schools and hospitals could ever be made woman-friendly. Only within an alternative health system, they argued, could women safely receive health care that was neither sexist nor demeaning.[8] Thus, although medical men frequently did not recognize the distinction, most women students of the 1970s were not radical feminists. Rather, most were liberal feminists who entered medical school buoyed by the hope that feminist activism could transform medicine from within as well as without.

When these courageous young women arrived in medical school during the 1970s—the numbers jumped from 1,256 (11% of first year classes) in 1970 to 4,966 (29%) by 1980[9]—they were prepared for much of what they found there. They had heard horror stories about gynecologists, obstetricians, and surgeons and were ready to face a hostile environment. But they were not prepared for the epistemological and cultural convictions shared by their teachers and peers that sanctioned lecture slides of women's breasts and genitalia and comments denigrating women's skills and mental abilities, sometimes expressed through humor, always disparagingly. Jokes, belittling, and harassment were hardly new experiences for American women entering medical school. But they became a more vitriolic part of medical schools' culture of exclusion in reaction to newly vocal students, outside activists charging the medical establishment with elitism and racism, and federal authorities enforcing new laws restricting gender discrimination in higher education.

To make the experiences of women medical students integral to the history of American medicine is to force a gendering of medical education history. Our histories must bring gender into the stories we tell about the making of professional identity, health care, patient experiences, and the structure of medical practice, not relegate them to a separate chapter.[10] Gender is everywhere embedded in the experiences and self-definitions of physicians and other health professionals, both women and men. Medical training has always endeavored to inculcate particular values that exemplify what it means to be a good professional. As Robert Nye shows, the concept of "character"—an essential element for defining the ethical and moral respectability of the nineteenth-century gentleman-physician—was profoundly gendered. In the 1970s, as feminists disparaged the connotations of "lady" and "gentleman" alike, women physicians struggled similarly to challenge public and professional models of the "good" doctor. Women in the 1970s did not bring gender awareness to the medical school; they challenged what was already there.

Boys in White

In a revealing section in *Boys in White* (1961), one of the most influential analyses of postwar medical school training, sociologists Howard Becker, Everett Hughes, and their colleagues acknowledged that to prepare for their intensive study of the University of Kansas Medical School they read many popular novels about students and doctors. When they asked the sixty-two students in their study what each had read about medicine before coming to medical school, "with one exception, if they had read anything, it was these novels."[11] In one of the few tongue-in-cheek moments in this formal text, the researchers described trying to "test" these fictional insights. The novels, they wrote, pictured "students hardening themselves to face suffering and death . . . [as they] parade their lack of feeling by continual obscene joking . . . harass the girls in the class by gags with the sex organs of cadavers or profane the lab by throwing around the detritus of dissection . . . and use obscene rhymes as mnemonic devices." But at Kansas, the researchers noted ruefully, "we observed no instances of harassed girl students . . . [and] only one instance of 'profaning' the lab."[12] They even "tried to stir things up a bit ourselves by telling dirty jokes," but they were disappointed by the results. Most students seemed uncomfortable with the "off-color" remarks that the researchers had offered in hopes of invoking more of the same.[13] The authors noted that the Kansas students "remain male enough to look at and comment on their female patients" with remarks like "I sure wouldn't mind doing an examination on someone like her," but the researchers saw such comments as funny and not needing further explanation.[14]

American medical culture by the mid-twentieth century was a thoroughly masculinized world, hostile to women and to anyone not Anglo-American and Protestant. Few schools were known for accepting and hiring Jews, Catholics, and African Americans, and the Woman's Medical College in Philadelphia provided a safe space for white women and for some women of color.[15] Sociologists who explored the training of physicians—and medical sociology was a leading field in the 1950s and 1960s—tended to accept this homogeneity uncritically.[16] Thus, the *Boys in White* researchers assumed the objectivity of the admission process that had so clearly structured the student population they investigated. All sixty-two of their "random sample" were white men, and about half were

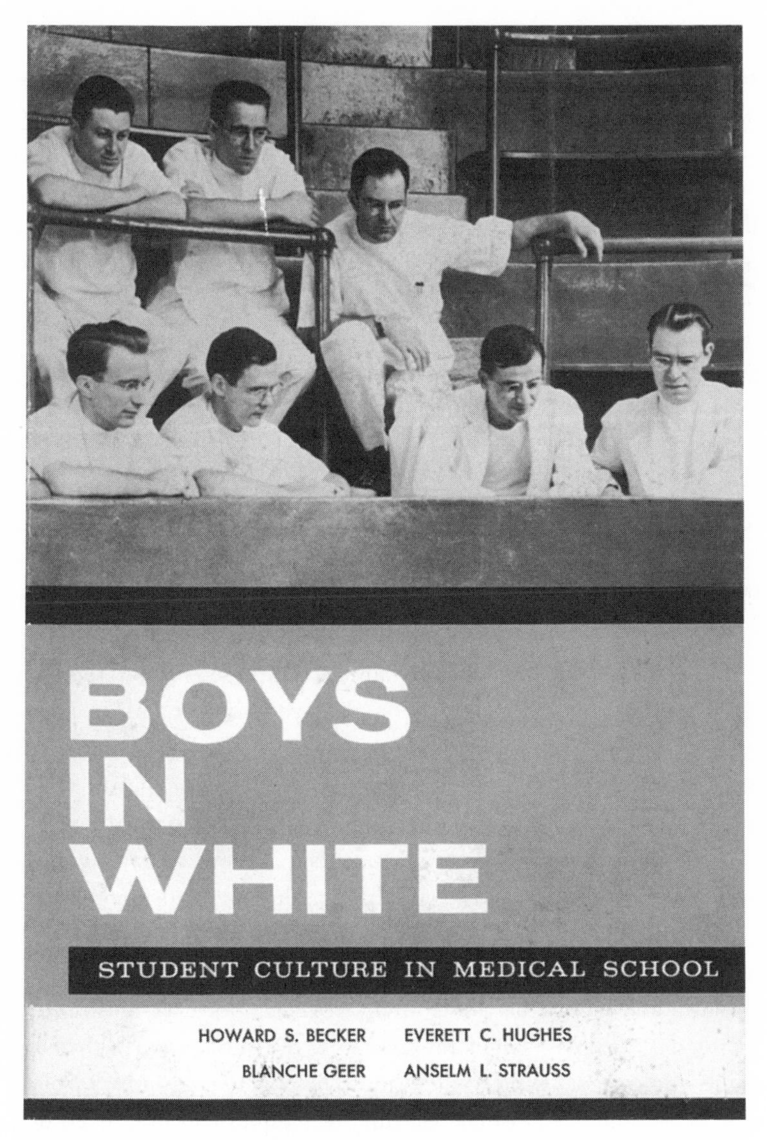

Cover of *Boys in White* by Howard S. Becker, Blanche Geer, Everett C. Hughes, and Anselm L. Strauss (Chicago: University of Chicago Press, 1961). © University of Chicago, 1961.

the sons of professionals.[17] To explain this demography—clearly drawing on a local medical administrator's words—they observed, "the small numbers of women and Negroes do not reflect any intent to discriminate. The school gets very few applicants of either category."[18] Compare this with the recollection of Marjorie Sirridge, who had confronted "discouraging" interviewers at Kansas's medical school about a decade earlier. The members of her committee asked her repeatedly, " 'You know, this isn't really—you don't really think this is a field for women, do you?' And I'd say, 'Well, yes, I really do.' " One of the men on the admission committee even told her, "I don't know why you women want to keep coming to medical school; then the men can't tell all the jokes and dirty stories they want to."[19]

Following the example of their teachers, male students often harassed the small numbers of women in their classes. Some historians have suggested that "obscene pranks" must be understood within a broader system of student esprit de corps, but such an analysis reads closer to the picture offered by medical novelists of the time.[20] Warren Bennett, a student in the 1940s when Harvard's medical school first admitted women, recalled that the Nu Sigma Nu members were "really against the idea of women in the medical school, and they did make things uncomfortable for them." On one occasion, they organized a dance and invited a woman student to sing with their band. "Then someone handed her a little glass of wine which was totally laced with chlorohydrate [*sic*]. Within about twenty minutes she passed out completely and had to be carried from the place."[21] Teachers impressed on their students that they had gained entrance to a special world, where public conventions regarding touching and talking about body parts did not apply. When professors would "flash a picture of a nude woman" or tell "dirty jokes," women and men alike saw this as a form of shocking socialization.[22] Their harassing intent was clear, however, and women were frequently discouraged from manly specialties like surgery when their teachers asked the whole class: "Who would want to be operated upon by a woman?"[23]

The use of humor and other belittling techniques had long been defended by educators as necessary or at least appropriate in medical training. Both the subject matter and pedagogy of many preclinical subjects, educators admitted, needed "sexing" up, and throughout the 1950s and 1960s, they debated the problems of disaffected, bored students.[24] The "well-prepared first year medical students show signs of boredom," argued a 1966 Commonwealth Fund study on *The Crisis in Medical Services and Medical Education*, "not only from the repetition of material with which they are already familiar . . . [but because] they find ordinary medical school teaching inferior to the teaching of science in the better undergraduate institutions where learning is regarded as an active process for the students . . . rather than a spongelike passivity which seems to be expected in many medical schools."[25] One woman who had graduated from Bryn Mawr recalled that she suffered "culture shock when I went to medical school . . . it was like going back to high school—the learning by rote, the didactic approach."[26]

Even before the feminist movement began to target doctors, hospitals, and medical training, other political and social forces had begun to challenge assertions by medical educators that their selection of students and faculty fairly balanced merit with civic benevolence.[27] During the 1960s, civil rights activists scornfully dismissed the idea that the practice and the intent of medical schools' admission and hiring policies were anything but discriminatory. Medical training, they declared further, created arrogant physicians who neglected the needs of the poor and communities of color.[28] By the late 1960s, individual medical deans as well as the Association of American Medical Colleges began to develop well-publicized policies of race-based affirmative action. Trying to assuage angry local communities tired of being used as janitors and research subjects but never hired as administrators or medical professionals, deans sought out "qualified" students of color and tried to expand the pool by setting up small programs to help minority high school students get laboratory experience and build their resumes.[29] These efforts were an early wedge in breaking down the notion of scientific objectivity and neutrality underlying the culture of postwar medical education. Training more doctors of color, though, at the same time, could be justified in terms of expanding access to the products of scientific medicine to the ghetto and to other medically underserved groups.

During the 1970s, the appealing option of simply adjusting academic affirmative action policies to include women was made more difficult as feminists demanded a transformation of medical culture and the values embodied in its practices. While many professors could ignore the abortion protesters outside their hospitals, the health collectives set up in tiny shop fronts, and the pamphlets informing their increasingly vocal women patients, they were shocked when feminist arguments began to be expressed inside the citadel. Furthermore, these activists seemed strangely familiar. Feminist health activists during the 1970s were usually young, white, and middle class. They were "our own daughters," attacking the medical profession as dehumanizing and discriminatory even as they sought to gain entrance to its most eminent schools and hospitals.

Ladies in White

Before the sexual revolution of the 1960s, the treacherous politics of respectability structured the public and private lives of women medical students and physicians. In one way, Elizabeth Blackwell and her peers had an advantage. They had to convince the men of medicine as well as the general public that a middle-class woman could work as a health professional without losing her femininity or her standing as a "lady," but it was a struggle infused with moral certainty. They never were and never could have been "one of the boys."[30] Even as the term "lady" began to sound old-fashioned, its power lingered in the structure of professional gender relations. Thus, as late as 1968, one woman medical reformer argued that "no matter how much a girl may want to prove that she is 'one of the boys,' sooner or later 'dirty jokes' or rowdy demonstration will present a situation from which she has to withdraw as a lady—and some men are not really at ease until they have verified that the line exists."[31]

After 1920, with the disappearance of all but one of the nation's women's medical colleges, American women trained at coeducational institutions where administrators, as well as their teachers and peers, were frequently unwelcoming and dismissive. In arguing for the transformation of American medical education, reformer and suffrage supporter Abraham Flexner had believed that gender, like creed, would become irrelevant, for "science, once embraced, will conquer the whole."[32] But instead, reinforced by the cultural ideology of the new scientific medicine that feminized empathy and masculinized objectivity, American medical educators practiced discriminatory hiring and admissions policies.[33] Most medical schools imposed a strict quota on women applicants; in the words of one administrator, "We do not want the one woman we take to be lonesome, so we take two per class."[34] The women who chose this life were considered odd by definition. In her 1925 autobiography, Lilian Welsh recalled the "gossip" in Baltimore that a woman choosing to study medicine must have had "some profound emotional disturbance," and quoted her teacher William Osler saying "jocularly" that "human kind might be divided into three groups—men, women and women physicians."[35] More harshly, one woman was told by a fellow student in the 1920s: "You are not a man, you are not a woman. You are an unsexed thing, studying medicine out of morbid curiosity."[36]

Uncomfortable male students vacillated between harassment and exaggerated chivalry. The 110 men of one class in the early 1960s sent their ten female peers roses on Valentine's Day, showing, as one of the women recalled, that "they didn't know what to do with us."[37] The constant pressure to maintain ladylike behavior in environments where being a "gentleman" was part of a professional identity added to this strain. "A girl who wants to become a physician," one woman doctor warned in the 1940s, "must realize that everything is stacked against her" and that even after she has been admitted, "the medical schools are constantly trying to get rid of her, if she doesn't toe the mark both scholastically and personally."[38]

Racism deepened the difficulties for women physicians of color, who debated whether race discrimination trumped gender discrimination. New Jersey obstetrician Lena Edwards, a 1924 graduate of Howard University's medical school, commented in *Women Doctors Today*, a 1947 collection of essays targeted at high school and college students: "I do not feel that being a Negro doctor has ever been so much of a handicap as being a woman doctor," for every woman doctor "is bound to encounter snippy patients and male doctors who belittle her."[39] The racism encountered by Virginia Alexander, a 1925 graduate of the Woman's Medical College, lingered larger in her recollections. One of her professors "took especial delight and pains in retelling to his classes, where there were colored students, every discreditable, dirty and insulting story about colored people he could think of."[40]

Many of the women doctors of the early and mid-twentieth century practicing specialties considered fitting for their gender—such as pediatrics, public health, and psychiatry—deliberately cultivated a stiff and off-putting persona, at least to colleagues who might potentially mock or harass them. Rita Finkler, practicing in Newark, New Jersey, in the 1920s and 1930s, chose outfits in somber brown and black with severe horn-rimmed glasses, no jewelry, and hats without feathers or flowers.[41] A stereotypically "feminine" aspiring physician was easily judged flippant and not serious. Trying to assess whether female applicants were normal or "abnormal," one dean admitted that "when a very feminine-looking gal comes in, I might wonder if she's fit for medicine."[42]

Demonstrating acceptable signs of femininity had long been a code for social and psychological normality. Yet women professionals were also warned that it was impossible to combine a career and a family. Thus, the numerous biographies of Elizabeth Blackwell written in the 1940s and 1950s suggested that she was and had to be *A Doctor Alone*.[43] In books advising young people about a medical career, these warnings were clear in both illustrations and text. In the 1949 student guide *How to Become a Doctor*, the chapter on "Success and Failure in Medical School" opens with a sketch of a pretty wife hushing her daughter

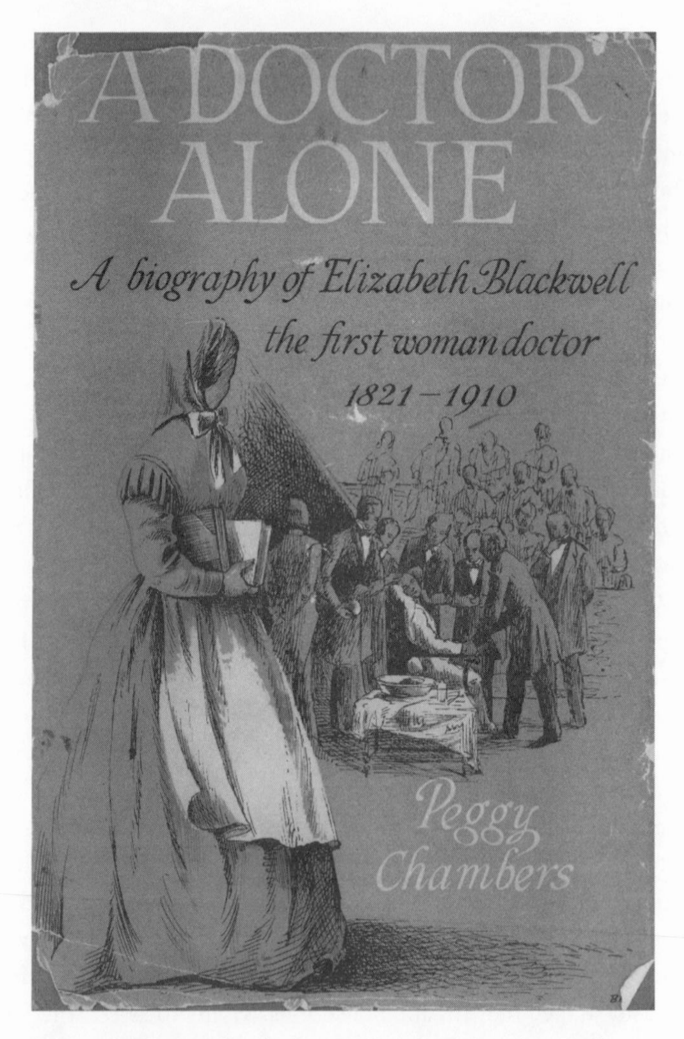

Cover of *A Doctor Alone* by Peggy Chambers (London: Bodley Head, 1956). Reproduced with permission from Random House Archive & Library.

while the son watches his father study. The chapter on "Women and the Study of Medicine" shows a rather unkempt, older woman alone looking through a microscope.[44]

In an era when homophobia entangled with a reinvigorated ideal of domesticity, many women physicians felt they needed to assert their femininity, and the lay press agreed.[45] "In spite of the fierceness of the fray, the outstanding

SUCCESS AND FAILURE IN MEDICAL SCHOOL

CHAPTER EIGHT

THE APPLICATION FOR ADMISSION

CHAPTER FOUR

WOMEN AND THE STUDY OF MEDICINE

CHAPTER SEVEN

Illustrations in *How to Become a Doctor* by George R. Moon (Philadelphia: Blackiston, 1949). Reproduced with permission of the McGraw-Hill Companies.

women doctors don't seem to have sacrificed femininity for success," a *New York Times* reporter reassured readers in 1961, and "most are unusually gentle, charming and attractive." New York City Health Commissioner Leona Baumgartner, for example, was "entirely feminine" and "indulges a taste for colored hats that set off her blond good looks."[46] Women physicians able to manipulate such stereotypes could make them into calculated techniques for professional success. "If you're feminine and let them think *they're* making the decisions," a successful physician advised her younger peers, "you'll get along very well. Men don't like aggressive, crusading women."[47]

In their efforts to "get along," many women physicians refused to join professional groups that set them apart by gender. Organizers in the 1940s and 1950s who "fought for equal rights or special arrangements" encountered a "spirit of complacency and secure smugness" among their peers.[48] When Bertha Offenbach urged women physicians on the staff of the Massachusetts General Hospital to join the American Medical Women's Association during the 1950s, each woman told her: "I have no interest in joining . . . a women's medical association . . . there is no sex in medicine."[49] In *Child of Destiny*, a 1956 biography of Blackwell, the aspiring physician struggled for professional respect, readers were assured, "not with bluster, noise, or self-assertion, but quietly, with a gentle manner and a will of iron." When Elizabeth and her sister Emily established their own clinic, they "shied away from beating the drums for women's rights," for "they were desperately anxious to establish high medical standards, and they knew that nothing would wreck their work so quickly as a reputation for eccentricity."[50] During the Cold War years, when the corporate "organization man" was the epitome of success, conformity to gender norms added a sharper edge to the construction of a woman's professional identity. When she was in medical school in the 1950s, psychiatrist Jean Carlin recalled, all the women students "believed the quieter we were the more likely we were to graduate."[51]

No Longer Ladies

As activists in the nascent feminist movement in the early 1970s began to point out sexism in the workplace and the family, the power of an educational system that made *Dick and Jane* the basis for reading and thinking about proper gender roles loomed large.[52] Turning political outrage into legal pressure, liberal feminists in the National Organization of Women and the Women's Equity Action League used a massive lawsuit against American medical schools to enforce federal rules against sex discrimination. In 1972, Congress passed the Higher Education Amendments Act whose Title IX threatened universities with the loss of federal funding if they practiced sex discrimination in hiring or student training.[53] As the result of what one feminist identified as "both the strengthened self-esteem of young women—one effect of the feminist movement—and the laws prohibiting discrimination," the numbers of women applying to medical schools in the United States rose tremendously, tripling between 1970 (2,734) and 1974 (8,712).[54] The expanding size of medical school classes and the twenty-two new schools opened during the 1960s meant that the percentage of men accepted stayed steady.[55]

In the early 1970s, most medical administrators believed they could respond to legal rules simply by accepting more women students. Few thought that adopting affirmative action would significantly alter the culture or pedagogy of their schools. Members of admission, internship, and residency committees continued to ignore legal strictures on discriminatory questioning, but, cognizant of new mores among young people, they began to pose more pointed questions to women applicants who in this age of *Hair* and *Godspell* would presumably not be shocked. Married applicant Susan Benes recalled that the nine men sitting around the table at her medical school interview "grilled me the entire time about what form of contraception I was using. Would I promise not to get pregnant while I was in medical school? How could I possibly hope to be both a doctor and a mother?"[56] Most of these were familiar questions, but the assumption that pregnancy could now be predictably controlled gave them a nastier edge. Women who became pregnant during training or residency faced harassment and humiliation. Male students and professors at one California school informed a pregnant third-year student that "you have no commitment to medicine, you're not really interested," and one of her teachers added "that I had better be thinking real hard about having an abortion."[57]

Outside the medical establishment, the alternative clinics established by feminist health activists sought to "completely redefine the patient and provider roles." The physicians who worked at these clinics were often relegated to minor, legally necessary roles.[58] Convinced that medical socialization created dehumanized professionals, irrespective of their gender, feminists were especially intolerant of women doctors. In a directory that rated doctors for gender sensitivity and bedside manner, one New York feminist group noted that "some female physicians can be worse than their male colleagues."[59] To illustrate the arrogance of physicians toward their women patients worried about the potential dangers of the pill, science journalist Barbara Seaman's exposé *The Doctors' Case against the Pill* quoted gynecologist Elizabeth Connell defending physicians who dismissed their patients' fears, for otherwise "they would all come running to me with a headache."[60] "Doctors are doctor chauvinists as well as male chauvinists," early editions of *Our Bodies, Ourselves* concluded; "most women doctors are no exception to this, having taken a role of 'honorary men.'"[61]

By the mid-1970s, American medical schools were filled with feisty and intrepid women students conscious of themselves as activists. In 1968, Carol Lopate could still argue that "the woman medical student often finds herself in the position of pretending not to have seen some little piece of joking, or of acknowledging it with the faintest of smiles."[62] But that time was fast disappearing. Rather than blushing at "anatomical jokes and girlie pin-ups among the lecture slides," women students took their protests to the professor himself and to administrators now uncomfortably torn between faculty traditionalists and legal pressures to provide a "nonsexist" educational environment.[63] Students complained about "the cardiologist who continuously refers to the women in his physical diagnosis class as future pediatricians (even after they have informed him differently)"; "the gynecologist who has the gall to inform the women in the class that they shouldn't be in medical school (they should be home having babies)"; and "the doctor who comes in and begins his lecture, 'My, we have some lovely young ladies here.'"[64] If professors did not know or would not follow the federal guidelines against discrimination, courageous individuals reminded them. A premed undergraduate from Brooklyn College "became so exasperated" during one of her admission interviews that she asked her interviewer "whether this question was asked of me because I was a female. He fumbled a little and his face turned a crimson red."[65]

Activists and policy analysts began to build up extensive evidence showing widespread, structural discrimination in medical school admission policies.[66] Studies in the early 1970s showed that medical school interviewers told 61 percent of the women applying (compared with 8% of male applicants) that medicine was too demanding a career to allow for a family.[67] A survey of medical schools in 1971 found that, of the seventy-three institutions responding to the survey, 45 percent had not a single "minority member" (no man of color or woman whatever her race) on their admission committees.[68]

More slowly, as women students began to confront what sociologist Judith Lorber termed "the masculine mystique of American medicine," they began to identify and reject stereotypes that had led many women before them to play

Cartoon from feminist newspaper *Sister*, July 1973.

down their own skills and intelligence. Phrases like "make a conscious effort not to appear to show up the men in their class"; "you should be good, but not too good"; and "it isn't right for a woman to have her hand up first with the right answer" were repeated and stripped of their power.[69] Instead, students sought to construct a new kind of professional identity that recognized the special needs of women as potential mothers and family members, but nonetheless demanded equality in respect and opportunity, and even asserted the right for women to display pride in their abilities without losing their femininity.

Beyond publicizing breaches in rules forbidding sex discrimination, women students drew attention to informal methods of harassment inherent in medical socialization. Repeating the jokes and put-downs in public began to break apart any sense of a secret medical fraternity (with all its implications). Some medical administrators continued to protest weakly that these comments were "harmless," but the evidence both qualitative and quantitative showed that harassment and discrimination clearly occurred at all levels: student peers, lecturers, clinical teachers, and senior administrators.[70] "Because of you, a man probably went into chiropractic school," taunted one male student.[71] "A woman doesn't belong in the OR except as a nurse," said the chief of an OBGYN unit.[72] Drawing on the stereotype, combated in numerous policy studies, that married women doctors stopped practicing and became housewives, a dean told incoming students that "he wanted to keep down the numbers of women doctors—in the interest of the taxpayer."[73] A Wayne State University medical school lecturer had sought to break the ice with his students with "this little joke," the *New York Times* reported with relish: "Question: Which word doesn't belong—egg, sex, woman, rug. Answer—Sex doesn't belong, because you can beat an egg, a rug and a woman, but you can't beat sex."[74]

Nor were students content just to speak up about this behavior. They took notes, taped it when they could, and reported it to senior administrators. In one of the first formal complaints made under Title IX to the Office of Civil Rights at the Department of Health, Education, and Welfare, women students at the University of Wisconsin recorded the comments of their pathology professor, who told a woman student that she couldn't make an announcement because she had "too many clothes on" and then said to all the women in the class, "We have to give you equal rights and the first thing you know . . . women will dominate society. Thank God I'll be dead before you do though." After the complaint was filed, he apologized to the class.[75]

Responding to attacks on both the left and the right that they were prudish and had no sense of humor, feminists stopped blushing and began to create humor of their own. Feminist cartoons and jokes appeared in alternative women's journals and in collections like *Titters* and *Pulling Our Own Strings*.[76] In 1972, when feminist endocrinologist Estelle Ramey, a professor at Georgetown's medical school, condemned a new anatomy textbook illustrated by photographs of *Playboy*-type models, she was called "a sour old maid who couldn't take a joke." But Ramey's campaign, aided by her position as president of the newly established American Women in Science, led the publisher to withdraw the book.[77] Ramey's skill in medical politics was to turn harassing humor against her opponents. A widely quoted interview with her Georgetown students, reprinted in *Perspectives in Biol-*

ogy and Medicine, began with Ramey's disclaimer: "I really believe that men are just as capable as women in medicine."[78] Gently mocking the familiar stereotypes of women physicians, Ramey said, "Women in the medical sciences, like women in harems, range from gentle beauties to castrating witches . . . Most of us, like our unscientific sisters, are somewhere in between."[79] Older women physicians were now recognizing structural discrimination, Ramey admitted, adding with a twist: "There is a strong bias at the top even among men who happily entertain the notion of a woman Assistant Professor. The reasons are standard: women are too prone to emotion, verbosity, pettiness, and pregnancy. Even when they try to act like men, they just aren't gentlemen."[80] Proud of this combination of activism and satire, one woman student said in 1973: "We need more women in medicine with the mind and mouth of Estelle Ramey."[81]

Why Would a Girl Go into Medicine?

In 1972, the Harvard Medical School, under pressure to demonstrate a commitment to gender equality, hired pediatrician Mary Howell in the newly created position of associate dean for student affairs. She became the most senior woman administrator at Harvard's medical school. Howell was extraordinarily qualified. She had a PhD in psychology, an MD, and had been at Harvard since 1969 as an instructor in pediatrics and as chief of the Behavior Unit in the Children's Service of the Massachusetts General Hospital. A feminist, Howell had worked with the Boston Women's Health Book Collective and was part of the group, along with Barbara Seaman and Belita Cowan, which later became the National Women's Health Network.[82]

Both Howell and feminists outside Harvard saw her appointment as an opportunity to challenge medical sexism across the country. In 1973, she completed a survey based on questionnaires filled out by 146 women students from forty-one American medical schools. Reproduced first as a mimeographed report and then published by the Feminist Press under the pseudonym Margaret Campbell, *Why Would a Girl Go into Medicine?* was circulated "in an underground manner among medical school counselors, deans and interested feminists."[83] "All medical schools," Howell stated bluntly, "exhibit some degree of discrimination against women students."[84] Her most powerful evidence were concrete examples, both horrifying and easily repeatable, of the jokes, insults, and other acts of exclusion and harassment students had sent her. Women students, Howell explained, were extremely good informants, for they "are novices in the health care system, not yet fully initiated and therefore not yet hardened, and they can be sensitive observers and recorders of the denigrating assumptions about women that influence the conduct of health care."[85] Her examples included the resident who told students: "While you're in the hospital you're a doctor and you have to cuss like the rest of us. You can be a lady when you leave"; the woman psychiatrist introduced to the freshman class as "the wife of one of [the city's] best radiologists"; and, one of the most widely repeated examples, the lecturer who said, "the only significant difference between a woman and a cow is that a cow has more spigots."[86] Howell's evidence also undermined the "anatomy room defense" (that humor was a way of "distancing the student from the attraction and repulsion of dealing with human anatomy"), for, commentators pointed out, "the continual mockery of women's bodies" was not matched by any analogous mocking of men's bodies.[87]

According to Howell, her study was intended to document "discrimination, coping, change, and concern for the health care of women . . . It is not about discouragement or defeat."[88] She saw this evidence as part of a broader claim that physicians' attitudes demeaning women "affect a large proportion of the patient population," an argument she laid out in her 1974 article, "What Medical Schools Teach about Women," in the *New England Journal of Medicine* (based on the work of "Margaret Campbell").[89] Students, she argued, were "widely taught both explicitly and implicitly, that women patients (when they receive notice at all) have uninteresting illnesses, are unreliable historians, and are beset by such emotionality that their symptoms are unlikely to reflect 'real' disease."[90]

Established women physicians responding to Howell's article were not convinced that harassment had harmed them or their patients. One woman wrote to the *New England Journal of Medicine* protesting that she had experienced "the usual amount of anti-feminist difficulty," but it "may well have developed in me work habits and a seriousness of purpose that I might not otherwise have had" and that "a few good humored gibes along the way" were hardly demeaning.[91] In reply, Howell commented that "the price of character building by way of a steady attack on one's self-esteem seems unduly high."[92] A number of women physicians did share Howell's assessment. When the journal's editor remarked that women doctors at a recent Harvard Medical School Alumni Day were "handsome in countenance, dressed with elegant simplicity, far from shapeless, and . . . married," readers attacked him for thoughtless sexism.[93]

By 1975, Howell had publicly acknowledged that she was the author of *Why*

Would a Girl Go into Medicine? That year she organized a national conference on women's health at Harvard, where, amid equally provocative papers, Howell proposed that feminists set up a "women's health school" where students would learn a new style of medical decision making based on teamwork rather than hierarchy, and where they would not need to "act like men, identify with men, and adopt the characteristics of competitive individualism that have been traditionally ascribed to men."[94] Howell later said that as she became recognized as a prominent feminist critic of medical education, Harvard administrators and faculty began to use her and other feminist advocates at the medical school as "student pacifiers rather than student advocates."[95] Characterizing her work at Harvard as a struggle to stay "sane in an insane place," Howell wrote an "Open Letter to the Women's Health Movement" in May 1975 to explain her resignation. The school's administrators had used her presence at Harvard, she declared, "to ward off complaints about their unwillingness to recognize the needs of the underrepresented in health affairs"; yet, those struggling against "disadvantage and deprivation of privilege" needed not "token appointments" but "a real voice" in decision making.[96] She moved to Maine and set up a private practice. Howell's successor at Harvard, psychiatrist Carola Eisenberg, was appointed as dean, not associate dean, of student affairs, and continued the tradition of feminist educator-activist.[97]

Conscious that some administrators were behind them and that they now made up a visible proportion of their classes, women medical students began to act as a group using strategies feminists had developed outside the halls of medicine to confront legal and social discrimination.[98] When the head of the University of Pennsylvania's Department of Medicine announced to students that "medicine is the profession that separates the men from the boys," Vanessa Gamble recalled, "he got hissed and could not understand why."[99] Medical students in California reported that "women now boo and hiss like hell in classes if a professor says or alludes to women inappropriately."[100] Forcing reluctant teachers to accept analogies between race and sex discrimination, one group of women students confronted a professor of pathology who defended his use of "cheesecake pictures" and jokes because they made his lectures "less gruesome" to the class, which was, after all, mostly male. When the women asked him why he "didn't also tell 'nigger jokes' since the majority of the class was white," he became "quite belligerent" and later complained to the chair of pathology.[101] By the 1970s, behavior that was explicitly racist was no longer acceptable in most academic settings, but making sexist jokes was still considered both entertaining and a kind of provocative socializing, reinforcing the idea that women students needed to have a "sense of humor."

Activism, Change, and Backlash

Despite the exhilaration of challenging sexism within the medical citadel, women students continued to battle assumptions that their chosen career would leave them masculinized and unhappy. Being an activist in medical school by the late 1970s became, in some ways, more difficult, as students and teachers alike grew conscious of the power of harassing comments and actions.[102] According to one report, a dean conducted an active campaign against admitting women "by corresponding with his friends on admissions committees throughout the country."[103] A number of medical schools began to try to select students who were "compared to the entire population of applicants . . . quite conservative." That such efforts were not fully successful was the result, feminist commentators believed, of diverse voices on admission committees and "the well-known ability of student applicants, chameleon[-]like, to appear to be whatever admission committees seem to want them to be."[104]

Some elements of feminist thinking, though, were becoming widely accepted, especially the idea that professional women did not make "bad mothers." In a clever inversion of this stereotype, Carola Eisenberg told the readers of the *New England Journal of Medicine* that in her work as a child psychiatrist at three different medical schools, "I became all too well aware of the extent to which many male physicians fail to meet their responsibility as fathers and husbands."[105]

The depiction of the older woman physician as embittered, hierarchical, and antifeminist was also contested. Although some feminist activists outside medicine continued to argue that women doctors were "only too ready to become 'honorary men' and take their places in the male establishment, enjoying privilege and abandoning their sisters," many became aware of the stereotyping they had reinforced.[106] Women medical students, the 1976 edition of *Our Bodies, Ourselves* admitted, "undergo considerable stress and harassment" and had to battle "contemptuous remarks about women as patients, nudie slides in lectures, and dirty jokes."[107]

The feminist movement also radicalized women professionals who had previously considered their own personal success or failure an individual rather than a structural matter. Some women physicians set up their own consciousness-raising groups; others joined health collectives and feminist clinics. In the mid-1970s, a group of feminist physicians in Boston gained control of the regional chapter of the American Medical Women's Association and began to enlarge its system of female mentors, address family/career conflicts, and develop strategies for "how to advance your career in the 'male-buddy' system."[108] Boosted by an invigorating sense that a feminist professional could fashion a medical career that was responsive to the demands of family life, activists condemned "the organization of work in medicine," and the ways that "key professional norms" assumed a "male biological clock."[109] Such women had "moved beyond reform and came to advocate revolutionary changes in the health care system and in American society itself," reflected activist pediatrician Helen Rodriguez-Trias in 1984. Rodriguez-Trias, a 1960 medical graduate of the University of Puerto Rico, had organized protests against sterilization abuse and helped to make the concerns of poor women and women of color central to feminist health activism. Growing numbers of women in "medical schools and other health institutions" had "placed more women in positions of increased power," and Rodriguez-Trias believed they "in turn are providing support for others who enter now."[110]

In 1979, Mary Howell praised the young women students entering medical school as "remarkably sophisticated in their political analysis and understanding," but she noted that the broader public backlash against feminism epitomized by the rise of the New Right and the Moral Majority was being felt in medical schools. Sexist remarks, she reflected, used to be made "from a position of unthinking and naïve ignorance of the effect of such pejoratives," but now such comments "are justified by disparaging remarks about the trouble-making personalities of those who would object."[111] Nor had medical faculty members worked out how to maintain a jocular medical culture within the strictures against sexist behavior. In the 1980s, one woman complained that a Philadelphia administrator had assured new women students that "the faculty was very supportive and willing to offer help to students whenever we needed it. To illustrate the point, he showed a slide of a woman (student) on the lap of a man in a white coat (professor). There were giggles from the audience."[112] These giggles suggested that students remained uneasy about the implications of such behavior from their teachers and wondered if women in medical school still had to pretend to "get along."

As more women entered the American medical profession as activists as well as students, they wondered whether they would have to choose between feminism and professionalism. How powerful were the dangers of being co-opted into "elitist professionalism"?[113] At a conference on women and medicine in 1976, Barbara Ehrenreich, a leading health feminist and the co-author of *Witches, Midwives, and Nurses* and *Complaints and Disorders: The Sexual Politics of Sickness*, warned women medical students that they must decide to whom they would be accountable—"to the institutions that employ you, to the medical schools or to yourselves and all the women who helped put you where you are now." She challenged students to become "rebels and troublemakers" but warned that if they chose that path they would not become deans of medical schools, famous clinicians, or well-funded researchers, for "you're going to get some crummy recommendations along the way from your professors . . . You'll probably not get the best residency." Such personal sacrifice, however, would be worth it for "there will be sisterhood and security in whatever solidarity you can develop with the other people who are struggling for change."[114] Ehrenreich believed that a feminist commitment could undermine the powerful lessons of medical socialization, but other feminists continued to be "suspicious of the overpowering process by which ordinary people are transformed into physicians." "On the basis of experience," Mary Howell reflected as she confronted these challenges in her own life, "they doubt that one can be both a feminist and a professional medical expert."[115]

Different Doctors? Better Doctors?

In an essay published in the *New York Times Magazine* in 1988, pediatrician Perri Klass asked, "Are Women Better Doctors?" Klass had published a regular column on her experience as a medical student at Harvard during the 1980s, expanded in *A Not Entirely Benign Procedure: Four Years as a Medical Student* (1987).[116] Drawing on her own experiences and conversations with other women physicians and students, Klass concluded, "Yes, women are different as doctors—they're better." Internist Nevada Mitchell of Kansas City told her, "There's a world of difference. The women I come into contact with are less aggressive, more likely to have one-on-one relationships with patients, less likely to go for a high volume of patients." Gynecologist Marilyn Richardson disagreed, complaining about patients who held "a misconception that has evolved with consumer awareness, an erroneous belief that women doctors are more compassionate, more understanding." Klass's informants did agree that "what would be taken as normal behavior in men"—showing temper in the operating room or treating nurses arrogantly—was "considered aggressive and obnoxious in women." If women

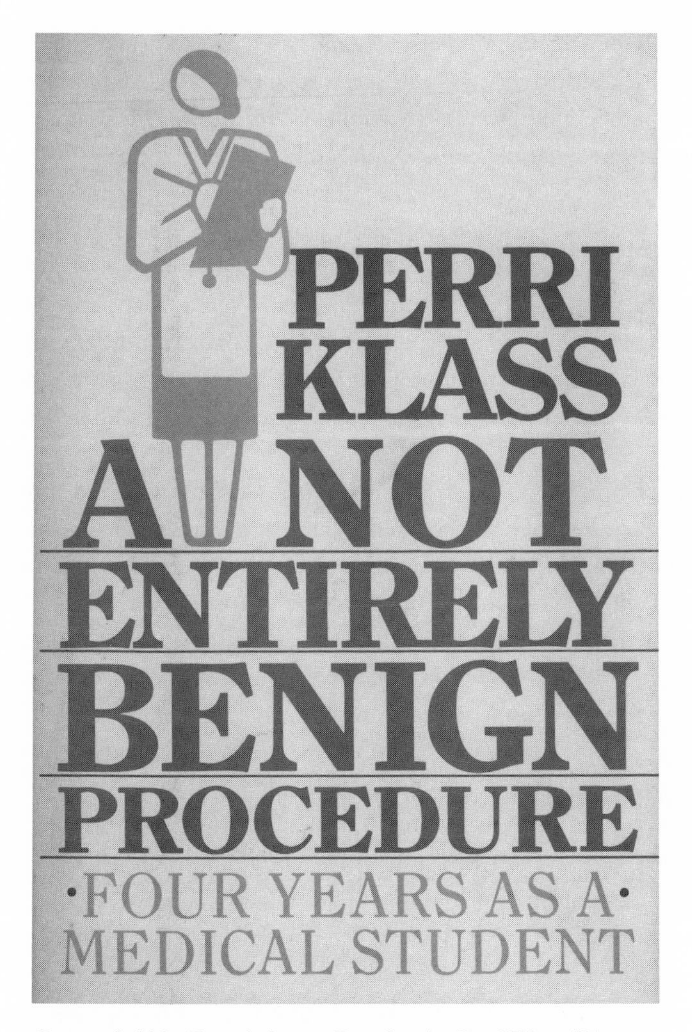

Cover of *A Not Entirely Benign Procedure* by Perri Klass (New York: Putnam, 1987). Illustration reproduced with permission from Paul Gamarello.

doctors were different and more humane, Klass concluded, then "the influx of women into medicine, we can hope, will help us to design medical careers that will enable all doctors to lead more integrated lives."[117]

Her article sparked a fury of discussion. One woman mocked "the well-publicized accomplishments of the ubiquitous Dr. Klass" and noted that "her name alone evoked a 20-second raspberry from a recent round table of seasoned careerists including several mothers."[118] Among a wider audience, though, her essay led many health professionals, educators, and potential and former patients, to reflect on the implications of a "feminized" medicine. One writer to the *New York Times* argued that the empathy question—whether it was a quality distinct to a single gender, innate, or the result of unconscious socialization—was "sexist and irrelevant," for "sensitivity, empathy and sincerity are not female qualities; they are human ones." Another writer argued that "the correct conclusion, I think, is that good *people* make better doctors."[119] New York pediatrician Ruth Kaminer was not sure whether women doctors practiced differently, but she did admit that "some residual prejudice against giving women power still exists" and that women physicians "have not acquired the habit of using power comfortably."[120]

Klass's comments on the trials of being a working wife and mother produced less fury. Few of her respondents felt that a woman who chose a medical career harmed her children or her marriage.[121] It was a powerful change from the frantic efforts by women physicians of an earlier generation to live up to notions of femininity and normality. Klass was even able to conclude her article with a Ramey-like joke about her efforts to soothe her four-year-old who had discovered that his new pediatrician was a man: "Benjamin, I have to tell you something . . . Boys can be doctors too, if they want to. If they go to school and learn how, boys can be very good doctors, really."[122]

When cultural feminism came to the citadel, it came with great force. If there was, as Perri Klass and others had suggested, a "feminization" of American medicine, was that healthy or dangerous? The women physicians of the 1940s and 1950s had been certain of the answer: There should be, they had said, "no sex in medicine." But they knew better than anyone how deeply and harshly gendered were the traditions and structure of the American medical profession. Their hopes that individual merit, perseverance, and quiescence would be rewarded with professional respect and recognition had so often been disappointed.

By the early 1980s, it was clear that, although affirmative action policies were formally accepted by medical schools, by deans across the country, and by the Association of American Medical Colleges, Title IX was an empty threat. Unlike its effect on high school and college sports programs, no federal funds were withheld from medical schools for sex discrimination. Legal pressures had spurred the creation of a larger academic bureaucracy with skilled administrators who had developed a safe, middle ground in response to complaints of sex discrimination. Indeed, many medical administrators and faculty had come to accept some of the demands of their feminist students but had integrated them in ways less disruptive to medical practice and pedagogy. There were now more women doctors as members of hospital staff and medical school teachers but still few in the higher echelons of medical power.[123] New ethics and "professionalism" classes sought to reinforce antiracist and antisexist values, and, even before federally mandated guidelines, many hospitals and universities established maternity leave policies and on-site childcare facilities. But it was not clear whether larger numbers of women health professionals had led to improved health care for women patients.[124]

In response to such ambivalent developments, new feminist voices such as Catherine McKinnon and Andrea Dworkin argued that the basic causes of sexual harassment and violence against women had to be addressed by special social and legal measures for which the vocabulary of equality and the denial of difference would not suffice. Their arguments about the special gendered qualities of women created a new feminist politics of difference. Simultaneously, feminist professionals drew attention to the discriminatory content and funding of medical research, and in the early 1990s, after pressure from both Congress and health advocacy groups, the National Institutes of Health and other federal agencies established new policies that demanded that the health of women and underserved minorities be made research priorities.[125] With federal research money at the heart of the economy of American medical schools, this renewed attention to women's health could certainly be considered an achievement of some of the feminist goals of 1970s activists. The blossoming of so-called women's health centers in major hospitals and medical centers, often using the insignia and slogans of the feminist health movement and proudly featuring their staff of women professionals, however, pointed to a disturbing co-optation of health feminism.[126]

Feminists and other progressive women entering medical schools in the 1970s hoped to transform training in ways that would allow both women and men students to liberate themselves from social stereotypes. This new medical culture would create and nurture nonsexist, empathetic men and women physicians who would treat their patients and research subjects with humanity and dignity.[127] Yet, as feminist critics of women physicians argued throughout the 1970s, creating a woman-friendly environment in medical schools and hospitals was supposed to be just the start of a revolution that would transform health care and the doctor-patient relationship and make gender as crucial as race and class in shaping progressive health care and research policies. The women who graduated from medical school in the 1970s are now senior members in the medical establishment, and some have indeed become, as Barbara Ehrenreich had urged them in 1976, "rebels and troublemakers." To listen again to their experiences is to be shocked, to be drawn into a world that needs to be recalled, and, in the telling of their stories, to learn new ways to interrogate the "boys in white."

NOTES

My thanks to the participants of the National Library of Medicine symposium on "Women Physicians, Women's Politics and Women's Health" and to the faculty and student members of the Yale Holmes Workshop where this paper was first presented; to Elizabeth Fee, Ellen S. More, and Manon Parry, editors of this volume, for their insight and patience; and especially to Gina Feldberg, Meg Hyre, and John Harley Warner.

Epigraphs. First epigraph quoted in Howard S. Becker, Blanche Geer, Everett C. Hughes, and Anselm L. Strauss, *Boys in White: Student Culture in Medical School* (Chicago: University of Chicago Press, 1961), 9; second epigraph quoted by Sonia Bauer, Letter to Editor, *New England Journal of Medicine*, 1974, *291*: 1142.

1. Donald Pollock, "Training Tales: U.S. Medical Autobiography," *Cultural Anthropology*, 1996, *11*: 339–361.

2. For jokes produced within the medical profession for other physicians, see, for example, *It Only Hurts When I Laugh: Cartoon Classics from Medical Economics Magazine* (Greenwich, CT: Fawcett, 1968).

3. Margaret A. Campbell [Mary A. Howell], *Why Would a Girl Go into Medicine? Medical Education in the United States: A Guide for Women* (Old Westbury, NY: Feminist Press, 1973), 41.

4. Florence Haseltine and Yvonne Yaw, *Woman Doctor* (Boston: Houghton Mifflin, 1976), 16. Phyllis Donnan was admitted to the internship only after the intervention of the hospital's chief of medicine, one of her former medical school professors.

5. Patricia Gerald Bourne and Norma Juliet Wikler, "Commitment and the Cultural Mandate: Women in Medicine," *Social Problems*, 1978, *25*: 432.

6. Oswald Hall, "The Stages of a Medical Career," *American Journal of Sociology*, 1948, 53: 327–336. See Robert Nye's Chapter 6 in this volume for a nuanced analysis of the strategies of exclusionary and aggressive practices among male physicians.

7. Helen Rodriguez-Trias, "The Women's Health Movement: Women Take Power," in *Reforming Medicine: Lessons of the Last Quarter Century*, ed. Victor Sidel and Ruth Sidel (New York: Pantheon Books, 1984), 113. My focus is on coeducational schools, so I am excluding the experiences of Woman's Medical College graduates; but see Steven J. Peitzman, *A New and Untried Course: Woman's Medical College and Medical College of Pennsylvania, 1850–1998* (New Brunswick, NJ: Rutgers University Press, 2000).

8. On the feminist health movement of the 1970s, see the essays by Susan Wells (Chapter 9) and Sandra Morgen (Chapter 7) in this volume; see also Sheryl Burt Ruzek, *The Women's Health Movement: Feminist Alternatives to Medical Control* (New York: Praeger, 1978); Ellen Frankfort, *Vaginal Politics* (New York: Quadrangle, 1972); Gina Corea, *The Hidden Malpractice: How American Medicine Treats Women as Patients and Professionals*

(New York: William Morrow, 1977); Sandra Morgen, *Into Our Own Hands: The Women's Health Movement in the United States, 1969–1990* (New Brunswick, NJ: Rutgers University Press, 2002); and Susan M. Reverby, "Thinking through the Body and the Body Politic: Feminism, History, and Health-Care Policy in the United States," in *Women, Health, and Nation: Canada and the United States since 1945*, ed. Georgina Feldberg, Molly Ladd-Taylor, Alison Li, and Kathryn McPherson (Montreal: McGill-Queen's University Press, 2003), 404–420.

9. Council on Graduate Medical Education, *Fifth Report: Women and Medicine* (Washington, D.C.: U.S. Dept of Health and Human Services, 1995), figures from table 14, "Female Medical School First-Year and Total Enrollment and Graduates, Selected Years," p. 33. In 1971, women were 9.6% of graduates; in 1981, 25% of graduates.

10. Here, I am drawing on work of Regina Morantz-Sanchez, Ellen S. More, and Judith Leavitt, especially Leavitt's expansive analysis of gender; see Judith Walzer Leavitt, "'A Worrying Profession': The Domestic Environment of Medical Practice in Mid-Nineteenth-Century America," *Bulletin of the History of Medicine*, 1995, 69: 1–29.

11. Howard S. Becker, Blanche Geer, Everett C. Hughes, and Anselm L. Strauss, *Boys in White: Student Culture in Medical School* (Chicago: University of Chicago Press, 1961), 103–104. Unfortunately, the researchers did not list the novels they read.

12. Ibid., 103. For a critical analysis by a feminist sociologist of this and other studies of medical school socialization that left women out, see Judith Lorber, "Women and Medical Sociology: Invisible Professionals and Ubiquitous Patients," in *Another Voice: Feminist Perspectives on Social Life and Social Science*, ed. Marcia Millman and Rosabeth Moss Kanter (Garden City, NY: Anchor Books, 1975), 75–105.

13. Ibid., 104–105, for a discussion of jokes made around schoolwork.

14. Ibid., 324. For more on jokes and humor, see Mahadev L. Apte, *Humor and Laughter: An Anthropological Approach* (Ithaca, NY: Cornell University Press, 1985); Jerry Palmer, *Taking Humor Seriously* (New York: Routledge, 1994); and especially Susan E. Lederer, "'Porto Ricochet': Joking about Germs, Cancer, and Race Extermination in the 1930s," *American Literary History*, 2002, *14*: 720–746.

15. See, for example, Kenneth M. Ludmerer, *Time to Heal: American Medical Education from the Turn of the Century to the Era of Managed Care* (New York: Oxford University Press, 1999); Peitzman, *A New and Untried Course*; Dan A. Oren, *Joining the Club: A History of Jews and Yale* (New Haven, CT: Yale University Press, 1985); and Gert H. Brieger, "Getting into Medical School in the Good Old Days: Good for Whom?" *Annals of Internal Medicine*, 1993, *119*: 1138–1143.

16. For reflections on gender and Cold War sociology, see Steven C. Martin, Robert M. Arnold, and Ruth M. Parker, "Gender and Medical Socialization," *Journal of Health and Social Behavior*, 1988, *29*: 333–343; Samuel W. Bloom, *The Word as Scalpel: A History of Medical Sociology* (New York: Oxford University Press, 2002); and especially Judith Lorber, *Women Physicians: Careers, Status, and Power* (New York: Tavistock, 1984).

17. Becker et al., *Boys in White*, table 2, p. 61; 12 out of 62 fathers were doctors, and 32 out of 62 fathers were professional men.

18. Ibid., 60.

19. "Marjorie Sirridge: Latecomer to Academic Medicine," in *In Her Own Words: Oral Histories of Women Physicians*, ed. Regina Markell Morantz, Cynthia Stodola Pomerleau, and Carol Hansen Fenichel (New Haven, CT: Yale University Press, 1982), 157. Later, she became one of his prize pupils and found that "he never really told such sto-

ries." "I suspect he felt he was expected to say something about the inappropriateness of women in medical school" (p. 157). On local law societies whose members refused to admit women because "it's not the same with a woman around. They aren't free to express themselves, to tell off-color stories," see Cynthia F. Epstein, "Encountering the Male Establishment: Sex-Status Limits on Women's Careers in the Profession," *American Journal of Sociology*, 1970, 75: 977.

20. Ludmerer, *Time to Heal*, 75.

21. Chloral hydrate is a hypnotic and a sedative. Warren Bennett, Class of 1947, in Nora N. Nercessian, *Worthy of Honor: A Brief History of Women at Harvard Medical School* (Boston: Committee on the Celebration of 50 Years of Women at Harvard Medical School, 1995), 72. Bennett later married one of these women students.

22. See, for example, Nercessian, *Worthy of Honor*, 63, 75.

23. Chester d'Autremont, Class of 1944, in Nercessian, *Worthy of Honor*, 63. D'Autremont felt such slides were "unwarranted, unnecessary" (p. 63). For further examples in the 1960s and 1970s, see Corea, *Hidden Malpractice*, 31–44.

24. Ludmerer, *Time to Heal*, 196–204.

25. *The Crisis in Medical Services and Medical Education* (1966), a report sponsored by the Commonwealth Fund and Carnegie Corporation, quoted in Carol Lopate, *Women in Medicine* (Baltimore: Johns Hopkins Press for the Josiah Macy, Jr. Foundation, 1968), 62.

26. "Marjorie Wilson: Career Administrator," in Morantz et al., *In Her Own Words*, 179.

27. On the Health New Left, see Naomi Rogers, "'Caution: The AMA May Be Dangerous to Your Health': The Student Health Organizations (SHO) and American Medicine, 1965–1970," *Radical History Review*, 2001, *80*: 5–34; and Lily M. Hoffman, *The Politics of Knowledge: Activist Movements in Medicine and Planning* (Albany: State University of New York Press, 1989).

28. For two influential critiques of medical socialization, see Samuel W. Bloom, *Power and Dissent in the Medical School* (New York: Free Press, 1973), and Robert H. Coombs, *Mastering Medicine: Professional Socialization in Medical School* (New York: Free Press, 1978).

29. See Rogers, "'Caution: The AMA May Be Dangerous to Your Health,'" and Ludmerer, *Time to Heal*, 249–256.

30. On the few occasions when nineteenth-century women doctors considered cross-dressing (to pass as a male physician or student), see John Harley Warner, *Against the Spirit of System: The French Impulse in Nineteenth-Century American Medicine* (Princeton, NJ: Princeton University Press, 1998), 49–53.

31. Lopate, *Women in Medicine*, 82. For similar examples, see Morantz-Sanchez, *Sympathy and Science*; Marjorie Sirridge and Brenda R. Pfannenstiel, "Daughters of Aesculapius: A Selected Bibliography of Autobiographies of Women Medical School Graduates, 1849–1920," *Literature and Medicine*, 1996, *15*: 200–216; and Arleen Marcia Tuchman, *Science Has No Sex: The Life of Marie Zakrzewska, M.D.* (Chapel Hill: University of North Carolina Press, 2006).

32. Abraham Flexner, *Report on Medical Education in the United States and Canada: A Report to the Carnegie Foundation for the Advancement of Teaching* (New York: Carnegie Foundation, 1910), 161. This quote was in reference to "sectarian" medical training. On the question of race in medicine, however, Flexner conceded that black communi-

ties might continue to need physicians of color. On Flexner and women physicians, see Thomas Neville Bonner, *Iconoclast: Abraham Flexner and a Life in Learning* (Baltimore: Johns Hopkins University Press, 2002), 191–221.

33. See Ellen S. More and Maureen A. Milligan, eds., *The Empathic Practitioner: Empathy, Gender, and Medicine* (New Brunswick, NJ: Rutgers University Press, 1994). On science, masculinity, and women physicians in earlier eras, see chapters by Regina Morantz-Sanchez (Chapter 3) and Carla Bittel (Chapter 1) in this volume.

34. Unnamed medical dean, cited in Walsh, *Doctors Wanted*, 243–244.

35. Lilian Welsh, *Reminiscences of Thirty Years in Baltimore* (Baltimore: Norman, Remington, 1925), 46, 37. On harassment experienced by the early classes of women at Johns Hopkins, see Michael Bliss, *William Osler: A Life in Medicine* (New York: Oxford University Press, 1999), 232–234.

36. Lopate, *Women in Medicine*, 26.

37. "Molly McCoy, M.D.," in *Life and Death: The Story of a Hospital*, ed. Ina Yalof (New York: Fawcett Crest, 1988), 122–127, quotation p. 123. On Harvard students during the 1940s holding doors and helping women students on with coats, see Edith Schwartz Taylor, in Nercessian, *Worthy of Honor*, 49.

38. Sally Knapp, *Women Doctors Today* (New York: Thomas Y. Crowell, 1947), 22, 30–31.

39. Ibid., 59, 62. Edwards's daughter Marie Metoyer graduated from Cornell's medical school and became a psychiatrist.

40. Virginia May Alexander, "Can a Colored Woman Be a Physician?" *Crisis*, 1933, quoted in Wilbur H. Watson, *Against the Odds: Blacks in the Profession of Medicine in the United States* (New Brunswick, NJ: Transaction Publishers, 1999), 51.

41. Knapp, *Women Doctors Today*, 179, 180. "She even had a tuxedo shirt and black suit for medical conferences." On crucial issues of masculinity, heteronormality, and medical professionalism, see Judy Tzu-Chun Wu's essay (Chapter 4 in this volume).

42. Lopate, *Women in Medicine*, 73.

43. Peggy Chambers, *A Doctor Alone: A Biography of Elizabeth Blackwell, the First Woman Doctor, 1821–1910* (London: Bodley Head, 1956); see also Dorothy Clarke Wilson, *Lone Woman: The Story of Elizabeth Blackwell, the First Woman Doctor* (New York: Little, Brown, 1970).

44. George Moon, *How to Become a Doctor* (Philadelphia: Blakiston, 1949). See also a 1949 poll of 100 hospital chiefs of staff whose comments included "if she is married and childless she is frustrated" and "if she raises a family she is neglecting her practice," cited in Walsh, *Doctors Wanted*, 245.

45. On gender and the Cold War, see Elaine Tyler May, *Homeward Bound: American Families in the Cold War Era* (New York: Basic Books, 1988); Joanne Myerowitz, *How Sex Changed: A History of Transsexuality in the United States* (Cambridge, MA: Harvard University Press, 2002); and David K. Johnson, *The Lavender Scare: The Cold War Persecution of Gays and Lesbians in the Federal Government* (Chicago: University of Chicago Press, 2005).

46. Jacqueline Seaver, "Women Doctors in Spite of Everything," *New York Times*, March 26, 1961.

47. Lopate, *Women in Medicine*, 111.

48. Mabel Gardner to Catherine MacFarlane, quoted in Morantz-Sanchez, *Sympathy and Science*, 338; Lopate, *Women in Medicine*, 20. For the argument that women physicians

in the early twentieth century "fought unabashedly for the rights of that exceptional minority," yet were suspicious of "militant separatism," see Regina Morantz-Sanchez, "So Honored, So Loved? The Women's Medical Movement in Decline," in *"Send Us a Lady Physician": Women Doctors in America, 1835–1920*, ed. Ruth J. Abram (New York: Norton, 1985), 235, 236.

49. More, *Restoring the Balance*, 192.

50. Ishbel Ross, *Child of Destiny: The Life Story of the First Woman Doctor* (New York: Harper & Brothers, 1949), 2, 209. Later, "driven by loneliness," Blackwell adopts a child (p. 186).

51. Lorraine Bennett, "Women in Medicine: Rx for Pain," *Los Angeles Times*, February 19, 1976. Carlin was then the associate dean of the University of California at Irvine's College of Medicine.

52. For the analogies of boys being given "Dr. Dan, the Bandage Man" and girls, a nurse doll, see Walsh, *Doctors Wanted*, 246–247.

53. See More, *Restoring the Balance*, 217–218; Carolyn S. Pincock, "New Horizons for the American Medical Women's Association," *Journal of the American Medical Women's Association*, 1975, *30*: 10; and see Deborah Shapely, "Medical Education: Those Sexist Putdowns May Be Illegal," *Science*, 1974, *184*: 450. Nixon's Executive Order No. 11478 issued in 1969 had urged federal agencies "to establish and maintain an affirmative action program of equal employment opportunity for all civilian employees," and during his administration the U.S. Equal Employment Opportunity Commission conducted the first compliance reviews of hiring policies toward women by institutions of higher education receiving federal grants.

54. Mary Howell, "What Medical Schools Teach about Women," *New England Journal of Medicine*, 1974, *291*: 304; More, *Restoring the Balance*, 218–219.

55. At some elite medical schools, more women were admitted: at Harvard in 1971, there were 11 women out of 139 students (almost 8%), and in 1974, 55 women in a class of 185 (one-third); Evan Jenkins, "Women in Medicine Up Sharply," *New York Times*, July 17, 1974.

56. "Susan Benes: Ophthalmologist in Training," in Morantz et al., *In Her Own Words*, 233–234.

57. Quoted in Robert S. Broadhead, *The Private Lives and Professional Identity of Medical Students* (New Brunswick, NJ: Transaction Books, 1983), 72.

58. Helen I. Marieskind and Barbara Ehrenreich, "Toward Socialist Medicine: The Women's Health Movement," *Social Policy*, 1975, *6*: 39; and for one example of a feminist antiphysician guide, see the Federation of Feminist Women's Health Centers, *How to Stay Out of the Gynecologist's Office* (Culver City, CA: Woman to Woman Publications, 1981). On tensions within health feminism, see Reverby, "Thinking through the Body and the Body Politic," 404–420; and Ruzek, *The Women's Health Movement*, 108–124.

59. Ruzek, *The Women's Health Movement*, 163.

60. See Barbara Seaman, *The Doctor's Case against the Pill* (New York: Avon, 1969), 15–16; and see Elizabeth Watkins, *On the Pill: A Social History of Oral Contraceptives, 1950–1970* (Baltimore: Johns Hopkins University Press, 1998).

61. Boston Women's Health Course Collective, *Our Bodies, Our Selves: A Course for and by Women* (Boston: New England Free Press, 1971), 128; and see Ruzek, *The Women's Health Movement*, 87. For a close analysis of *Our Bodies, Ourselves* as feminist narrative, see Susan Wells's essay in Chapter 8 in this volume.

62. Lopate, *Women in Medicine*, 83.

63. Georgia Dullea, "More Women Now in Law and Medicine," *New York Times*, January 15, 1975. On the widespread ignorance among women students about affirmative action plans or policies, see Campbell, *Why Would a Girl?* 19–20. On the continuing reluctance for physicians of color to tell administrators or peers about their experiences of racial harassment, see Vanessa Northington Gamble, "Subcutaneous Scars: A Black Physician Shares What It Feels Like to Be on the Received End of Racial Prejudice, Despite a Successful Career" [2000], in *Women, Health, and Nation*, Feldberg et al., 305–310.

64. D. X. Fenton, *Ms.-M.D.* (Philadelphia: Westminster Press, 1973), 20.

65. Esther Jacobowitz, letter to editor, *New Physician*, 1973, *22*: 606. See also Elizabeth R. McAnarney, "The Impact of Medical Women in United States Medical Schools," in *Women in Medicine 1976: Report of a Macy Conference*, ed. Carolyn Spieler (New York: Josiah Macy, 1977): "A female applicant in the mid-1970s, faced with an intrusive comment or question, would be more likely to tell the interviewee she thought his remark was inappropriate than her early counterpart," and was "less likely to allow offensive attitudes to go unheeded" (p. 19).

66. More, *Restoring the Balance*, 217–218; and Ludmerer, *Time to Heal*, 256–259.

67. *Access to the Medical Profession in Colorado by Minorities and Women: Colorado Advisory Committee to the U.S. Commission on Civil Rights* (Washington, D.C.: U.S. Commission on Civil Rights, 1976), 16. One in five women reported that they had been advised to give up career ambitions altogether and devote themselves to raising a family.

68. Ibid., 28.

69. Lorber, "Women and Medical Sociology," 89; Patricia H. Beshiri, *The Woman Doctor: Her Career in Modern Medicine* (New York: Cowles Book Company, 1969), 119; Lopate, *Women in Medicine*, 110, 83.

70. For one feminist response to humor as harassment, see Naomi Weisstein, "Why We Aren't Laughing . . . Anymore," *Ms*, 1973 (November): 49–51; and see Nancy A. Walker, *A Very Serious Thing: Women's Humor and American Culture* (Minneapolis: University of Minnesota Press, 1988), on "the freedom to laugh at oneself and also make ridiculous the cultural sources of women's oppression" (p. 180).

71. Campbell, *Why Would a Girl?* 23.

72. Ibid., 25.

73. Ibid., 28. For the conflicting evidence around this idea, see More, *Restoring the Balance*; Lopate, *Women in Medicine*; and McAnarney, "The Impact of Medical Women," 23.

74. Jenkins, "Women in Medicine Up Sharply."

75. Shapely, "Medical Education," 449–451. On note-taking, see Dullea, "More Women Now in Law and Medicine."

76. Gloria Kaufman and Mary Kay Blakely, eds., *Pulling Our Own Strings: Feminist Humor and Satire* (Bloomington: Indiana University Press, 1980); see also Gloria Kaufman, ed., *In Stitches: A Patchwork of Feminist Humor and Satire* (Bloomington: Indiana University Press, 1991); Deanne Stillman and Anne Beatts, eds., *Titters: The First Collection of Humor by Women* (New York: Macmillan, 1976). For examples of the explosion of attention to women not as the objects but as the instigators of humor, see Regina Barreca, ed., *Last Laughs: Perspectives on Women and Comedy* (New York: Gordon and Breach, 1988); Lisa Merrill, "Feminist Humor: Rebellious and Self-Affirming," *Women's Studies*, 1988, *15*: 271–280; Walker, *A Very Serious Thing*; Nancy Walker and Zita Dresner, *Re-*

dressing the Balance: American Women's Literary Humor from Colonial Times to the 1980s (Jackson: University Press of Mississippi, 1988); Alice Sheppard, "From Kate Sanborn to Feminist Psychology: The Social Context of Women's Humor, 1885–1985," *Psychology of Women Quarterly*, 1986, *10*: 155–170; June Sochen, ed., *Women's Comic Visions* (Detroit: Wayne State University Press, 1991); and Linda Morris, ed., *American Women Humorists: Critical Essays* (New York: Garland, 1994).

77. Ramey, quoted in Corea, *The Hidden Malpractice*, 84; and see Lee Katterman, "AWIS, Marking Its 25th Anniversary, Eyes Changed but Unfinished Tasks," *The Scientist*, 1995, *9*: 8; Patricia Sullivan, "Estelle R. Ramey: Used Wit in Women's Advocacy," *Washington Post*, September 10, 2006; and "Estelle R. Ramey, 89; Used Medical Training to Rebut Sexism," *New York Times*, September 12, 2006. For a revealing debate between male and female medical students and young physicians over this text in the progressive *New Physician*, see letters, *New Physician*, 1972, *21*: 349–350.

78. Estelle Ramey, "An Interview with Dr. Estelle Ramey," *Perspectives in Biology and Medicine*, 1971, *15*: 424–431, quotation p. 424.

79. Ibid., 427.

80. Ibid., 428.

81. Campbell, *Why Would a Girl?* 56.

82. On Howell, see Ruzek, *The Women's Health Movement*, 84–87; More, *Restoring the Balance*, 232–235; Pauline B. Bart, "Mary Howell (1932–1998)," *Off Our Backs*, 1998, *25*: 15; Wolfgang Saxon, "Mary Howell, a Leader in Medicine, Dies at 65," *New York Times*, February 6, 1998; and Sandra Morgen, Chapter 7 in this volume.

83. Shapely, "Medical Education,"449.

84. Campbell, *Why Would a Girl?* 1. One of the blurbs on the cover of the second edition was by Norma Swenson, identified as "co-author, *Our Bodies, Ourselves*": "For every woman who has ever doubted that male physicians are systematically conditioned to feel contempt for women—as patients and as doctors—here then is the unpleasant truth from an authoritative source."

85. Campbell, *Why Would a Girl?* 3. On the argument comparing medical school to a men's club, see Howell, "What Medical Schools Teach about Women," 306; and see Bourne and Wikler, "Commitment and the Cultural Mandate," 430–431.

86. Campbell, *Why Would a Girl?* 37, 39, 41.

87. Shapely, "Medical Education," 450, 451.

88. Campbell, *Why Would a Girl?* 3.

89. Howell, "What Medical Schools Teach about Women," 304, 306. Howell speculated that most "medical humor" was the result of the institutionalized "stoic repression of anxiety"; "What Medical Schools Teach," 306.

90. Howell, "What Medical Schools Teach about Women," 305.

91. See, for example, Janet B. Hardy, letter to the editor, *New England Journal of Medicine*, 1974, *291*: 1140; and "Correspondence: Women in Medicine," *New England Journal of Medicine*, 1974, *291*: 1138–1142.

92. Howell, response, *New England Journal of Medicine*, 1974, *291*: 1141. On descriptions of letters of recommendation for internships, see Sonia Bauer, letter to editor, *New England Journal of Medicine*, 1974, *291*: 1141–1142: "Some were grossly sexist," such as "her intelligence and aggressiveness are such that she stands out among, but does not antagonize, her male colleagues" and "she was warm and friendly, has a nice smile, and I

found it a pleasant hour even though I felt that I was dealing with a mind that has twice the capacity of mine" (Bauer italicized the words "even though").

93. Franz J. Ingelfinger, "Doctor Women," *New England Journal of Medicine*, 1974, *291*: 304. For one response, see Barbara H. Roberts (a physician at the Peter Bent Brigham Hospital), letter to the editor, *New England Journal of Medicine*, 1974, *291*: 1141, on the editor's "blatant sexism." The piece was "unbelievably disparaging" and "I seriously doubt if any report on a panel of male physicians would include comments on their physical attractiveness, dress, or body build, not to mention their marital status."

94. Mary C. Howell, "A Women's Health School?" *Social Policy*, 1975, *6*: 50, 53.

95. Howell quoted in More, *Restoring the Balance*, 234; see also Corea, *Hidden Malpractice*, 33–35; Morgen, *Into Her Own Hands*, 133–145.

96. Mary C. Howell, "An Open Letter to the Woman's Health Movement," quoted in Walsh, *Doctors Wanted*, 281, and Corea, *Hidden Malpractice*, 42.

97. See, for example, Carola Eisenberg, "Medicine Is No Longer a Man's Profession, or, When the Men's Club Goes Coed It's Time to Change the Regs" (1989), reprinted in *The Social Medicine Reader*, ed. Gail Henderson et al. (Durham, NC: Duke University Press, 1997), 266–270; and Judith Lorber, "Why Women Physicians Will Never Be True Equals in the American Medical Profession," in *Gender, Work and Medicine: Women and the Medical Division of Labour*, ed. Elianne Riska and Katarina Wegar (London: Sage Publications, 1993), 62–76.

98. Broadhead, *Private Lives and Professional Identity*, 97.

99. "Vanessa Gamble: Tomorrow's Physician, Tomorrow's Policy Maker," in Morantz et al., *In Her Own Words*, 214–215.

100. Broadhead, *Private Lives and Professional Identity*, 93. On women who demanded that their teachers stop addressing classes as "gentlemen" or including nude "girlie pictures" among lecture slides, see Jenkins, "Women in Medicine Up Sharply." At UCLA's medical school, the last slide in a lecture was sometimes "a picture of a naked female rump with 'The End' across it"; Corea, *Hidden Malpractice*, 83.

101. Campbell, *Why Would a Girl?* On the distinctive history of health feminists of color, see, for example, Jennifer Nelson, *Women of Color and the Reproductive Rights Movement* (New York: New York University Press, 2003); and Jael Silliman et al., *Undivided Rights: Women of Color Organize for Reproductive Justice* (Cambridge, MA: South End Press, 2004).

102. On reports by the mid-1970s on improvement in fewer intrusive admission questions and less "chauvinistic behavior" in clinical rotations, see McAnarney, "The Impact of Medical Women," 19, 29.

103. Beshiri, *The Woman Doctor*, 75.

104. Mary C. Howell, "The New Feminism and the Medical School Milieu," *Annals of the New York Academy of Sciences*, 1979, *323*: 210–214.

105. Eisenberg, "Medicine Is No Longer a Man's Profession," 269.

106. Howell, "A Women's Health School?" 53; see also Sandra Morgen's essay, Chapter 7 in this volume.

107. Boston Women's Health Book Collective, *Our Bodies, Ourselves* (New York: Simon and Schuster, 1976), 341.

108. New England Branch of the American Medical Women's Association, flyer (1975) cited by Walsh, *Doctors Wanted*, 270, 272.

109. Bourne and Wikler, "Commitment and the Cultural Mandate," 431.

110. Rodriguez-Trias, "Women's Health Movement," 125. See Joyce Wilcox, "The Face of Women's Health: Helen Rodriguez-Trias," *American Journal of Public Health*, 2002, *92*: 566–569; and Morgen's essay, Chapter 7 in this volume.

111. Howell, "The New Feminism," 212. For a discussion of the backlash against health feminism in the 1980s and 1990s, see Morgen, *Into Our Own Hands*, 145–205; and Sue V. Rosser, *Women's Health: Missing From U.S. Medicine* (Bloomington: Indiana University Press, 1994), 161–174.

112. Deborah L. Jones, "Father Knows Best?" in *Women in Medical Education: An Anthology of Experience*, ed. Delese Wear (Albany: SUNY Press, 1996), 66.

113. Howell, "A Women's Health School?" 51.

114. Bennett, "Women in Medicine."

115. Howell, "The New Feminism," 212.

116. Perri Klass, *A Not Entirely Benign Procedure: Four Years as a Medical Student* (New York: Putnam, 1987).

117. Perri Klass, "Are Women Better Doctors?" *The New York Times Magazine*, April 10, 1988. She began at Harvard in 1982, when there were 53 women in a class of 165, and graduated in 1986.

118. Polly Flounder, letter to the editor, *New York Times*, May 3, 1988.

119. David M. Lans, letter to the editor, *New York Times*, May 8, 1988; Jeff Ecker, letter to editor, *New York Times*, May 9, 1988. For an important analysis of gender and empathy, see More and Milligan, *The Empathic Practitioner*.

120. Ruth Kahan Kaminer, letter to the editor, *New York Times*, May 9, 1988.

121. On the argument that autobiographies by women physicians published in the early 1980s less often focused on "patriarchy in medicine" and that "the genre has become more or less focused on the challenges of work," see Pollock, "Training Tales," 344.

122. Klass, "Are Women Better Doctors?"

123. See Janet Bickel, "Women in Medical Education: A Status Report," *New England Journal of Medicine*, 1988, *319*: 1579–1584; Sheryl Burt Ruzek and Julie Becker, "The Women's Health Movement in the United States: From Grass-Roots Activism to Professional Agendas," *Journal of the American Medical Women's Association*, 1999, *54*: 4–8, 40; and Susan W. Hinze, "'Am I Being Over-Sensitive?' Women's Experience of Sexual Harassment during Medical Training," *Health*, 2004, *8*: 101–127.

124. For important examples of the blossoming analyses of gender, health care, and the medical profession in the 1990s, see Delese Wear, *Privilege in the Medical Academy: A Feminist Examines Gender, Race, and Power* (New York: Teachers College Press, 1997); Kate L. Moss, ed., *Man-Made Medicine: Women's Health, Public Policy, and Reform* (Durham, NC: Duke University Press, 1996); Sue V. Rosser, *Women's Health—Missing from U.S. Medicine* (Bloomington: Indiana University Press, 1994); Susan Sherwin, et al., *The Politics of Women's Health: Exploring Agency and Autonomy* (Philadelphia: Temple University Press, 1998); Sheryl Burt Ruzek, Virginia L. Olesen, and Adele E. Clarke, eds., *Women's Health: Complexities and Differences* (Columbus: Ohio State University Press, 1997); and Rosemary Pringle, *Sex and Medicine: Gender, Power and Authority in the Medical Profession* (Cambridge: Cambridge University Press, 1998).

125. Florence P. Haseltine and Beverly Greenberg Jacobson, eds., *Women's Health Research: A Medical and Policy Primer* (Washington, D.C.: Health Press International, 1997); Adele E. Clarke and Virginia L. Olesen, eds., *Revisioning Women, Health and Heal-*

ing: Feminist, Cultural, and Technoscience Perspectives (New York: Routledge, 1999); Carol S. Weisman, *Women's Health Care: Activist Traditions and Institutional Change* (Baltimore: Johns Hopkins University Press, 1998); and Beryl Lieff Benderly, *In Her Own Right: The Institute of Medicine's Guide to Women's Health Issues* (Washington, D.C.: National Academy Press, 1997).

126. Morgen, *Into Our Own Hands*; Susan E. Bell and Susan M. Reverby, "Vaginal Politics: Tensions and Possibilities in *The Vagina Monologues*," *Women's Studies International Forum*, 2005, *28*: 430–444.

127. Nursing schools, similarly, saw an onslaught of feminist students—both women and a small number of men—who envisioned new models for the nurse-doctor, nurse-patient relationships; see Corea, *Hidden Malpractice*, 63–72.

Part III / Expanding the Boundaries

CHAPTER TEN

Women Physicians and Medical Sects in Nineteenth-Century Chicago

Eve Fine

In 1891, Lucy Waite, an 1883 graduate of Chicago's Hahnemann Medical College, published a novel titled *Doctor Helen Rand*. Writing under the pseudonym of "Lois Wright," Waite told of two women physicians who lived together. Helen Rand was a regular physician. Kate Summerville was a homeopath. Though Waite's novel was not about the medical practices of these two women, she did provide an explanation for how these practitioners from opposing schools were able to live and work together amicably. She did so by having another character, Rand's long-lost brother, ask them how they could "work together in harmony and belong to opposite schools." Summerville explained, "Our amiability . . . is all due to our ignorance . . . After long and exhaustive discussions, we found that our education in all branches of medicine had been the same with the exception of therapeutics, and we effected a compromise on that point. Helen acknowledged that my drugs did not always kill and that her's [*sic*] did not always cure, and upon that we raised the flag of truce."[1]

This fictional account reflects one of the major themes of this chapter: recognition that homeopathic and regular women physicians could and did establish close friendships and collaborative relationships. As historians, we have not examined interactions between regular and sectarian women physicians. Indeed, we have not focused much attention on the history of sectarian women physicians. As Naomi Rogers noted in 1990, "sectarian women have been all but invisible to historians."[2] With the exception of Anne Taylor Kirschmann, few historians have answered Rogers's call to examine the history of sectarian women.[3] Our understanding of the history of women physicians in the nineteenth century is shaped primarily by research on regular women doctors.[4] We consequently view the nineteenth century as a period characterized by separatism in medicine—as a time when women, barred from entering most regular male medical institutions, acquired medical training and experience by establishing their own separate, all-female medical schools and hospitals.

In Chicago, as in many other major U.S. cities, the founding of a women's medical school in 1871 provided women with the opportunity to become physicians. Women in Chicago, however, had several additional options for becoming physicians. Like many of their male counterparts, some women simply entered practice without medical degrees as no requirements for entering medical practice existed in Illinois until 1878.[5] Substantially more women graduated from coeducational sectarian schools. Recognizing that significant numbers of nineteenth-century women physicians were sectarians, some historians have recently acknowledged that sectarian medical schools helped provide women with access to medical education before women's medical schools existed, but they continue to regard separatist women's medical colleges as primarily responsible for the education of women physicians in the nineteenth century.[6] For example, Thomas Bonner wrote, "more important in the long run than the sectarian schools in accounting for the growth and acceptance of women doctors in the United States was the drive to establish regular schools of medicine for women."[7] Similarly, Ellen S. More points out that "as late as 1917 a high percentage of the women physicians who had been practicing more than twenty years were graduates of sectarian medical colleges" but later concludes that "until nearly the end of the nineteenth century . . . the majority of women physicians still received their medical education in all-female colleges."[8] The majority of sectarian women physicians, however, received their medical education in coeducational schools.[9]

This chapter considers the history of the entire community of women doctors, both regular and sectarian, practicing in Chicago. It examines women's varied options for entering medical practice, explores the reasons for the choices they made, and provides evidence showing that throughout the nineteenth century

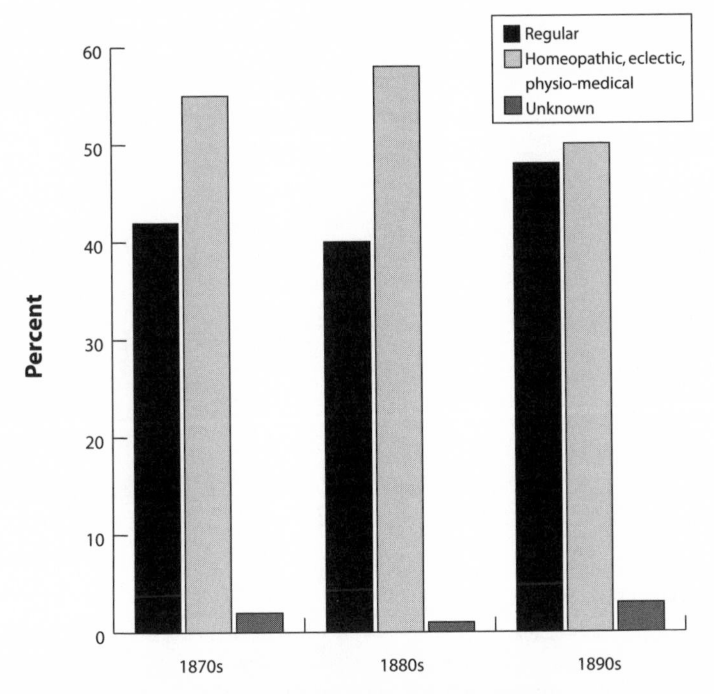

Registered Chicago women physicians: percentage regular and sectarian practitioners, 1870–1900.

Sources: Figures and percentages hand-counted from the Illinois State Board of Health, *Official Register of Physicians and Midwives* (Springfield: Weber & Co., State Printer, 1880); Illinois State Board of Health, *3rd and 4th Annual Report*, 1881; Illinois State Board of Health, *6th Annual Report*, 1884; Illinois State Board of Health, *Official Register of Physicians and Midwives* (Springfield: H. W. Bokker, State Printer, 1886); Illinois State Board of Health, *10th Annual Report*, 1890; Illinois State Board of Health, *11th Annual Report*, 1892; Illinois State Board of Health, *Official Register of Legally Qualified Physicians*, 1903.

the majority of Chicago's women doctors were sectarian practitioners, primarily homeopaths, who trained in coeducational schools. This finding is consistent with Anne Kirschmann's work showing that "at different points in time and in particular locations, women homeopaths were equal to or more numerous than women regular practitioners."[10] This chapter calls into question our focus on separatist women's medical colleges as largely responsible for the education of women doctors in the nineteenth century and suggests that coeducational sectarian medical schools played an even more significant role in the education of

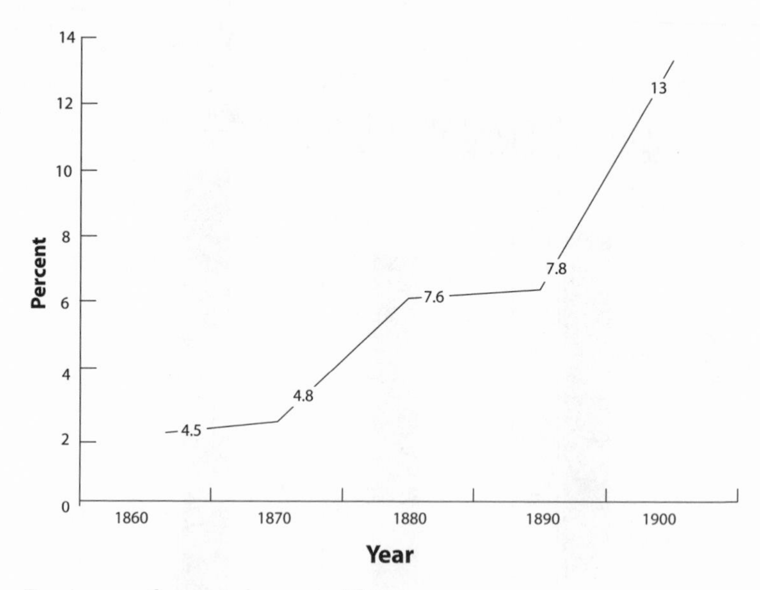

Percentage of women doctors in Chicago.

Sources: For 1860—Deborah L. Haines, "City Doctor City Lawyer: The Learned Professions in Chicago, 1833–1860" (Ph.D. diss., University of Chicago, 1986). Percentages are based on figures from Haines's hand-counting from the 1860 U.S. Census. For 1870—Percentages are based on figures hand-counted from the *Edwards Directory of the City of Chicago*, 1870. For 1880–1890, percentages are from Walsh, *Doctors Wanted: No Women Need Apply* (New Haven, CT: Yale University Press, 1977), 185. Walsh's percentages are based on figures from the U.S. Census.

women physicians than we have recognized thus far. As the fictional account of Helen Rand and Kate Summerville suggests, the chapter also shows that sectarian and regular women physicians interacted with one another in both social and medical settings. Finally, I consider how these collaborations influence our understanding of the history of women in medicine.

Women's Options for Medical Education in Chicago

Women physicians, graduates of both homeopathic schools and regular women's schools, began practicing medicine in Chicago well before 1870. The numbers and percentages of women physicians practicing in Chicago rose substantially when the city began producing its own women practitioners. Throughout the nineteenth century, most women practitioners in Chicago graduated from Chicago medical schools. Women in Chicago first gained the opportunity to obtain

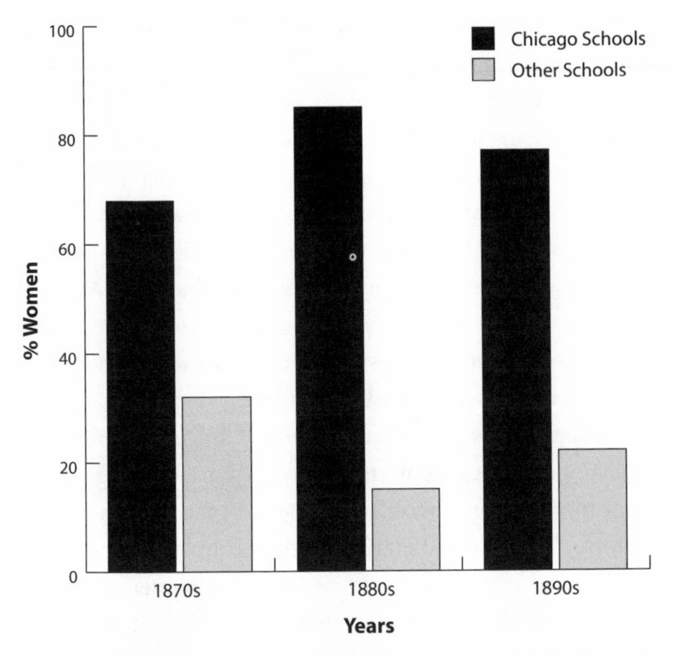

Registered Chicago women physicians' educational background by geographical location.

Sources: Figures and percentages hand-counted from the Illinois State Board of Health, *Official Register of Physicians and Midwives* (Springfield: Weber & Co., State Printer, 1880); Illinois State Board of Health, *3rd and 4th Annual Report*, 1881; Illinois State Board of Health, *6th Annual Report*, 1884; Illinois State Board of Health, *Official Register of Physicians and Midwives* (Springfield: H. W. Bokker, State Printer, 1886); Illinois State Board of Health, *10th Annual Report*, 1890; Illinois State Board of Health, *11th Annual Report*, 1892; Illinois State Board of Health, *Official Register of Legally Qualified Physicians*, 1903.

medical education when the Bennett College of Eclectic Medicine opened as a coeducational institution in 1869.[11] Shortly thereafter, Chicago's Hahnemann Medical College, followed by the regular Chicago Medical College, opened to women students.[12] Mary Harris Thompson, an 1863 graduate of the New England Female Medical College who subsequently settled in Chicago and established the Chicago Hospital for Women and Children in 1865, was among the small group of women admitted to the Chicago Medical College in 1869. After only one year, however, the Chicago Medical College discontinued its "experiment" in coeducation.[13] Frustrated and angered by the abrupt closing of the Chicago Medical College to women, Thompson, together with supportive male physicians, including William H. Byford, a faculty member of the Chicago Medical College, established the Chicago Woman's Medical College as a regular school of medicine for the exclusive benefit of women students.[14]

In 1874, the Hahnemann Medical College also reversed its policy of admitting women students.[15] Laywomen members of Hahnemann Hospital's Ladies' Aid Society vocally protested this action and threatened to withdraw essential financial support from the Hahnemann Hospital. Their actions convinced Hahnemann's faculty and trustees to reconsider, and Hahnemann Medical College remained open to women.[16] In 1876, a disagreement among the faculty of Hahnemann Medical College led to the establishment of a second homeopathic college, the Chicago Homeopathic Medical College, which also accepted women students.[17] In the late 1880s and 1890s, women's choices expanded to include a regular coeducational night school, the Harvey Medical College; Jenner Medical College, another regular school; a physio-medical college; and two new homeopathic schools, National and Hering, both devoted to strict Hahnemannian principles. In 1897, women gained admission to the regular College of Physicians and Surgeons, which became coeducational when it affiliated with the University of Illinois.

Chicago thus serves as an interesting example of a city in which women seeking a medical education could attend one of several medical schools. Examining the choices women in Chicago made provides new insights into the history of women physicians and raises important questions. Women's choices are revealed in the records of the Illinois State Board of Health's official registers of physicians and midwives.[18] These records show that sectarian medical schools, particularly the Hahnemann Medical College, played a critical role in providing women with opportunities to become physicians in Chicago, a role that at the very least equaled the importance of the Woman's Medical College. In the 1870s, the Woman's Medical College contributed the most graduates, twenty-one (42%), to Chicago's population of women physicians. The Hahnemann Medical College and the Chicago Homeopathic Medical College together contributed an almost equal number of women physicians, twenty-two (44%), to the city of Chicago. Most of these homeopathic graduates (17) attended Hahnemann Medical College. Adding in women graduates of the Bennett College of Eclectic Medicine, six (12%), brought the total number of women practitioners in Chicago who graduated from local sectarian colleges to twenty-eight (56%).

In the 1880s, even more women (62%) chose to attend sectarian medical colleges. Again, most of these women attended the Hahnemann College. Indeed,

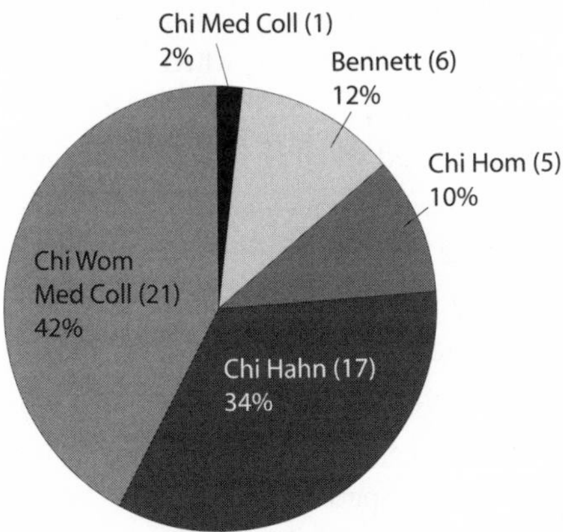

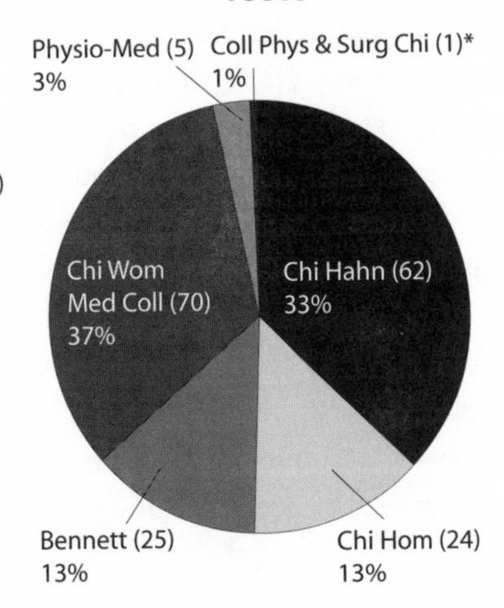

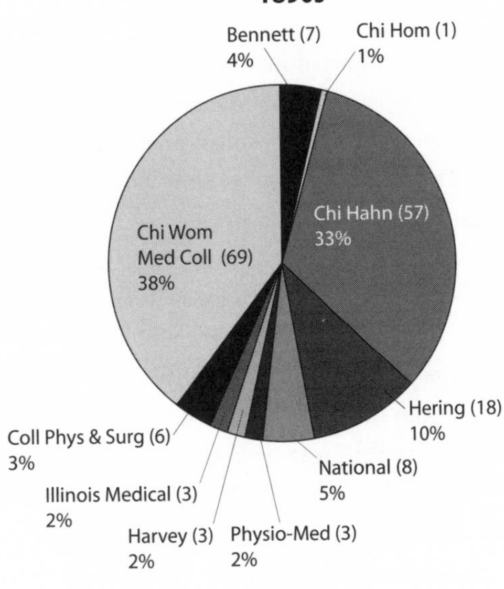

Chicago medical schools granting degrees to women physicians practicing in Chicago in the 1870s, 1880s, and 1890s. An asterisk indicates that for the 1880s one woman, Helen M. Goodsmith, is listed as an 1888 graduate of the College of Physicians and Surgeons, which did not officially become coeducational until 1897. See Illinois State Board of Health, *10th Annual Report*, 1890.

Sources: Figures and percentages hand-counted from the Illinois State Board of Health, *Official Register of Physicians and Midwives* (Springfield: Weber & Co., State Printer, 1880); Illinois State Board of Health, *3rd and 4th Annual Report*, 1881; Illinois State Board of Health, *6th Annual Report*, 1884; Illinois State Board of Health, *Official Register of Physicians and Midwives* (Springfield: H. W. Bokker, State Printer, 1886); Illinois State Board of Health, *10th Annual Report*, 1890; Illinois State Board of Health, *11th Annual Report*; Illinois State Board of Health, *Official Register of Legally Qualified Physicians*, 1903.

Hahnemann contributed sixty-two (33%) to Chicago's growing medical profession, only slightly fewer women than the Woman's Medical College, which contributed seventy (37%). This pattern continued in the 1890s, when 55 percent of the city's women medical practitioners received their medical educations in local sectarian schools. Sixty-nine (38%) graduated from the Woman's Medical College, only slightly more women than the fifty-seven (33%) from Hahnemann. As these figures show, substantial and significant numbers of women chose to attend the Woman's Medical College. Even more women, however, chose to attend sectarian schools, particularly the Hahnemann Medical College. Kirschmann's research shows that Chicago may not have been exceptional. In Boston, for example, in 1900, where approximately 18 percent of all physicians were women, 51 percent of the women were homeopaths.[19]

Gender and the Choice of a Sectarian Medical Education

Given the available options, why did a majority of women in Chicago choose to attend sectarian medical schools? Gender may have contributed to women's decisions because the percentage of women in Chicago who obtained sectarian educations differs substantially from the percentage of men who did so. As the data from the Illinois State Board of Health show in 1878, 1884, and 1890, slightly more than 70 percent of all male practitioners were regulars and approximately 25 percent were sectarian.[20]

For 1878 and 1884, approximately 40 percent of all women practitioners were regulars and approximately 60 percent were sectarian. By 1890, the percentage of women who identified themselves as regular practitioners increased slightly to almost 45 percent, while the percentage of women sectarian practitioners fell to just over 55 percent. Still the pattern for women physicians is markedly different than for men. What explains this difference? Why did more Chicago women than men choose sectarian medical educations?

Was it easier for women to gain admission to sectarian schools? Was the course of study less rigorous in sectarian schools? Did it cost less to attend sectarian schools? Did women who entered sectarian schools simply prefer coeducation and was this preference strong enough to override sectarian differences? Were women more likely than men to prefer sectarian medicine?

No evidence exists to suggest that Chicago's regular schools were any more expensive, more selective in admissions, or more rigorous in their academic requirements than its sectarian schools. Indeed, if they were, it is quite likely

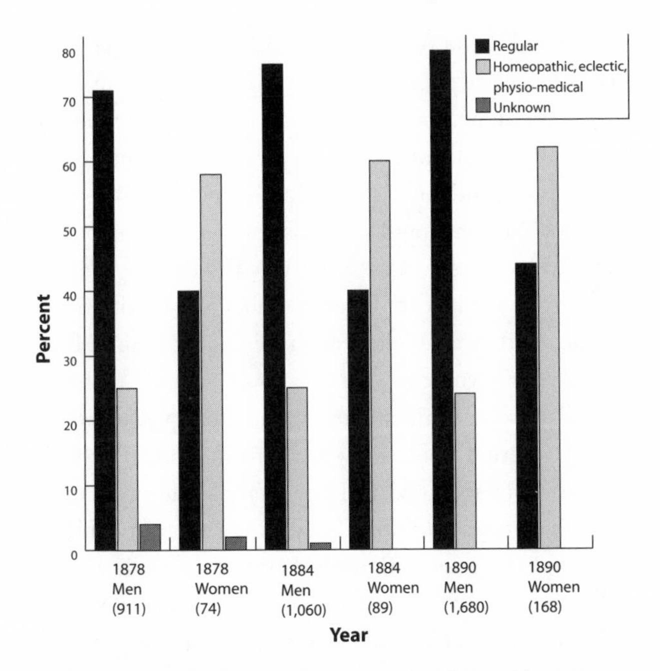

Chicago men and women practitioners, by choice of practice, 1878, 1884, and 1890.

Sources: Figures and percentages hand-counted from the Illinois State Board of Health, *Official Register of Physicians and Midwives* (Springfield: Weber & Co., State Printer, 1880); Illinois State Board of Health, *3rd and 4th Annual Report*, 1881; Illinois State Board of Health, *6th Annual Report*, 1884; Illinois State Board of Health, *Official Register of Physicians and Midwives* (Springfield: H. W. Bokker, State Printer, 1886); Illinois State Board of Health, *10th Annual Report*, 1890; Illinois State Board of Health, *11th Annual Report*, 1892.

that men would have flocked in greater numbers to these sectarian schools. For most of the nineteenth century, however, women could not enter Chicago's regular medical colleges. Because one of the goals of women's medical schools was to prove that women could successfully practice medicine, it is possible that Chicago's Woman's Medical College was more selective and more rigorous than both Chicago's sectarian schools and its regular male schools. This explanation, however, does not sufficiently explain women's choices, for many of the women who entered Chicago's sectarian medical schools were highly qualified, well educated, and capable of entering and successfully completing the medical course offered by the Chicago Woman's Medical College.

Lucy Waite provides one such example. A longtime Chicagoan, Waite possessed excellent credentials for entering any medical school in the city. By name and paternal heritage alone, she had a place in the Chicago medical community. Her paternal grandfather, Daniel D. Waite, was one of Chicago's pioneer physicians and served as president of the Chicago Medical Society in 1859. Lucy Waite's father, an abolitionist and suffragist, was a prominent Chicago judge, Charles B. Waite. Her mother, Catherine Van Valkenburg Waite, a prominent suffragist, was an 1852 graduate of Oberlin College. In addition to her extensive work for suffrage, Catherine Waite established and ran the Hyde Park Seminary, a school for young women.[21]

Charles and Catherine Waite's commitment to equal rights and education for women and Catherine Waite's involvement in women's club activities and suffrage work clearly contributed to Lucy Waite's education and to her decision to enter medicine. In addition, Lucy Waite was probably influenced by her mother's example because Catherine Waite had actually attempted to enter medical school. In 1866, just one year after Mary Thompson had established the Hospital for Women and Children, Catherine Waite applied to Rush Medical College but was rejected.[22]

Lucy Waite, the second child in Charles and Catherine's family of five children, graduated with a bachelor of arts degree from the "old" University of Chicago.[23] With this degree and her family pedigree, Waite could easily have gained admission to the Woman's Medical College of Chicago. Instead, she enrolled in the Hahnemann Medical College and graduated in 1883. Though no sources explaining Waite's decision to attend the Hahnemann Medical College exist, her educational background and qualifications make it highly unlikely that academic concerns played a role. Perhaps Waite preferred to obtain her medical education in a coeducational school or perhaps she preferred to study homeopathic therapeutics. Indeed, because qualified men could enter any medical school of their choosing, we assume that the minority of men who entered homeopathic or eclectic schools did so out of their preference for homeopathic or eclectic therapeutics.

Still, it is possible that Waite and other women placed a higher priority on coeducation than on therapeutic distinctions. As historian Ellen S. More suggests, Sarah Adamson (later Dolley), the third woman in the United States to obtain a medical degree, chose to enter an eclectic medical school, the Central Medical College rather than the Female Medical College of Pennsylvania because she preferred to attend a coeducational school. Although More notes that the Female Medical College was not yet open in 1849 when Adamson entered the Central Medical College, she suggests that Adamson was aware of plans for its establishment but chose not to wait for it to open because coeducation "was a significant issue for her."[24] It is also possible that Adamson's decision to attend the Central Medical College was based on her reluctance to wait and see whether the plans to open an all-female medical college in Philadelphia would bear fruit.

Another early woman medical graduate, Emily Blackwell, did exhibit a strong preference for coeducation. She did not even consider entering the Female Medical College of Pennsylvania and, in fact, regarded it as an obstacle in her quest to gain admission to a regular male medical college. "I fear this stupid Philadelphia College may make it difficult for me to enter any college here," she wrote.[25] Emily Blackwell's preference for coeducation, however, did not lead her to consider entering a sectarian school. Influenced perhaps by her older sister Elizabeth Blackwell's disdain for sectarian medicine, she persisted in her efforts to attend a regular medical school until she finally gained admission to Chicago's Rush Medical College in 1852.[26] Emily Blackwell's example suggests that it is unlikely that a preference for coeducation would override a preference for regular medical training. Blackwell, however, may well have been an exception and other women physicians may have been less committed to regular medicine and less aware or concerned about sectarian allegiances. Indeed, Sarah Dolley (nee Adamson) claimed that she was initially unaware of the sectarian nature of her medical school.[27] We need more examples of sectarian women physicians and their reasons for choosing sectarian medicine before concluding whether a preference for coeducation played a substantial role in their choices. Kirschmann provides several such examples in her discussion of homeopathic women and their reasons for choosing homeopathy but does not identify a preference for coeducation as a major factor in their decisions. Though she does note that women may have been attracted to homeopathy by "liberal opportunities to integrate . . . homeopathic organizations and institutions," she points to several others factors as being significant in their choice of homeopathy.[28]

Prime among the factors that led women to pursue a homeopathic education, Kirschmann argues, were their personal experiences with disease, particularly with successful treatment by a homeopathic physician, or family connections to homeopathic physicians.[29] Personal or social connections with acquaintances who patronized or had family ties to homeopathic physicians also may have played a role in women's choice of a medical school. Julia Holmes Smith, for example, chose to enter a homeopathic medical school because her friends referred her to a homeopathic physician for medical care. Holmes had recently moved to Boston and had not yet engaged the services of a physician. When she fell ill, several of her acquaintances recommended Mercy B. Jackson, a faculty member of the Boston University Medical School. "While in Boston," Smith wrote,

> I saw for the first time in my life a woman physician, and a homeopathic physician. I hadn't been very well, and one day when I had a headache, I sent for Dr. Jackson, who was a professor in the Boston University. She said there was nothing the matter with me and that I had headaches because I had no definite work to do. She advised my studying something—German, Latin, Italian, or music—all of which I had already studied. Then she suggested that I study medicine. She invited me to hear her lecture. I did hear her, and was interested. On the second day I listened to two more professors, and on the third day, I took out a perpetual ticket and began the study of medicine.

Smith left Boston in the midst of her studies and completed her education at the Chicago Homeopathic Medical College. Smith may well be the only woman who entered medicine on the basis of a medical prescription! Her decision to study and practice homeopathic medicine, however, did not result from a considered choice between homeopathic and regular medicine. Instead, her referral to and acquaintance with Mercy B. Jackson led her to enroll in Boston University. Though personal and family connections certainly may have led some women to enter sectarian schools, such connections probably played an equal role in men's decisions to enter sectarian schools. Such connections consequently cannot fully explain why proportionately more women than men entered sectarian schools. A final possibility to consider is that women were simply more likely than men to prefer sectarian medicine. To understand why, we must return to the reasons women, both sectarian and regular, sought to become physicians in the first place. As historian Regina Morantz-Sanchez convincingly shows, the movement to educate women as physicians developed from a vigorous critique of the harsh treatments characteristic of regular medicine and from the rise of a popular health-reform movement that focused on preventing disease by teaching and observing the laws of physiology and hygiene.[30] Advocates of educating women as physicians and many women physicians themselves believed that women had a special role to play in medicine—that they would improve the health and medical treatment of women and children by focusing more on prevention than on cure and by practicing gentler and milder forms of medicine.[31] As Morantz-Sanchez and several authors of chapters in this book have argued, this notion that women had a "special role" in medicine resulted from Victorian beliefs in women's nature.[32] Victorians believed that qualities such as nurturance, altruism, and morality were innately female and that these qualities made women particularly well suited to raising and educating their children and providing peaceful homes for their husbands.[33] Women physicians and their supporters argued that the same inherent qualities that made women good mothers would also make women good doctors.[34] Certainly some women physicians, like Marie Zakrzewska, rejected such notions of innate gender differences. As Arleen Tuchman acknowledges, however, Zakrzewska's position represented a distinctly minority point of view.[35] Most women physicians shared their culture's views about women's nature and believed that as women physicians they were uniquely positioned to protect, advocate for, and improve the health and health care of women and children.

Whether women actually practiced medicine differently than men is a matter of historical debate. Though historians have examined the obstetrical practice of regular women in the New England Hospital for Women and Children to determine whether it differed from the obstetric care men provided in Massachusetts General and the Boston Lying-In Hospital, they have reached conflicting conclusions.[36] Taking a broader approach to the question of the nature of women's medical practices, More concludes that the practices of male and female physicians "did differ in important ways." She argues that "the settings in which women practiced medicine, their deep empathy for and interest in the problems of women and children, and the meaning they and society at large accorded their activities bracketed women's medicine as a special domain."[37]

Expanding the analysis of differences in women's medical practices even further, I suggest that we consider the settings in which women chose to pursue their medical educations within the context of the broader criticisms of regular medicine and women's motivations to improve health care for women and children. Given this context, women were possibly more likely than men to prefer sectarian medicine because it promised alternative treatments more consistent with their ideals. As Susan Cayleff argues, "women did not become hydropathic practitioners because they were barred from regular medical schools . . . Instead, bad experiences as patients, a reformist nature, . . . and the positive gender consciousness of the water-cure movement make *choice* and a compatible belief system a far more likely explanation than rejection."[38]

Similarly, Kirschmann points to women's "expectations and assumptions of therapeutic difference" as one of several factors explaining "women's choice of homeopathy."[39] In a city such as Chicago, where women could choose to attend the regular women's medical college or one of several coeducational sectarian schools, the fact that most chose sectarian schools suggests that one way women acted on their desire to reform the practice of medicine and improve the health of women and children was to study sectarian medicine. This suggestion does not imply that women who entered regular medicine were not also motivated by a desire to improve the medical care of women and children. Indeed, this motivation inspired many women who entered medicine, regardless of sectarian concerns. Women who entered regular medicine, however, believed their gender—their womanhood—would transform medical practice. Though women who entered sectarian medicine undoubtedly shared the widely held belief that gender would make a difference in their medical practice, they also relied on the promise of alternative therapeutics to transform their practice of medicine. Women's interests in health reform, their criticisms of regular medical practice, and their desire to improve the medical treatment of women and children thus provide a viable explanation for why women were more likely than men to pursue sectarian medicine.

Collaboration between Regular and Sectarian Women Physicians in Chicago

The data from Chicago clearly show that sectarian medical schools provided women with medical education throughout the nineteenth century. The history of women physicians in Chicago also shows that women graduates of both sectarian and regular schools formed close friendships and collaborated with one another in medical practice. Lucy Waite again provides an example.

In 1895, the Chicago Hospital for Women and Children underwent a major transition after the death of Mary Thompson, its head and founder. Waite served temporarily in Thompson's stead as head of the hospital and eventually became its gynecologist and head physician and surgeon. How did a homeopathic physician come to succeed as such a preeminent regular woman physician and to lead the pioneering institution for women physicians in Chicago? Surprisingly, Waite's leadership of the Chicago Hospital for Women and Children came at Thompson's request. When Thompson fell ill in May 1895, she turned to her friend and colleague, Lucy Waite, for help. According to a letter from the board of trustees to the medical staff of the Hospital for Women and Children, Thompson asked Waite to "take charge of the Hospital work during her illness."[40] The medical staff honored this request and Waite managed the hospital until November 1895.

Why would Thompson select a homeopathic physician to manage the institution to which she had devoted her life? Why did she not select one of the women already serving on the hospital's attending staff? Why did she not select one of her esteemed colleagues from the Northwestern University Woman's Medical School? Though we will never know exactly why Thompson selected Waite, she clearly valued and respected Waite and her expertise and trusted her to continue her life's work.

Thompson's choice is surprising in part because our understanding of the history of the relationships between regular and homeopathic physicians is characterized by hostility rather than by friendship, trust, and collaboration. The history of medicine in Chicago provides several examples of hostility between homeopathic and regular physicians. The argument between homeopaths and regulars over representation of homeopaths on the medical staff of Cook County Hospital and its influence in delaying the opening of the hospital are well documented.[41] Homeopaths succeeded in winning appointments to Cook County Hospital's medical staff in 1879, though eclectic physicians only gained this privilege in 1889.[42] The inclusion of sectarian physicians on Cook County's medical staff, however, did not signify that regular physicians finally, if grudgingly, accepted the medical qualifications and competency of their sectarian brethren. Rather, it was a decision enforced by the Cook County Board of Commissioners who were subject to the political influence of sectarian physicians and their supporters. Similarly, the appointment of homeopathic and eclectic physicians to the hospital's medical staff does not suggest that regular physicians began working collaboratively with sectarians. Instead regulars, homeopaths, and eclectics each received a proportional allotment of patients and treated them separately.[43]

The American Medical Association (AMA), well known for its lengthy opposition to association and consultation with sectarian practitioners, provides another example of antagonism between regulars and sectarians in Chicago. One source of this antagonism was the AMA's Code of Ethics forbidding regular practitioners from consulting with and, according to some practitioners, from socializing with homeopathic, eclectic, and other sectarian practitioners.[44] Though not all physicians upheld this consultation clause, the AMA's views and policies may have significantly influenced relations between Chicago homeopaths, eclectics, and regulars because the AMA was headquartered in Chicago. Conversely, the strength and popularity of homeopaths in Chicago may have increased the AMA's tendency to regard homeopaths as a threat to their professional authority and solidified their opposition to collaboration with sectarians.[45] In either case, relations between the sectarians and regulars in Chicago remained hostile for most of the nineteenth century. In 1893, for example, efforts to hold a joint medical congress in conjunction with the Columbian Exposition failed and though the national homeopathic and eclectic medical societies met together, the AMA refused to participate and held a separate congress. Editorials in Chicago's homeopathic and eclectic medical journals also frequently reflected the animosity between regular and sectarian practitioners.[46]

The example of Lucy Waite and Mary Thompson, however, suggests that animosity and hostility did not characterize all relationships between homeopathic and regular physicians. Possibly their common interest in and practice of surgery drew them together. It is also possible that Thompson was acquainted and friendly with Waite's mother, Catherine Van Valkenburg Waite, since they both shared the experience of attempting, unsuccessfully, to open the doors of Rush Medical College to women in the 1860s. If so, the relationship between Thompson and Waite's mother may have fostered her interest in Waite's career.

Despite Thompson's wish to have Waite serve in her stead, the board of trustees and the medical staff of the hospital did not initially select Waite as Thompson's permanent replacement. Indeed, her place in the hospital generated considerable controversy. Though Waite served as head of the hospital until November 1895, the board of trustees chose Marie Mergler "to fill the vacancy left by the death of Dr. Thompson."[47] The board of trustees, however, did want to show Waite their appreciation for the work she had done while leading the hospital, and consequently asked the medical staff of the hospital to "appoint her as one of the Consulting Staff" and to request that she "continue to send her private patients to the hospital."[48] Arguing that there was no need to increase the number of consulting physicians, the medical staff refused to include Waite. The board of trustees voiced displeasure with this "misunderstanding between the Staff and the Board of Trustees regarding the not carrying out of their wishes in respect to Dr. Waite."[49] Though the medical staff appointed a committee of two to confer with the board of trustees about this "misunderstanding," it was not referred to again and Waite never did receive an invitation to join the consulting staff.[50]

What were the objections of the medical staff? The stated reason, that it was not advisable to add to the number of the consulting staff at that time, cannot be the real reason because one month later, on December 28, 1895, the medical staff added E. Fletcher Ingals and Frank Billings as consultants.[51] On May 2, 1896, they added Rosa Engert as a consulting physician in Obstetrics and Gynecology.[52] Even though Lucy Waite's homeopathic training did not influence the hospital's founder, Mary Harris Thompson, Waite's status as a homeopath may have bothered Marie Mergler and the rest of the attending staff of the Hospital for Woman and Children. Indeed, Bertha Van Hoosen, in unpublished manuscript drafts of her autobiography later published as *Petticoat Surgeon*, wrote: "Dr. Lucy Waite was a homeopath and the constitution of the hospital expressly stated that physicians on the staff of the hospital must be women physicians graduated from a *regular* school of medicine [emphasis original]."[53] Curiously, Mergler and the medical staff of the hospital never referred to Waite's homeopathic training and by-laws of the hospital in explaining their refusal to appoint her to the staff—or if they did, it is not recorded in the hospital's official minutes.

Despite the medical staff's initial rejection of Thompson's choice of Waite as her successor, the hospital's board of trustees turned to Waite two years later when Mergler resigned her position.[54] By this time, Frances Dickinson, a graduate of the Chicago Woman's Medical College and president of the Harvey Medical College, had provided Waite with the means of overcoming the obstacle posed by the hospital's constitution by awarding her a regular degree from the Harvey Medical College. The time Waite had spent studying in Europe between 1895 and 1897 may also have contributed to her success. Waite became a successful manager of the Chicago Hospital for Women and Children. "During her administration," Van Hoosen wrote, "Dr. Waite not only built, and presented to the hospital a large nurses' home, but of even greater importance, kept the hospital free from indebtedness."[55] Prominent male Chicago physicians and surgeons also expressed their admiration for Waite's surgical skill and hospital administration.[56]

Many may argue that Mary Thompson, Frances Dickinson, Bertha Van Hoosen, and others accepted Lucy Waite because she was a surgeon, an area of medicine in which the therapeutic differences between homeopaths and regulars had little relevance. By taking a second degree from the Harvey Medical College, they also may argue, Waite became a regular physician. Indeed, Waite used this degree to claim status as a regular physician when she updated her Illinois state registration. However, several other examples of close associations between women regular and sectarian practitioners in Chicago suggest that sectarian and regular women could and did work together as physicians despite strong allegiances to competing therapeutic systems.

Chicago's clubs for women fostered cooperation and collegiality between sectarian and regular women by providing opportunities for women doctors of various sects to become acquainted with one another. The relationships they forged through membership in such organizations as the Fortnightly, the Chicago Woman's Club, and the Queen Isabella Association undoubtedly fostered their interest in working together as physicians. The homeopathic physician, Julia Holmes Smith, as Muriel Beadle wrote in her history of the Fortnightly Club, "was the first of a group of women doctors who joined not only the Fortnightly but many other groups as well."[57] The Fortnightly was a literary and self-education club organized by Kate Newell Dodgett. Smith joined the Fortnightly in 1876. The following year, Sarah Hackett Stevenson, a graduate and faculty member of the regular Chicago Woman's Medical College, became the next woman physician to join the Fortnightly. As physicians and as women's club members, Sarah Hackett Stevenson's and Julia Holmes Smith's shared interests in health and medicine frequently led them to participate in the same activities.

Stevenson and Smith, for example, both supported and worked in the Illinois Training School for Nurses, a school organized by several members of the Fortnightly Club. Stevenson was among the small group of Fortnightly Club members who met in the summer of 1880 to organize the nursing school. They not only aimed to train and employ young women as nurses, thus helping to establish "a new and dignified profession for women," but also to supply trained nurses to the public and to improve the nursing care in Cook County Hospital, the training ground for the new school of nursing.[58] When the Illinois Training School for Nurses opened in 1881, Stevenson served on its board of directors, chaired its hospital committee, and was a member of the faculty. Stevenson remained a member of the board of directors until 1907.[59] Smith served on the board of directors from 1886 to 1920 and was a lecturer in the school.[60] Smith and Stevenson also joined the Woman's Christian Temperance Union. In 1886, when the Frances E. Willard National Temperance Hospital opened, Stevenson became president of its medical staff and Smith later joined the staff as a consulting physician.

In addition to their membership in the Fortnightly, Stevenson and Smith became members of the Chicago Woman's Club, an offshoot of the Fortnightly formed by members who wanted to engage not only in the literary discussions, intellectual pursuits, and self-improvement that characterized the Fortnightly but also to use their knowledge and social position to foster social reform in the city.[61] Smith and Stevenson were active members of the Chicago Woman's Club and served several terms as its president. Julia Holmes Smith served as the club's president from 1879 to 1881 and from 1883 to 1884. Sarah Hackett Stevenson was president of the Chicago Woman's Club from 1892 to 1894. Another homeopathic physician, Leila G. Bedell, an 1878 graduate of the Boston University School of Medicine who subsequently moved to Chicago, served as president from 1885 to 1887. Bedell and Stevenson encouraged and sponsored Frances Dickinson's membership in the Chicago Woman's Club. Several other Chicago physicians, too, were members of the Chicago Woman's Club. They included Julia Ross Low, homeopath; Marie Reasner, eclectic; Harriet B. Alexander, regular; Rosa Englemann, regular; and in later years, Bertha Van Hoosen, regular.[62]

With so many physicians as members, it is not surprising that the Chicago Woman's Club embarked on several projects related to health and medicine. Indeed, one historian has suggested that the dedicated involvement of medical women in the Chicago Woman's Club influenced the direction of the club's work.[63] One of the club's first accomplishments, for example, was to organize the Woman's Physiological Institute of Chicago. This institute's purpose "was to disseminate among women greater knowledge of the laws of life and health."[64] Physician members of the Chicago Woman's Club and the Woman's Physiological Institute gave lectures in various locations, including Hull House, on topics such as "Our Babies, Hygiene for Mother and Infant," "The Germ Theory of Disease," and "Moral Hygiene, Physiologically Considered."[65]

In addition to the Woman's Physiological Institute of Chicago, physician members of the Chicago Woman's Club established or belonged to several new organizations for medical women. In 1887, for example, Frances Dickinson together with Lucy Waite established the Medical Woman's Sanitary Association.[66] As Dickinson noted in her reminiscences, "Among the woman doctors who had found each other in civic work in that early day, were representatives of the three schools [regular, homeopathic, and eclectic]. After much deliberation, many of us united under the banner of the Medical Woman's Sanitary Association."[67] Because no records of this organization exist, we do not know precisely what goals the Medical Woman's Sanitary Association aimed to achieve. We do know from Dickinson's reminiscences that the Medical Woman's Sanitary Association successfully lobbied the Chicago Board of Health to employ a woman physician. In 1889, this organization sent several of its members—Kate Bushnell (regular), Alice Ewing (homeopath), and Rachel Hickey (regular)—to help victims of a devastating flood in Johnstown, Pennsylvania.[68]

In 1890, women physicians representing the regular, homeopathic, and eclectic branches of medicine began to plan and implement joint programs to highlight the work of women physicians for the 1893 Columbian Exposition. Women physicians' involvement with the planning began when Dickinson became a member of the Exposition's Board of Lady Managers. This board, headed by Bertha Honore Palmer, the wife of wealthy Chicago real estate developer and hotelier, Potter Palmer, planned to erect a separate Woman's Building on the fairgrounds to showcase women's work. Dickinson, along with several other women physicians and professional women, established a rival organization, the Queen Isabella Association, which strongly opposed this plan. The "Isabellas" argued that women's achievements should be integrated into general displays, not segregated in a woman's building.[69] Though the "Isabellas" lost this argument, its physician members worked with the Illinois Women's Exposition Board and the Illinois Training School for Nurses to build and run a model hospital and an emergency clinic on the fairgrounds. Sarah Hackett Stevenson, as a regular physician, Julia Holmes Smith, as a homeopath, and Marie Reasner, as an eclectic physician, recruited women doctors from all three branches of medicine to staff this clinic, which treated more than three thousand patients during the exposition.[70] This hospital not only provided several women with the opportunity to practice medicine at the World's Fair and served as a showcase for women's achievements in medical practice, it also provided a unique example of an institution in which women from all three branches of medicine—regular, homeopathic, and eclectic—practiced side by side in a joint effort to provide medical services to the masses of people attending the Columbian Exposition. Indeed, Chicago's eclectic medical journal, *The Medical Times*, praised Chicago's women doctors writing, "The women in the profession are setting the men an excellent example in forgetting school differences. It is a notable and highly praiseworthy fact that the lady physicians of Chicago are a harmonious body of intelligent workers."[71]

Chicago women doctors, thus, appear to have been considerably more successful than their male counterparts in overcoming sectarian differences. Although, as in the case of Lucy Waite, the ability of women from different schools of medical practice to work cooperatively may have been fostered by the lack of therapeutic conflict in fields such as surgery, or by her subsequent acquisition of a regular medical degree, this does not explain all collaborations. Julia Holmes Smith, for example, was a strict adherent of homeopathy, even becoming a faculty member and then dean of the National Homeopathic Medical College, a school founded to teach strict Hahnemannian principles.[72]

The collaboration between regular and sectarian women physicians in Chicago clearly violated the AMA policy barring consultation between regular and sectarian physicians. More importantly, it contradicts our vision of the relations between regular and sectarian physicians in the nineteenth century as being characterized more by hostility and contempt than by friendship and cooperation. Since relations between male regular and sectarian physicians in Chicago were acrimonious, my findings suggest that, at least in Chicago, there was a gender difference in physicians' willingness to tolerate, to befriend, and to work with practitioners from competing branches of medicine. Their membership in the Chicago Woman's Club, their mutual interests in social reform and in fostering improved health for women and children, and their shared experiences as women physicians may have helped Chicago's medical women overcome sectarian divisions. Thus they could cooperate as physicians to train nurses and to work side by side, providing medical care to victims of the Johnstown flood, the patients of the Frances Willard National Temperance Hospital, and ailing visitors to the Columbian Exposition.

This friendship and collaboration between women physicians from various schools of medicine raise a host of new questions about the experiences of women physicians in the nineteenth century. What, for example, did women physicians from different schools of medicine talk about when they gathered? It is hard to imagine that they did not share stories about the various institutions in which they gained educational and clinical experience. Did regular women, for example, realize that even though sectarian medical schools embraced the education of women and that women at times accounted for 20 percent or more of their medical classes, Chicago's sectarian medical schools did not include women on their faculties until late in the 1890s—and then only in token positions?[73] Did this knowledge give regular women greater appreciation for the Chicago Woman's Medical College, which boasted a substantial core of women among its faculty and leadership? Did this knowledge prevent the Chicago Woman's Medical College from following the example of the New York Infirmary's Medical School? Unlike the latter school, it did not close its doors when a major regular medical school in the city, the College of Physicians and Surgeons began admitting women after affiliating with the University of Illinois. Similarly, did sectarian women's awareness of the role regular women played in the faculty and leadership of the Woman's Medical College cause them to advocate for an increased role for women on their schools' faculties or to complain about marginalization and unequal treatment as Lucy Waite did in a letter to the editor of *The Medical Visitor*, a local homeopathic journal?[74] Despite their absence from the faculty ranks of their respective medical schools, sectarian women were far more integrated than were regular women in their professional organizations. Sectarian women played much larger roles and served in leadership positions in local and statewide medical societies and were much better represented in local medical publications than were regular women.[75] Did this influence women's engagement with regular medical societies? Did it contribute in any way to the substantial involvement of Chicago women in the creation and early activities of the Medical Women's National Association (later renamed the American Medical Women's Association)? These are just some of the questions raised by recognizing the interactions between regular and sectarian women physicians.

Was Chicago unusual, or was it typical of other cities across the nation? Certainly, in Chicago and in other cities, there were regular women who would not associate with sectarians. Indeed, Marie Mergler and other members of the staff of the Chicago Hospital for Women and Children are prime examples. Elizabeth Blackwell was vehemently antisectarian and, as Carla Bittel notes, Mary Putnam Jacobi was "committed to . . . weeding out homeopathic practices."[76] In addition, Ellen S. More shows that Sarah Dolley expressed considerable animosity toward homeopathy. Dolley, a founding member of the Practitioners' Society of Rochester (later renamed the Blackwell Medical Society), which restricted its membership to "regularly qualified physicians," clashed with her local medical society when it voted to allow consultation with homeopathic physicians and attended the 1889 annual meeting of the AMA to show support for adhering to AMA's policy forbidding consultation with homeopaths.[77] Yet other aspects of Dolley's private life and medical practice pose interesting contradictions given her public and organizational stand on collaborating with sectarians. Ironically, much like the fictional Helen Rand, discussed at the beginning of this chapter, Dolley lived and practiced medicine together with a homeopathic physician, Anna Searing.[78] How did Dolley justify this? Did Searing, like Dolley, repudiate her sectarian origins? Dolley also maintained lifelong friendships and correspondence with her classmates from the eclectic Central Medical College: Lydia Folger Fowler, phrenologist and health reformer; and Rachel Brooks Gleason, hydropathist. Though these friendships do not constitute "consultation," they certainly suggest that Dolley had some degree of respect for these individual sectarian practitioners.

Other historians provide additional suggestions of cooperation between regular and sectarian women. Gloria Moldow, for example, notes that Washington, D.C., physician Caroline Brown Winslow, a graduate of both an eclectic and a homeopathic school, founded numerous women's groups, including the D.C. chapter of the Women's Educational and Industrial Union (WEIU) that included both regular and sectarian physician members.[79] Similarly, Kirschmann shows that in Boston in the 1890s, homeopathic and regular women both joined a variety of women's organizations, such as the Daughters of the American Revolution and the Association of Collegiate Alumnae, and served as physician consultants to the Boston chapter of the WEIU.[80] These organizations, however, were primarily political or educational and may not have initiated substantial medical or health-related projects. Though these examples do not provide evidence of medical collaborations as extensive as those engaged in by Chicago medical women, they do indicate that we have much to gain by including sectarian women in our studies of women physicians and by examining the interactions between sectarian and regular women physicians.

Conclusion

This essay provides evidence that most Chicago women physicians in the nineteenth century were educated in coeducational institutions, albeit sectarian institutions, and that Chicago's women physicians interacted and collaborated with one another. They were not separated by sectarian allegiances. The history of women physicians in Chicago suggests that we may need to revise our understanding of nineteenth-century medical education for women as being characterized by sex segregation and acknowledge that this characterization applies only to regular women practitioners. If regular women represented most female practitioners in the country, our understanding of the nineteenth century as dominated by separatism would be justified. The example of Chicago, however, suggests that significant numbers of women physicians were sectarian practitioners. Local studies of other communities, particularly communities that offered women a choice of medical schools, will help us obtain a fuller picture of women physicians in the nineteenth century, a history that includes all types of women practitioners.

NOTES

1. Lois Wright, *Doctor Helen Rand* (Chicago: Physician's Publishing Company, 1891), 31.

2. Naomi Rogers, "Women and Sectarian Medicine," in *Women, Health, and Medicine in America: A Historical Handbook*, ed. Rima D. Apple (1990; New Brunswick, NJ: Rutgers University Press, 1992), 282. Rogers recognizes the work of Susan Cayleff and Jane Donegan on women hydropathists as exceptions to the historical neglect of sectarian women. She also notes that Gloria Moldow includes homeopathic women in her work on physicians in Washington, D.C. Because homeopathic women constituted such a small minority of women medical practitioners in Washington, D.C., however, Moldow concentrates on the separatist organizations in which the majority of Washington, D.C.'s women physicians worked. See: Susan E. Cayleff, *"Wash and Be Healed": The Water-Cure Movement and Women's Health* (Philadelphia: Temple University Press, 1987); Jane B. Donegan, *"Hydropathic Highway to Health": Women and Water-Cure in Antebellum America* (Westport, CT: Greenwood Press, 1986); Gloria Moldow, *Women Doctors in Gilded-Age Washington: Race, Gender, and Professionalization* (Urbana: University of Illinois Press, 1987).

3. Anne Taylor Kirschmann, *A Vital Force: Women in American Homeopathy* (New Brunswick, NJ: Rutgers University Press, 2004).

4. The pioneering and influential works that have inspired most recent work on the history of women physicians focus on regular women practitioners and on their strategy of establishing separate medical institutions to gain access to medical education and clinical practice. See, for example, Mary Roth Walsh, *"Doctors Wanted: No Women Need Apply": Sexual Barriers in the Medical Profession, 1835–1975* (New Haven, CT: Yale University Press, 1977); Virginia G. Drachman, *Hospital with a Heart: Women Doctors and the Paradox of Separatism at the New England Hospital, 1862–1969* (Ithaca, NY: Cornell University Press, 1984); Regina Markel Morantz-Sanchez, *Sympathy and Science: Women Physicians in American Medicine* (New York: Oxford University Press, 1985).

5. Eve Fine, "Doctors without Degrees: Women Physicians in Chicago, 1850–1900," paper presented at the American Association for the History of Medicine, May 2000.

6. Thomas N. Bonner, *To the Ends of the Earth: Women's Search for Education in Medicine* (Cambridge, MA: Harvard University Press, 1992), 14–16, 149; and Ellen S. More, *Restoring the Balance: Women Physicians and the Profession of Medicine, 1850–1995* (Cambridge, MA: Harvard University Press, 1999), 20–21, 42.

7. Bonner, *To the Ends of the Earth*, 16.

8. More, *Restoring the Balance*, 21, 42.

9. A few all-female homeopathic medical schools operated in the nineteenth century, but most were short-lived. The exception was the New York Medical College and Hospital for Women, which graduated 370 women homeopaths between 1863 and 1914. See Kirschmann, *A Vital Force*, 56, 58, and n. 15, p. 188. Chicago's Hahnemann Medical College graduated 426 women. See William Harvey King, *History of Homoeopathy and Its Institutions in America*, 4 vols. (New York: Lewis Publishing Company, 1905), 2: 356–379.

10. Kirschmann, *A Vital Force*, 46.

11. Editorial, "Bennett College," *Chicago Medical Times*, 1869, *1*: 5.

12. Minutes of the Board of Trustees Meeting, July 24, 1869, "Copy of the Original Minute Book of the Hahnemann Medical College and Hahnemann Hospital of the City

of Chicago," pp. 25–26, Northwestern University Memorial Hospital Archives, Chicago. Records of the Chicago Medical College were destroyed in the 1871 Chicago fire. Consequently, we do not know precisely when the board of trustees or faculty members of the Chicago Medical College decided to admit women students. Women began attending the college early in October 1869.

13. Charles Warrington Earle, *The Demand for a Woman's Medical College in the West* (Waukegan, IL: 1879). Reprinted in James Nevin Hyde, *Early Medical Chicago* (Chicago: Fergus Printing, 1879), 47.

14. "Dr. Mary Harris Thompson," *The Woman's Medical Journal*, July 4, 1895, 7: 192; Minutes, August 2, 1870, "Minutebooks," Northwestern University Woman's Medical School Records, Archives and Special Collections on Women in Medicine, Drexel University College of Medicine, Philadelphia, Pennsylvania.

15. Minutes of the Board of Trustees, April 22, 1874, "Copy of the Original Minute Book of the Hahnemann Medical College and Hahnemann Hospital of the City of Chicago," pp. 61–62, Records of the Hahnemann Medical College and Hahnemann Hospital of Chicago, Archives of the Northwestern University Memorial Hospital, Chicago.

16. Minutes of the Board of Trustees, June 24, 1874, and June 27, 1874, "Copy of the Original Minute Book of the Hahnemann Medical College and Hahnemann Hospital of the City of Chicago," pp. 63–69.

17. Allen C. Cowperthwaite, "Chicago Homeopathic Medical College," in *History of Homeopathy and Its Institutions in America*, 2: 401–406; J. S. Mitchell, "History of the Chicago Homeopathic College," in *Medical and Dental Colleges of the West*, ed. H. G. Cutler (Chicago: Oxford Publishing, 1896), 291–297. The Chicago Homeopathic Medical College remained coeducational until 1883 when it decided to discontinue educating women. It did not readmit women until 1901. See *The Annual Announcement of the Chicago Homeopathic College for the Collegiate Year 1883–84* (Chicago: Blakely, Marsh, Printers, 1883), 7–8, and "Items of Interest," *The Woman's Medical Journal*, 1901, *11*: 274.

18. Illinois became one of the earliest states in the nation to reinstitute medical licensing when it passed the 1877 act to "Regulate the Practice of Medicine in the State of Illinois." A copy of this act is found in Illinois State Board of Health, *Official Register of Physicians and Midwives* (Springfield: Weber & Co., 1880), 1–2. After July 1, 1877, the state required all medical practitioners to submit a diploma or pass an examination in order to obtain a certificate to practice. Practicing without certification was punishable by a fine or a jail term. The State Board of Health published its lists of certified physicians in *Official Registers* and in its *Annual Reports.*

19. Walsh, "*Doctors Wanted*," 186; Kirschmann, *A Vital Force*, 176.

20. Using data from the American Medical Association on medical school graduates, William Rothstein noted that nationally in 1890 about 8 percent of medical school graduates were homeopaths and about 4 percent were eclectic. See William G. Rothstein, *American Physicians in the Nineteenth Century: From Sects to Science* (Baltimore: Johns Hopkins University Press, 1985), 287, 345. This suggests that sectarian men were particularly well represented in Chicago.

21. Carole Dianne Smith, "Waite, Catherine Van Valkenburg," in *Women Building Chicago, 1790–1990: A Biographical Dictionary*, ed. Rima Lunin Schultz and Adele Hast (Bloomington: Indiana University Press, 2001), 922–923.

22. Smith, "Waite, Catherine Van Valkenburg," 922.

23. William K. Beatty, "Lucy Waite," in *Biographies of Physicians and Surgeons*, ed.

F. M. Sperry (Chicago: J. H. Beers, 1904), 62; and William K. Beatty, "Lucy Clapp Waite," in *Women Building Chicago, 1790–1990: A Biographical Dictionary*, 924. For more information on the education of women at the University of Chicago, see Rosalind Rosenberg, *Beyond Separate Spheres: Intellectual Roots of Modern Feminism* (New Haven, CT: Yale University Press, 1983), 31; and Barbara Miller Solomon, *In the Company of Educated Women* (New Haven, CT: Yale University Press, 1985), 57.

24. More, *Restoring the Balance*, 19.

25. Emily Blackwell Journal, August 1851, p. 22, Blackwell Family Papers, Schlesinger Library, Radcliffe Institute for Advanced Study, Harvard University.

26. Nancy Ann Sahli, "Elizabeth Blackwell, MD (1821–1910): A Biography" (Ph.D. diss., University of Pennsylvania, 1974), 85–86; Emily Blackwell Journal, October 1852, p. 46.

27. More, *Restoring the Balance*, 21.

28. Kirschmann, *A Vital Force*, 4, 35–73.

29. Kirschmann, *A Vital Force*, 35.

30. Regina Markell Morantz, "The 'Connecting Link': The Case for the Woman Doctor in 19th-Century America," in *Sickness and Health in America: Readings in the History of Medicine and Public Health*, ed. Judith Walzer Leavitt and Ronald L. Numbers (Madison: University of Wisconsin Press, 1978); and Morantz-Sanchez, *Sympathy and Science*, 28–49.

31. Morantz-Sanchez, *Sympathy and Science*, 59–60. See also Drachman, *Hospital with a Heart*, 71–72; More, *Restoring the Balance*, 42.

32. Morantz, "The 'Connecting Link': The Case for the Woman Doctor in 19th-Century America"; Morantz-Sanchez, *Sympathy and Science*; Arleen Marcia Tuchman, "Maternity and the Female Body in the Writings of Dr. Marie Zakrzewska, 1829–1902," and Judy Tzu-Chun Wu, "A Chinese Woman Doctor in Progressive Era Chicago," Chapter 4 in this volume.

33. For more on Victorian notions of women's nature and place in society, see Barbara Welter, "The Cult of True Womanhood: 1820–1860," *American Quarterly*, 1966, *18*: 151–174; Barbara J. Harris, *Beyond Her Sphere: Women and the Professions in American History* (Westport, CT: Greenwood Press, 1978), 32–72; Linda K. Kerber, "Separate Spheres, Female Worlds, Woman's Place: The Rhetoric of Woman's History," *Journal of American History*, 1988, 75: 9–39.

34. Morantz, "The 'Connecting Link': The Case for the Woman Doctor in 19th-Centuy America."

35. Tuchman, "Maternity and the Female Body in the Writings of Dr. Marie Zakrzewska, 1829–1902," 6.

36. Laurie Crumpacker, "Female Patients in Four Boston Hospitals of the 1890s," paper presented at the Berkshire Conference on the History of Women, October 1974. Available in the Berkshire Conferences on the History of Women, Papers (1974–2002), Schlesinger Library, Radcliffe Institute for Advanced Study, Harvard University, Cambridge, MA; Regina Markell Morantz and Sue Zschoche, "Professionalism, Feminism, and Gender Roles: A Comparative Study of Nineteenth-Century Medical Therapeutics," *Journal of American History*, 1980, *67*: 568–588; Drachman, *Hospital with a Heart*, 87–89.

37. More, *Restoring the Balance*, 42.

38. Cayleff, "*Wash and Be Healed*," 69.

39. Kirschmann, *A Vital Force*, 4.

40. May 25, 1895, "Minutes of the Staff of Attending Physicians of the Chicago Hospital for Women and Children," p. 107. These minutes were previously housed at the Mary Thompson Hospital, Chicago, Illinois, which is no longer in existence. (Hereafter cited as CHWC Minutes.)

41. Harry F. Dowling, *City Hospitals: The Undercare of the Underprivileged* (Cambridge, MA: Harvard University Press, 1982), 28, 46–47, 60–61; Thomas N. Bonner, *Medicine in Chicago, 1850–1950: A Chapter in the Social and Scientific Development of a City* (Madison, WI: American History Research Center, 1957), 41.

42. "Bennett Men in Cook County Hospital," *The Medical Times*, February 1889, *21*, 2: 88.

43. Bonner, *Medicine in Chicago*, 162; Dowling, *City Hospitals*, 61.

44. For more information on the AMA's consultation clause and various physicians' support or opposition to this clause, see Rothstein, *American Physicians in the Nineteenth Century*, 70–173, 301–305; More, *Restoring the Balance*, 40; and Paul Starr, *The Social Transformation of American Medicine* (New York: Basic Books, 1982), 101–102.

45. For discussions of the AMA's motivations for opposing consultation with sectarians see, John Harley Warner, "Orthodoxy and Otherness: Homeopathy and Regular Medicine in Nineteenth-Century America," in *Culture, Knowledge, and Healing: Historical Perspectives of Homeopathic Medicine in Europe and North America* (Sheffield, UK: European Association for the History of Medicine and Health, 1998), 5–29; Starr, *The Social Transformation of American Medicine*, 98–102; and Rothstein, *American Physicians in the Nineteenth Century*, 170–173, 298–305.

46. For examples see "Editorials: Exclusive Folly," *Chicago Medical Times*, June 1881, 3: 149–150; "Editorials," *Chicago Medical Times*, June 1883, *15*: 129–131; "Public Opinion of the Allopaths," *Chicago Medical Times*, Aug. 1887, *19*: 259; "Infanticide," *The Medical Visitor*, Feb. 1889, *5*: 38; "The Field," *The Medical Visitor*, Aug. 1889, *5*: 264; "Correspondence," *The Medical Visitor*, Oct 1889, *5*: 343–344; "Report of the Legislative Committee of the Illinois State Homeopathic Medical Association," *Clinique*, May 1902, *23*: 344–347; "Editorial: The American Medical Association," *The Medical Times*, June 1870, *6*; "Editorials: Exclusive Folly," *The Medical Times*, June 1881, *3*: 149–150; "Editorials," *The Medical Times*, June 1883, *15*: 129–131; "Public Opinion of the Allopaths," *The Medical Times*, Aug. 1887, *19*: 259; "Letters to the Editor: A Wail for Lost Supremacy," *The Medical Times*, March 1890, 22: 137–138; "Report of the Legislative Committee of the Illinois State Homeopathic Medical Association," *Clinique*, May 1902, *23*: 344–347.

47. November 8, 1895, CHWC Minutes, pp. 112–115.

48. Ibid.

49. January 18, 1896, CHWC Minutes, pp. 127–129.

50. Ibid.

51. December 28, 1895, CHWC Minutes, pp. 125–126.

52. May 2, 1896, CHWC Minutes, pp. 134–135.

53. Bertha Van Hoosen Collection, Box 2, Mss File 2, unpaginated; and Box 2, Mss File 1, p. 207, Michigan Historical Collections, Bentley Historical Library, University of Michigan.

54. Mergler announced her decision to resign at the January 4, 1897, meeting of the attending staff of the Hospital for Women and Children. See January 4, 1897, CHWC Minutes, p. 142.

55. Bertha Van Hoosen Collection, Box 2, Mss. File 1, p. 14.

56. F. M. Sperry, ed., "Lucy Waite, B.A., M.D.," *Biographies of Physicians and Surgeons* (Chicago: J. H. Beers, 1904), 63.

57. Muriel Beadle, *The Fortnightly of Chicago: The City and Its Women: 1873–1973* (Chicago: Regnery, 1973), 24.

58. Grace Fay Schryver, *A History of the Illinois Training School for Nurses, 1880–1929* (Chicago: Board of Directors of the Illinois Training School for Nurses, 1930), 1. See also Julia Wood Kramer, "Smith, Julia Holmes Abbot," in *Women Building Chicago, 1790–1990: A Biographical Dictionary*, 813.

59. Schryver, *A History of the Illinois Training School for Nurses*, 4–7, 25, 199. See also, Brigid Lusk, "Stevenson, Sarah Hackett," in *Women Building Chicago, 1790–1990: A Biographical Dictionary*, 845.

60. Schryver, *A History of the Illinois Training School for Nurses*, 199; Kramer, "Smith, Julia Holmes Abbot," 813; and William H. King, *History of Homeopathy*, 4: 228–229.

61. Lana Ruegamer, "The Paradise of Exceptional Women: Chicago Women Reformers, 1863–1893" (Ph.D. diss., Indiana University, June 1982), 15–16.

62. Henriette Greenebaum Frank and Amalie Hofer Jerome, *Annals of the Chicago Woman's Club for the First Forty Years of Its Organization, 1876–1916* (Chicago: Chicago Woman's Club, 1916), 97, 114, 126, 129.

63. Ruegamer, "The Paradise of Exceptional Women," 154. Lana Ruegamer also suggests that the membership of a significant number of homeopathic physicians' wives in the Chicago Woman's Club may have contributed to the Club's interest in issues relating to health and physiology.

64. Frank and Jerome, *Annals of the Chicago Woman's Club*, 37.

65. Ibid.

66. Julia Wood Kramer, "Frances Dickinson," in *Women Building Chicago, 1790–1990: A Biographical Dictionary*, 218.

67. "Reminiscences by Frances Dickinson, M.D., Jubilee Week, May 19, 1931," AMWA Records: "Women in Medicine" Scrapbook, pp. 436–492, Archives and Special Collections on Women in Medicine, Drexel University of the Health Sciences, Philadelphia, Pennsylvania.

68. Ibid.

69. The Queen Isabella Association's main aims were to commission a statue of Queen Isabella commemorating her role in Columbus's journey and to promote the interests of women, particularly working women, at the fair. See Jean Madeline Weimann, *The Fair Women* (Chicago: Academy Chicago, 1981), for more on the Board of Lady Managers of the Columbian Exposition and on the Queen Isabella Association.

70. Editorial, "The Nurses Display at the Fair," *The Medical Times*, August 1892, 24(8): 382; Sarah Hackett Stevenson, "Address to the Chicago Woman's Club by its president," Dr. Sarah Hackett Stevenson, May 19, 1894, in Frank and Jerome, *Annals of the Chicago Woman's Club*, 139; Kramer, "Smith, Julia Holmes Abbot," 813; Brigid Lusk, "Sarah Hackett Stevenson," in *Women Building Chicago, 1790–1990: A Biographical Dictionary*, 845–846.

71. "The Lady Physicians of Chicago," *The Medical Times*, June 1892, 24: 274.

72. William K. Beatty, "Julia Holmes Smith—Homeopath and Club-Woman," *Proceedings of the Institute of Medicine of Chicago*, 1984, 37: 97.

73. Women were not represented on the faculty of the Hahnemann Medical College until 1892 when Cornelia S. Stettler joined the "auxiliary" faculty as a clinical assistant to

the Chair of Gynecology. The first woman to join the faculty of the Bennett College of Eclectic Medicine was Jessie Forrester in 1895. See Reuben Ludlam, "Memoranda concerning the Hahnemann Medical College and the College building, on Cottage Grove Avenue, June 8, 1870, to that of laying the corner-stone of the new one, August 20, 1892," Hahnemann Medical College and Hahnemann Hospital Records, Box 1, 1855–1915, Northwestern University Memorial Hospital Archives, Chicago; Reuben Ludlam, "The Homeopathic School," in *The History of Chicago*, ed. John Moses and Joseph Kirkland (Chicago: Munsell, 1895), 277; Jessie G. Forrester, "Introductory Address to the Bennett Class of '95," *Medical Times*, 1895, *28*: 90–91.

74. Lucy Waite, "Surprising," *The Medical Visitor*, April 1889, *5* (4): 115.

75. Women's visibility and involvement in professional organizations is readily apparent in multiple volumes and issues of local sectarian journals including *The Medical Visitor*, *The Clinique*, and *The Medical Times*.

76. Carla Bittel, "Mary Putnam Jacobi and the Nineteenth-Century Politics of Women's Health Research," Chapter 1 in this volume. See also Carla Bittel, "The Science of Women's Rights: The Medical and Political Worlds of Mary Putnam Jacobi" (Ph.D. diss., Cornell University, 2003).

77. More, *Restoring the Balance*, 40, 45–46.

78. Ibid., 33.

79. Moldow, *Women Doctors in Gilded-Age Washington*, 135–156.

80. Kirschmann, *A Vital Force*, 49.

CHAPTER ELEVEN

Ruth A. Parmelee, Esther P. Lovejoy, and the Discourse of Motherhood in Asia Minor and Greece in the Early Twentieth Century

Virginia A. Metaxas

Esther Pohl Lovejoy arrived at the burning coastal Turkish city of Smyrna on September 24, 1922, and found herself to be the only American woman to witness the catastrophic exodus of the Greek people from Asia Minor. Lovejoy had been attending a meeting of the Medical Women's International Association in Geneva when she heard that a quarter of a million Christian minority refugees, driven from the countryside to the city by the Turkish Army, desperately awaited evacuation before certain death.[1] Having had experience with providing medical aid during wartime, Lovejoy immediately left for Turkey with two goals in mind. First, she wanted to help victims of the war between Greece and Turkey, just as she had helped civilians and soldiers during the First World War in France. Second, she expected to gather vital information that she would later share with the American public to raise support for medical aid for the Greek refugees.

Lovejoy was building on a tradition set in motion by many secular and missionary Americans who had served in various Near Eastern locations. These Americans witnessed many years of hostility and violence between the Turks, whom they often characterized as feared, despised, and uncivilized "Mohammedans" who followed the Islamic faith, and the Greeks and Armenians whom they considered to be "Christian martyrs," especially in the context of oppression and genocide under Ottoman rule. Indeed, late-nineteenth- and early-twentieth-century reports from American missionaries and diplomats had resulted in widespread public awareness of the plight of Armenians and Greeks in the Ottoman Empire. Many religious and secular grassroots organizations conducted massive public fund-raising campaigns so that American schoolchildren knew of and contributed to saving the thousands of "starving Armenians" displaced by the conflict in the area.[2] Similarly, the Greek genocide and forced population exchange of 1922 became part of the American consciousness through the efforts of American witnesses who sought to form public attitudes and possibly achieve humanitarian intervention. These unprecedented international human rights campaigns helped shape America's national and international identity[3] and, in some ways, laid the groundwork for an emerging American Empire.

Although much critical work has been written about women physicians who practiced medicine in the United States in the nineteenth and twentieth centuries, less has been done to analyze and contextualize the lives and careers of American medical women who served overseas in crisis situations, often as missionaries or in the military but also as lay physicians and nurses attached to international humanitarian organizations.[4] Such are the stories of Esther Pohl Lovejoy and Ruth A. Parmelee, who both provided medical aid in the context of war and growing nationalism in Greece and Turkey. Moreover, they served as conduits of information that fed the development of an early twentieth century gendered and racialized discourse about Armenians, Greeks, and Turks. "Near Eastern" Armenians, Greeks, and Turks were seen as "other" in the West in various ways; yet, the American public responded to crisis after crisis by supporting financial aid and relief personnel for the Christian populations of Turkey and Greece in part because of the way American medical women framed the story.

Uncritical works on women physicians working outside of the United States often present heroic "great doctor" narratives of individual physicians, such as *House of a Thousand Babies: Experiences of an American Woman Physician in China*,[5] or *Palace of Healing, The Story of Dr. Clara Swain, First Woman Missionary Doctor and the Hospital She Founded*.[6] Both these treatments of women practicing medicine overseas read as if they were written for an adolescent female audience, meant to inspire young readers for a life of adventure and service. Neither offers critical analysis of the imperialist activities of missions and medicine, nor do they adequately contextualize women's position in these projects. *Physician to the*

World, Esther Pohl Lovejoy, published in 1973, a semifictional biography, similarly lacks adequate historical background about the position of women in medicine.[7] In general, most books of this genre fail to provide adequate historical background on the actual conditions in the foreign country, and more importantly they fall short in problematizing the presence of Americans seeking to influence the internal affairs of such places.

Certain Samaritans,[8] Esther Pohl Lovejoy's well-known book outlining the work of the American Women's Hospitals in several countries during the twentieth century likewise uses a heroic narrative in content and tone, this time in the genre of group biography and institutional history. This book and Lovejoy's *Women Physicians and Surgeons, National and International Organizations*,[9] and Ruth Parmelee's autobiography *A Pioneer in the Euphrates Valley*[10] lack self-reflection in terms of questioning the expansion of American medical and missionary authority outside of the United States. Nonetheless, they are valuable sources for information about the activities of American physicians abroad, shedding light on the ways in which Lovejoy and Parmelee represented themselves, their clients, and their work. By the time Lovejoy and Parmelee entered their respective medical careers, American medical women had established their rightful place in providing care to women and children. Certainly, as authors, Lovejoy and Parmelee justified women's work in medicine and in addition they capitalized on the familiar trope of universal motherhood to fund particular medical projects in Turkey and Greece.

Esther Pohl Lovejoy is well known to medical historians because of her central and significant leadership role in the international women's medical movement. Ruth Parmelee is lesser known, partly because she spent most of her life outside of the United States. Comparing and contrasting the careers of these two physicians, both dedicated to the plight of the Armenians and Greeks in the first half of the twentieth century, shows how they played different but equally vital roles. Parmelee worked in the field, providing direct medical care and sending important information back to the sponsoring organizations in the United States. She carried the moral compass and served as a bridge between the missionary and the secular medical women's networks. Lovejoy, from her stateside office, managed the narrative and the fund-raising drives so that the American Women's Hospitals could provide the necessary ongoing relief. As Eve Fine demonstrates in her essay (Chapter 10 in this volume), regular and sectarian women physicians cooperated with each other even though they represented different traditions. Professional cooperation shaped the relationship between

Parmelee and Lovejoy as well. Parmelee approached her work as a medical missionary even while she worked within Lovejoy's secular organization, the American Women's Hospitals.

Missionary physician Ruth A. Parmelee served on the ground in Turkey and Greece from 1914 until after the Second World War. Born in Trebizond, Turkey, in 1885, of American missionary parents, she witnessed, throughout her childhood, the protracted hostilities between Turks and Armenians. After an American medical education completed at the University of Illinois in 1912, she spent the better part of the years between 1914 and 1922 in Harpoot, Turkey,[11] as a medical missionary, providing obstetrical services, training nurses, and caring for war orphans under the auspices of the American Board of Commissioners of Foreign Missions and in cooperation with the American humanitarian organization Near East Relief. In 1922, she fled to Greece leaving Asia Minor because of the internationally approved Lausanne Treaty, which forced an estimated one and a half million Christian Greeks and Armenians to leave Asia Minor. From 1922 to 1925, she provided immediate aid to the refugees in the American Women's Hospitals in Thessalonika, Greece, and from 1925 to 1933, she directed the American Women's Hospital in Kokkinia, near Athens, serving the large refugee community of seventy thousand people in that area. She worked in Greece for the rest of her professional career until she retired in the 1950s.[12]

Esther Pohl Lovejoy, a leader of the international women's health movement, began focusing her work on international populations during the First World War in France. Lovejoy was born in 1869 in Oregon, graduated from medical school in 1894, and had many years of medical experience, including public health work, before she organized women physicians' work during World War I in France. She turned her attention to aiding the Greek people after her experience in Smyrna, as president of the American Women's Hospitals after 1919.[13] Lovejoy convinced her colleagues that the exigencies of war, such as "famine and forced migrations—attended by starvation and pestilential diseases," wreaked havoc on civilian populations, particularly women and children. From her American Women's Hospitals office in the United States, she took a leadership role in justifying and ensuring decades of medical aid to the Anatolian Greek refugee population repatriated to Greece. Her record in spearheading dozens of other American Women's Hospitals projects all over the world is well known; she tirelessly worked to sustain women's international medical work until her death in 1967.[14]

This chapter focuses on the way that early-twentieth-century international medical women created a central place for themselves to provide aid to women and children overseas, often in dangerous and difficult circumstances. They shaped a narrative of motherhood to be shared with the American public in which childbirth and maternal responsibilities were described as relentless, even in the face of disaster, thus justifying their presence as helpers in sometimes horrific situations. The research is based on primary and secondary sources from three types of organizations—(1) Christian missionary, (2) secular humanitarian, and (3) medical. Specifically, the chapter relies on the records of the American Board of Commissioners of Foreign Missions (ABCFM), Near East Relief (NER), and the American Women's Hospitals (AWH) and on the memoirs, diaries, and correspondence of Ruth A. Parmelee and Esther Pohl Lovejoy.

Americans in the Near East

Before the First World War, over a thousand American humanitarians, mainly missionaries with an agenda to convert "natives" to American Protestantism, had served in the Near East, in various locations in the Ottoman Empire.[15] Beginning in the early nineteenth century in reaction to the zealousness generated by the religious revivals in America and in Britain, missionaries from the American Board of Commissioners for Foreign Missions and other societies from the United States and Britain were determined to establish missions in the Holy Land, in Palestine in particular, which was under the rule of the Ottoman Empire at the time. Driven by religious commitment, they endeavored to convert Palestinian Muslims, Jews, and non-Protestant Christians to Protestantism. Their efforts in Palestine were not successful; the religious and ethnic populations there proved resistant to their conversion efforts, thus they shifted their focus to other areas in the Ottoman Empire. They also encountered resistance in Greece and in cosmopolitan city-ports such as Smyrna or Constantinople, where Jews, Muslims, and Christians had lived side by side for centuries. Quickly the ABCFM abandoned its agenda to convert Jews and Muslims, concentrating instead on evangelical work among the Greek and Armenian Christians.[16] They set up a number of stations in Greece and in coastal cities of Asia Minor in the 1830s, hoping to convert the Christians there to Protestant religious convictions, drawing them away from what they considered "Oriental dogma."[17] Eventually, the missionaries spread into Central and Eastern Turkey, establishing contact and influence with many Armenian and Greek

Christian communities who lived under Ottoman rule. In general, missionary efforts to convert Greek and Armenian Christians proved fruitless. Believing in the superiority of their own social, political, religious, and economic ideas, the American and British missionaries strove to find other means by which to bring their cultural ways to Greek and Near Eastern peoples.[18] Soon missionaries established humanitarian projects and educational institutions to promulgate Western ideas. The work of the American missionaries was divided geographically in two ways: The ABCFM worked in Greece, European Turkey, and Asia Minor; and the Presbyterian Board covered the Arabic-speaking areas of the Ottoman Empire. British missionaries organized and established their own stronghold in the Arabic-speaking regions of the Near East.[19]

An additional agenda for American missionaries in the region was to assist war victims and to attempt to influence American government involvement. This began in Greece when American Philhellenes in the 1820s wished to aid the cause of Greek independence after four hundred years of domination by the Ottomans. Since at least the late eighteenth century, educated Americans in general had been fascinated by Classical Greek culture, idealizing in particular its democratic philosophical traditions. When the Greek War of Independence began in 1821, in the context of America's New Republican ideology, many citizens of the United States hoped that Greece would return to its democratic political origins. Some went so far in their belief in the cause as to fight in the war on the Greek side.[20] Other American supporters tried to convince the U.S. government to officially help the Greeks obtain their freedom in the war that lasted until 1830—when the Greeks finally established an independent Greek state—but they failed to do so given the establishment of the Monroe Doctrine of 1823, which prohibited U.S. involvement in areas that had been under European economic and political influence. Greece continued its irredentist struggle for the rest of the nineteenth century, but for the most part, the United States government continued, officially, to remain neutral in most instances because of American commercial interests in the growth of trade with Greece and the Near East.[21]

Nonetheless, many ordinary American citizens contributed money to assist civilians. Harvard-trained physician and reformer Samuel Gridley Howe, for example, raised $60,000 in American funds to establish a fifty-bed hospital at Poros after volunteering for six years in the Greek Army in support of independence.[22] Beginning in the 1830s, and lasting through the rest of the nineteenth century, missionary groups also encouraged and financed the education of Greek war orphans—most of whom were sent to the United States to receive higher education. Americans also established and ran elite schools for boys and girls in urban centers to which Greek bourgeois families sent their children. Realizing that few families would agree to convert to the Protestant faith, the schools offered a secular curriculum, emphasizing exposure to those Western (read American) social, political, economic, and cultural values the missionaries saw as appropriate for a modernizing society. The missionaries also set up projects introducing modern scientific methods in medicine and agriculture, believing that their Western knowledge was superior to the traditional ways of the people there.[23] As historian Dimitra Giannuli has argued in her description of missionary projects in the Ottoman Empire during the nineteenth century, "overall, the missionaries introduced modern and secular models of life aimed at spiritually enlightening the natives and raising their living standards."[24]

American missionaries first appeared in Turkey when the ABCFM established a mission in Constantinople in 1831.[25] By the early 1860s, the establishment of several schools, a press, and some medical projects clearly reflected these changing goals on the part of the ABCFM, which by then emphasized education rather than religious conversion. In the 1850s, there had been only a few hundred students in the ABCFM schools; however, by 1914, nearly twenty-five thousand Greek and Armenian Christian students attended several hundred primary, grammar, high, and normal schools scattered around Turkey.[26]

By the 1860s and 1870s, colleges had been established as well, including a school for women in Constantinople.[27] According to an official census of the ABCFM taken in 1919, American staff of the Protestant Missions in Turkey was made up of 50 ordained ministers, 71 un-ordained men, 61 wives, 104 unmarried American women, and more than 1,300 "native" personnel,[28] all of whom staffed the schools, hospitals, and presses run by the ABCFM. There were more women than men involved, both American and "native," which should come as no surprise, given the kinds of services offered—educational, social, religious, and medical.

Studying the life of physician Ruth Parmelee is useful in understanding the drive and motivation of many of the Americans in the region. Her devotion to humanitarian work had been internalized since childhood; it was almost unavoidable given her family background in which work and sacrifice were the norm. Both of her parents served as role models for the adult woman Ruth would become. Her father, Moses Payson Parmelee, M.D., graduated from Union Seminary in 1861, making his first trip to Erzrum, Turkey, to be a missionary in 1863.[29] In 1872, two years after the death of his first wife Nellie A. Frost Parmelee and daughter Jennie, Reverend Doctor Parmelee married Julia Farr. In the next decades, they lived and worked as missionaries in Erzrum and later Trebizond, Turkey, raising several children and serving in the medical and educational missionary projects there. Ruth, the last child of the family, born April 3, 1885, would follow her parents' calling as an adult by becoming both a physician and a missionary.[30]

Julia Farr and Ruth participated in expanding roles for women in missionary work in Turkey. Although in the early years of missionary presence in the Near East wives were expected to serve as "helpmates" to their husbands, by the end of the nineteenth century, wives of male missionaries, and single women as well, traveled to foreign lands to play a more public role by involving themselves in schools, orphanages, and other entrenched missionary institutions.[31] Reforms aiming to educate girls as well as boys in secular schools in Central and Eastern Turkey as well as in the European Ottoman territories encouraged the establishment of schools from kindergarten to high school and even colleges and seminaries. These schools needed teachers, and hundreds of educated women were recruited from America to serve. Mostly Christian Greek and Armenian girls and women attended these secular schools while most Turkish women attended Ottoman schools, where reforms, including the education of women, were also under way. By the end of the nineteenth century, the American missionary schools widened educational opportunities for young people throughout the Ottoman Empire in addition to establishing equal access to higher education for both women and men.[32]

Ruth Parmelee felt most at home identifying herself with missionary work, and even when she affiliated with the American Women's Hospitals, clearly a secular and in some ways a feminist organization, she would often assert her missionary identity.[33] Esther Pohl Lovejoy, however, unmistakably self-identified as a fiercely independent secular physician with the ambition to make the American Women's Hospitals a vehicle through which women could not only help others but also ensure professional autonomy and higher status for women in medicine throughout the world. Historian Ellen S. More characterizes Lovejoy as a "New Woman" who "stepped smartly to the rhythms of twentieth-century America" by unabashedly finding her calling in combining "medicine and politics."[34] When the First World War broke out, there were at least ten thousand women doctors in the United States, all barred from military service because of their sex. The Medical Women's National Association (MWNA) appointed a war service committee charged with collecting funds and providing opportunities for wartime service for women physicians.[35] Even though women could not conduct war service through government channels, they sought nonmilitary opportunities such as work for the American Red Cross. Finally, the MWNA established the American Women's Hospitals, modeled after the Scottish Women's Hospitals. Wrapped in military garb and patriotic rhetoric, and eager to help others, learn new medical techniques, and even to have an adventure or two, the American Women's Hospitals medical women provided relief work in France, Serbia, Turkey, and myriad other places during the war.[36] Lovejoy did not want to lose momentum as the war ended, so she found other projects for the AWH.

"Terrible Motherhood"

In late 1922, when Lovejoy first described the plight of Greek women refugees that she witnessed in Smyrna, she called it "terrible motherhood."[37] "Never were babies brought into the world in stranger and sadder surroundings," Lovejoy reported to the American public. One woman, she said, "gave birth to her first child as she stood in line [for a ship], fearing to give up her place." Women gave birth out in the open "on the quays, while streams of panicked humanity surged around us. Some babies were born alongside the gangplanks of departing ships." In her view, the sanctity of the family was torn apart as men were systematically separated from women and children by Turkish soldiers. The men, for the most part, were taken to the interior to be placed in labor gangs or killed. Panic-stricken children threw "their arms about the legs of their fathers and [shrieked] for mercy [while] wives clung to husbands in a last despairing embrace."[38] Some sons and husbands escaped if they were strong enough to swim to Greek ships in the harbor—and if they could avoid being shot by the Turkish soldiers ordered to stop them. At night, under the cover of darkness, Turkish soldiers raped the Greek women, and if the women survived to board a ship for Greece, they faced many unknowns, including permanent separation from loved ones and years of scraping a living in refugee camps. Lovejoy later lamented, in an article describing the scene at Smyrna, "I never believed such things could happen."[39] She devoted four chapters to the disaster in "the martyred city" of Smyrna in *Certain Samaritans*.[40] The women physicians Lovejoy described in her book were not just the biblical Good Samaritans who tried to help the war victims; these were "Certain Samaritans," medical women who served with drive and purpose to help women and children in dire need.

Nearly thirty thousand people perished in the Smyrna disaster, and within a few weeks, seven hundred fifty thousand fled from all areas of Asia Minor—most of whom were women and children without male support. They arrived in terrible condition at the Greek ports of Salonica and Athens, and on the islands of Crete, Mytilene, Chios, and Euboea. In each of these places, following Lovejoy's directives, American Women's Hospitals personnel aided the refugees. Hungry and sick, with few resources, the refugees suffered diseases such as typhoid and smallpox. The crowded and unsanitary conditions produced by the refugee trek left most infested with lice. Some were so depressed they committed suicide, unable to cope with their losses. The refugees were without shelter, blankets, or clothing and in desperate financial ruin.[41] Eventually, the refugees numbered more than one million, and Greece, a poor country of approximately five million people, had to find ways to aid and absorb this sick and impoverished population, their compatriots from Asia Minor. Although Lovejoy represented the secular American Women's Hospitals, she nonetheless characterized this movement of humanity as a "Christian exodus."[42]

Like Lovejoy, Ruth Parmelee was part of the flight from Turkey in 1922, and she too used religious terms to characterize the humanitarian catastrophe. Take, for example, the figure depicting Dr. Parmelee standing over a Greek refugee woman and her baby.[43] This is only one of many gendered representations of Christian Greeks that the AWH used to raise sympathy and financial support for relief aid in Greece during this time. What we see here is a typical Greek Orthodox Christian image, that of Madonna and Child, which implies religious connotation about the sanctity of the relationship between mother and child. This AWH publicity image was meant to stir the conscience of prospective Christian donors who would contribute funds to the medical projects needed in postdisaster Greece. In contrast, in the same image, notice the figure of another woman, in this case a Protestant missionary woman, Parmelee herself. For Parmelee and many other medical women involved in this relief work, saving the lives of women and children was particularly important. While creating a narrative of motherhood which would appeal to American donors and sympathizers was vital to continuing support for relief projects, it was also important to reinforce Parmelee's authority and responsibility in overseeing the well-being of mothers and children. Thus, the image also speaks about Parmelee's self-identity—that of doctor and child saver—that of scientist and humanitarian. The following section will discuss Parmelee's earlier work in Turkey during which time she witnessed the Armenian genocide and began to be aware of her role as a communicator of conditions in Turkey.

Dr. Parmelee in Turkey, 1914–1917

Ruth Parmelee also wrote reports on the violence she witnessed long before the culminating event in Smyrna, when the last of the surviving Christians left Turkey. Her work as a professional physician and missionary began in January 1914, just two years after she completed her medical education at the University of Illinois and immediately following her internship at the Philadelphia Women's Hospital in 1913. Having gained practical experience in laboratory work and in obstetrical and in gynecological practice during her internship, Parmelee was more than eager to get to work as soon as she could.[44] A few days after the ABCFM appointed her to go to Turkey, she wrote in her diary, "I am glad to be of service—that is what I need."[45]

But the Turkey to which Parmelee returned was much worse than she remembered from her childhood. As James A. Barton, leader of the humanitarian organization Near East Relief would say, "Turkey stood, in 1914, at the focal center of . . . conflicting international political forces."[46] As the Ottoman Empire underwent its final death throes, several of the European powers circled like vultures to gain economic and territorial pieces of it. Turkey also struggled with internal strife during this period; clashes between the majority population of Muslim Turks and the Christian Armenian minority flared up with frequency. A series of massacres of Armenians occurred in 1894–1896[47] and again in 1909, even as Turkey began to modernize under the Young Turk Party.

During the First World War, the Turkish military and government grew suspicious of Armenian loyalty and took the opportunity to solve the "Armenian problem" once and for all. In 1915, the Turkish military was given orders to eliminate Armenian community leaders, and the Turkish government gave a general order for Armenians all over the country to prepare for "deportation." Soon thousands were marched into the deserts of Syria, Mesopotamia, and Arabia; many were executed or died of starvation or exhaustion along the way. Thousands of Armenian women were raped by Turkish soldiers, and many women were abducted and taken into what Americans called "Turkish harems." These war crimes resulted in a genocide that left thousands of children or-

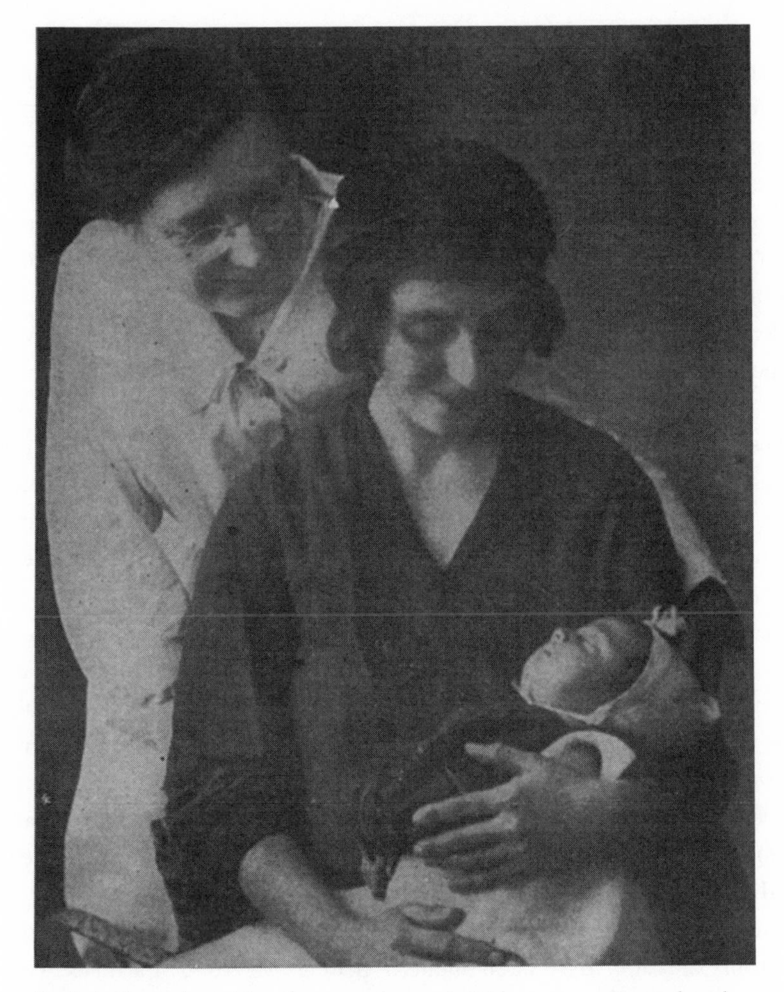

Ruth A. Parmelee with refugee mother and child, ca. 1925. Reproduced with permission from the Ruth Parmelee Papers, Hoover Institution, Stanford University.

phaned. At least one and a half million Armenians died in this horrendous genocide of 1915–1916.[48]

Parmelee left thorough records of her work in Turkey from 1914 to 1917, while she served as a missionary physician, and then again from 1919 to 1922, during which time she worked under the auspices of Near East Relief.[49] She recorded some of her accounts in diaries and correspondence with U.S. authorities in the missionary and relief agencies. In correspondence sent during the First World War, she described, in cautious terms, the interactions between the Turkish authorities and the missionaries or the Armenian people, fearing that mail or personal papers would be seized and read. She censored herself, yet made it clear to the reader that she was doing so. In one letter to a friend, for example, she managed to get information through about the Turkish atrocities against the Armenians by using code language that would signal the reader to lift the stamp and see her secret message. She did so by mentioning, at the end of the letter, "I am glad you are so good a fisherman," which alerted the reader to "fish" under the "pool" or stamp, where she found the words: "when the people were hauled to the government buildings, most let off, some exiled. That probably means death. Our summer camping region is now filled with bodies. Cruel Manner."[50] This communication may have been sent before the Turkish authorities uncovered a code that had been devised by the missionaries. Later, after the war ended, the papers and speeches that she wrote about the violence inflicted on the Armenians were more passionate and pointed. Clearly, she wanted continued American support for the Armenian cause.

In late May 1915, Turkish authorities collected all arms from the Armenians in the town where Parmelee resided, signaling danger. By June, rumors abounded that all Armenians would be exiled. Some tried to bribe officials or hide. Others promised to become what Parmelee called "Mohammedan"—a believer or follower of Islam. The missionaries saw fires on the horizon and realized that nearby villages were burning. By July, Parmelee concluded that the "attempt to extirminate [*sic*] a nation is diabolic, to the extreme," when she had heard that both men and women were being massacred *en masse*.[51]

During the summer of 1915, Parmelee and some of the missionary women visited an exile camp in Meszreh (modern spelling "Mezre").[52] Parmelee wrote a report on what she saw there, to be sent to the missionary authorities in Boston who subsequently publicized the facts. They found the people "in wretched condition," wearing filthy and ragged clothing. Hunger, illness, and emotional suffering among the Armenian refugees overwhelmed the camp, and as the American women entered, the inmates surrounded them, begging for food. Most of the camp refugees were women and children with a few old men from the towns north of Meszreh. The younger men had been killed along the way. In the center of the camp, a "square trench" had been built to hold dead bodies as the refugees died; each layer was covered with a small amount of "earth" until "further bodies" furnished the next layer. Bread, supplied by the Turkish guards, was scarce, and those who had a little bit of money left would send out to buy some. Parmelee was careful to recount stories about the women in the camp, saying that a few had told her that they used various ways to hide money. They placed gold pieces in the bottom of their water jugs, or swallowed coins. Sometimes, Dr. Parmelee said, the women inserted money into "the vaginal orifice." Unfortunately, the guards "would suspect these tricks and would pour out the drinking water onto the ground, and in some cases [go] so far as to examine the women internally or to search all their clothes." One woman told how she tricked the guards by hiding some money in her "monthly napkin." The "Mohammedan" guards followed their religious law and did not touch this "unclean woman."[53]

Parmelee and the rest of the missionary community were ordered to leave Turkey in the middle of May 1917, following the U.S. entrance into the war on the Allied side. They were filled with anxiety about what would happen to their Armenian charges—the thousands of men, women, and children who were ill and hungry in the wartime environment. Parmelee was determined to return to Turkey as soon as she was officially able, so she immediately threw herself into writing descriptions of what she had seen to raise funds for continued aid. She sent many of these accounts to Reverend James L. Barton at the ABCFM who would soon head Near East Relief, an organization that raised funds by depending on stereotypical images of "the Mohammedan Turk" and "the Christian Armenian." Parmelee had no problem vilifying Turks and exalting Armenians in her accounts of what she had seen, characterizing the Armenians as innocent victims and the Turks as cruel aggressors.

While she was in the United States, Parmelee gave several talks to the American people about what she had witnessed during the days of the "Armenian deportation."[54] She emphasized to her audiences the plight of the Armenian woman in particular, seeking to evoke sympathy for their cause. She spoke about how mothers "saw their girls kidnapped by Kurds or Turks" to be taken into the dreaded harem, where they would be sexually violated. Armenian women also saw their "husbands and sons taken away to be killed, or even shot down before their eyes." They often were forced to abandon "their dear ones left sick by the roadside." Sometimes, she said, in desperation, "women would throw their little babies into the river." Parmelee wondered how the women could survive such psychological loss, but then noted that they emotionally disassociated as a means of survival. "It was a mercy, perhaps, that [the Armenian women's] capacity to suffer had become somewhat dulled, or they could not have endured at all."[55]

Parmelee wanted to be sure that her audience understood the gravity of what she considered state-sanctioned genocide. She argued that there was "a definite plan to wipe out the whole nation" of Armenians. She quoted a conversation that allegedly took place between the Turkish Minister of War Enver Pasha and the American Ambassador Henry Morgenthau Sr. in which the former said, "It was the plan of the Turkish government to get rid of the Armenians, then the Greeks, then the foreigners, and have Turkey for the Turks."[56] She pleaded with her audience: "Because we are Americans and are in the war to give liberty to the oppressed peoples of the earth, we must send help to the martyr nations over in Asia Minor and the Caucasus." She equivocated slightly in her harsh characterization of Turks, distinguishing between the actions of the government and those of ordinary Turkish citizens. "We are hoping and praying that the Turkish Empire may never have the power it has had in the past. Not that the Turkish people must be annihilated—they need good government just as much as their subject races. But the rulers of Asia Minor . . . have proved themselves unfit to rule even their own people, let alone their oppressed races." Trying to win the American public's condemnation of the Turkish government's actions, she tried to reach their religious conscience, saying, "it will be a blot on Christendom, if when peace is made, the Sick Man of Turkey remains in power."[57]

Parmelee predicted that the Turkish government's plan to "Turkify" the nation would mean that the Greek minority would be the next to be attacked and removed—and she was right. By 1921, she characterized her third year of post–World War I relief work in "one word—deportations." She witnessed whole villages of marching Greek men, women, and children, in terrible condition. Parmelee estimated that twelve thousand of those who had started in forced marches died of starvation, exposure, or disease. Another eleven thousand were pushed towards Bitlis, "a wild region," and never heard from again. Eventually, the American relief workers received deportation orders as well. In Parmelee's mind, the Turkish authorities wanted to get rid of the NER workers to "counteract all efforts to aid the minority races and keep them alive."[58] She would later write, "what is the civilized world going to do about . . . [the Turkish government's] steady oppression of the Armenians and the deportation of so many of the Greeks?"[59]

Her rhetorical question was answered by the Treaty of Lausanne, signed in 1923 by the British Empire, France, Italy, Japan, Greece, Romania, and the Serb–Croat–Slovene state on one side and Turkey on the other. The treaty compelled Greece to relinquish all territory in Asia Minor, eastern Thrace, and two small islands off Turkey's northwest coast. It also crystallized the issues surrounding the largest single compulsory exchange of populations known in history. The treaty required all Muslims living in Greece, except for the Slavic Pomaks and Turkish Muslims in Thrace, and the Dodecanese, to be evacuated to Turkey. They numbered nearly four hundred thousand. In return, approximately 1.3 million Greeks were expelled to Greece. NER and AWH personnel would be directly involved in ameliorating the disastrous effects of this vast and rapid forced movement of dispossessed peoples.

By 1924, the AWH staff had created thirty-four hospital projects in Greek territory, including hospitals for Armenian orphans brought to Greece by NER. From January to June 1923, as the fleeing refugees swarmed over from Turkey, they set up a quarantine station and hospitals on Macronissi Island, "where pest ships from the Black Sea ports discharged their human Cargoes."[60] Four months after Smyrna, and upon her return to Greece after a fund-raising trip to the United States, Esther Lovejoy visited the quarantine island at Macronissi to see refugees coming ashore—"a tragic procession made up of women, children, the aged, and a small proportion of able-bodied men, struggling through the sand with their bundles on their backs." They were sent to the camp of the "unclean" until they were deloused and their belongings disinfected. Lovejoy described the women, whose husbands had probably been "detained," as strong and determined to survive. "There were no weaklings in this brood . . . Who was she? No one knew. What was she? Everyone knew. A strong mother, an honor and an asset to her people."[61]

Parmelee, on loan from NER but working for the American Women's Hospitals, focused her reports on maternity care: "Thousands of babies [were] being born on the road, on the steamer, or after arrival here [in Salonica] with no provision whatever for the comfort and necessities of the poor mother and infant." She reasoned that the new AWH hospital would "provide accommodations for women needing this care" as well as "give employment for a dozen nurses, and several doctors, to say nothing of the servants and attendants—all refugee, and in need of employment."[62] She said it was up to Americans to "care and help relieve the misery and suffering" indefinitely, until the "barbarities" that drove one and a half million people from their homes ended. She pledged, "We who are on the Front Line plan to remain on our post as long as God gives us strength for our work"—and she did just that, for decades after.[63]

Conclusion

Parmelee and Lovejoy were two of the many Americans who aided thousands of victims of war and conflict in Turkey and Greece in the early twentieth century, and they often put themselves in harm's way to do it. It is difficult to see these women as less than heroic figures, such as how society now views those who work for Doctors Without Borders. Ruth Parmelee may have seen herself in this heroic light when she looked back on her life's work in her book *A Pioneer in the Euphrates Valley.* In the preface, she said: "Never did I think I could be a pioneer, but one day, after beginning my work as a medical missionary, it suddenly came to me—'I *am* a pioneer, the first woman to practice medicine in the Harpoot region.'"[64]

The study of Parmelee and Lovejoy provides more than a heroic story. We can see how American medical women extended their well-established place in working with women and children beyond the geographic boundaries of the United States. In this case, interestingly, the narrative of motherhood they created used religious symbols and language because part of this discourse was borrowed from well-developed themes promoted by Protestant missionaries who had worked in Asia Minor and Greece well before the early twentieth century. By combining Christian imagery or language with the universal experience of motherhood and childhood, "exotic" Greek and Armenian women and children became more like Americans. American donors responded by funding relief efforts—Dr. Lovejoy raised more than $2 million in 1924 alone for the AWH hospitals in the Greece, the Balkans, and the Near East.[65]

Lovejoy added a feminist spin to the narrative. In her view, "being a refugee [did] not preclude the sacrifice and glory of motherhood. Neither hunger, nor thirst, pestilence nor cruelty, not even the spectacle of the waste men make of life, can deter woman from her profound function." Blaming men for the ways that they "warred for property and power," she reminded her readers that "women patiently renewed what they have destroyed." For Lovejoy, only women could understand the plight of other women in desperate straits—and she made sure that her organization would support a world "where motherhood transcends the destruction of men."[66] In time, the Greek government decorated both Lovejoy and Parmelee with medals of honor for this humanitarian work.

NOTES

I would like to acknowledge several colleagues for reading and commenting on earlier drafts of this chapter: Rosalyn Amenta, Steve Amerman, Polly Beals, Nikolaos Chrissidis, C. Jane Covell, Michael Horvath, Amy Kesselman, Tricia Lin, Nacia Miller, Byron Nakamura, Troy Paddock, Troy Rondinone, Michele Thompson, and Katherine Wiltshire. I also wish to thank Southern Connecticut State University for helping support this research through summer travel grants.

1. Esther Pohl Lovejoy, "Smyrna," *The Woman Citizen*, November 4, 1922, 8, in *Woman's Journal*, New York, vols. 1–16, no. 6; June 2, 1917–1931; "Smyrna! Where Motherhood Transcends the Destruction of Men," *The New Near East*, November 1922, p. 18. Papers of the American Relief Administration, European Unit, Box 698, Folder 1, "Donation," located at the Hoover Institution, Stanford University.

2. A recent treatment of the Armenian genocide and the response by the American public and government can be found in Peter Balakian, *The Burning Tigris: The Armenian Genocide and America's Response* (New York: HarperCollins, 2003).

3. Ibid., xiii.

4. See Barbara Reeves-Ellington's dissertation, "That Our Daughters May Be as Corner Stones: American Missionaries, Bulgarian Nationalists, and the Politics of Gender" (Ph.D. diss., Binghamton University State University of New York, 2001), for a critical study of the role of American women missionaries in shaping U.S. ambitions for empire during the nineteenth century in the Ottoman Empire.

5. Frances W. King-Salmon, *House of a Thousand Babies: Experiences of an American Woman Physician 1922–1940* (New York: Exposition Press, 1968).

6. Dorothy Clark Wilson, *Palace of Healing: The Story of Dr. Clara Swain, First Woman Missionary Doctor and the Hospital She Founded* (New York: McGraw-Hill, 1968).

7. Olive W. Burt, *Physician to the World, Esther Pohl Lovejoy* (New York: Julian Messner, 1973).

8. Esther Pohl Lovejoy, *Certain Samaritans* (New York: Macmillan, 1928).

9. Esther Pohl Lovejoy, *Women Physicians and Surgeons National and International Organizations: Book One, the American Medical Women's Association, the Medical Women's International Association, Book Two Twenty Years with American Women's Hospitals, a Review* (Livingston, NY: Livingston Press, 1939).

10. Ruth A. Parmelee, *A Pioneer in the Euphrates Valley* (n.p.: privately printed, 1967).

11. The modern spelling of Harpoot is Harput. The names and spellings of town and cities as they appear in the original sources will be used in this article.

12. "Outline of the Work of Dr. Ruth Parmelee," typescript ms., Papers of the American Women's Hospitals, Archives and Special Collections on Women and Medicine, Medical College of Pennsylvania, Philadelphia 19129.

13. Lovejoy, *Women Physicians and Surgeons*, 121.

14. For information about other American Women's Hospitals projects, see Esther Pohl Lovejoy, *Women Physicians and Surgeons National and International Organizations*. Direct quote taken from *Women Physicians and Surgeons*, p. 121.

15. Robert L. Daniel, *American Philanthropy in the Near East 1820–1960* (Athens: Ohio University Press, 1920), iv–xi, 1–70.

16. Dimitra Giannuli, "'Errand of Mercy': American Women Missionaries and Philanthropists in the Near East, 1820–1930," *Balkan Studies*, 1989, *39*: 223–262.

17. Ibid., 227.

18. Dimitra Giannuli, "American Philanthropy in the Near East Relief to the Ottoman Greek Refugees, 1922–1923" (Ph.D. diss., Kent State University, 1992; Ann Arbor: University Microfilms International, 1992), microfiche, pp. 1–30.

19. Giannuli, "'Errand of Mercy,'" 226.

20. Giannuli, "American Philanthropy," 1–2.

21. Daniel, 5–11; Giannuli, "American Philanthropy," 3.

22. Daniel, 10–11. Samuel Gridley Howe graduated from Brown University in 1821 and Harvard Medical School in 1824.

23. Giannuli, "American Philanthropy," 3, 27–33.

24. Ibid., 34.

25. Daniel, *American Philanthropy*, 42.

26. Joseph L. Grabill, "Missionary Influence on American Relations with the Near East," paper presented at Malone College at the First Annual Seminar on Christianity and History, March 31 and April 1, 1966, unpublished paper at Yale Divinity School.

27. Daniel, *American Philanthropy*, 71–92.

28. Edward Hale Bierstadt, *The Great Betrayal: A Survey of the Near East Problem* (New York: Robert M. McBride, 1924), 266.

29. John Vinton, et al., *Vinton Book, Volume III: Near East American Board of Missions to 1886 in the Near East ABCFM in Turkey* (n.p., n.d.), 80–81.

30. M. P. Parmelee, *Home and Work by the Rivers of Eden* (Philadelphia: American Sunday School Union, 1888); MS, "Outline of the work of Dr. Ruth Parmelee" found in the Parmelee Papers, Box 1, Folder "not applicable," Herbert Hoover Institution, Stanford University, Palo Alto, CA. Birth and death dates of the Moses P. Parmelee family can be found on a Parmelee Family website, www.geocities.com/mrjimwalters/index.html. I thank Jim Walters of Long Beach, California, who keeps this extensive database. Oddly, no information about Ruth is found in the database.

31. Giannuli, "'Errand of Mercy,'" 221–262.

32. Ibid., 234.

33. For example, note the caption on page 285 in which Parmelee is introduced as a member of the American Board of Foreign Missions.

34. Ellen S. More, *Restoring the Balance: Women Physicians and the Profession of Medicine, 1850–1995* (Cambridge, MA: Harvard University Press, 1999), 131–132.

35. Esther Pohl Lovejoy, *Women Doctors of the World* (New York: Macmillan, 1957), 301–303.

36. More, *Restoring the Balance*, 122–147; Kimberly Jensen, "Esther Pohl Lovejoy, M.D., the First World War, and a Feminist Critique of Wartime Violence," in *The Women's Movement in Wartime, International Perspectives, 1914–19*, ed. Alison S. Fell and Ingrid Sharp (London: Palgrave Macmillan, 2007), 175–193.

37. Lovejoy, "Smyrna!" 9.

38. Lovejoy as quoted in "Smyrna!"

39. Ibid.

40. Esther Pohl Lovejoy, *Certain Samaritans* (New York: Macmillan, 1928), 73–130.

41. Lovejoy, *Certain Samaritans*, 171–177, 190–198, 220–230.

42. Ibid., 269.

43. Photograph located in Ruth Parmelee Papers, Hoover Institution, Stanford University, Palo Alto, CA.

44. MS, "Outline of the Work of Dr. Ruth Parmelee"; "A Line a Day" diary, including the years 1913–1916, Ruth A. Parmelee Papers, Box 1, Herbert Hoover Institution.

45. "A Line a Day" diary. This quote was taken from an entry written on January 7, 1914, less than a week after the ABCFM had appointed her as a full-fledged medical missionary. Ruth A. Parmelee Papers, Box 1.

46. James A. Barton, *Story of Near East Relief (1915–1930): An Interpretation* (New York: Macmillan, 1930), 21.

47. Ruth Parmelee mentions witnessing a massacre of Armenians in 1895 when she lived with her missionary parents in Trebizond in her book *A Pioneer in the Euphrates Valley*, 15.

48. James L. Barton, *Story of Near East Relief*, 3–69; Robert L. Daniel, *American Philanthropy in the Near East 1820–1960* (Athens: Ohio University Press, 1920), iv–xi, 1–70; William Walker Rockwell, *The Deportation of the Armenians* (New York: American Committee for Armenian and Syrian Relief, 1916).

49. Parmelee left Turkey in 1917, upon the entry of the United States into the war, along with all of the missionaries ordered to evacuate, thus the break in her service there.

50. Letter dated January 16, 1916, "Personal Letter to Miss Daniels, Miss Ellen W. Catlin writes," in which a postcard from Ruth Parmelee, dated November 9 [1915] is quoted. Papers of the ABCFM, microfilm, Reel 714.

51. Parmelee, "A Line a Day," diary.

52. Ruth Parmelee, "A Visit to the Exiles in Mezereh" (n.d.), ABCFM Papers, Reel 714. This was probably written in 1917, shortly after Parmelee left Turkey. In a letter to Reverend J. L. Barton, who would soon become the head of Near East Relief, she promised to "try to remember data which would be helpful" to him "in the work you plan to undertake." Quote taken from a letter from Ruth A. Parmelee to Reverend J. L. Barton, September 14, 1917, Papers of ABCFM, Reel 714.

53. Parmelee, "A Visit to the Exiles in Mezereh."

54. Letter from Ruth Parmelee to Reverend J. L. Barton of the ABCFM, May 18, 1918. In this letter, Parmelee mentions making several speaking engagements in trips to Portland, Oregon, from her sister's home in Montana and on her cross-country trek to Chicago where she took a postgraduate course.

55. Ruth Parmelee, "The Armenian Deportations," in the Ruth A. Parmelee Papers, Box 2, File, "no title," p. 2.

56. Ibid., 3.

57. Ibid., 4.

58. Ruth Parmelee, "Relief Work at Harpoot, Turkey, 1919–1922," 5, in Ruth A. Parmelee Papers, Box 2, Folder, Correspondence, Selected Reports.

59. Ibid., 7.

60. Ruth A. Parmelee, untitled speech, folder "Address given presumably in 1926," p. 2.

61. Lovejoy, *Certain Samaritans*, 227–229.

62. Ruth A. Parmelee, "Refugee Work in Salonica, Greece, October 15–December 31, 1922," in Parmelee Papers, Box 2, Folder AWH Relief Work, 1.

63. Ibid.

64. Parmelee, *A Pioneer in the Euphrates Valley*, preface.

65. Ellen S. More, *Restoring the Balance*, 147.

66. Lovejoy, "Smyrna!"

CHAPTER TWELVE

Women Physicians and a New Agenda for College Health, 1920–1970

Heather Munro Prescott

At the twenty-first annual meeting of the American Student Health Association (ASHA) held in 1940, Joseph E. Raycroft, professor emeritus of Hygiene and Physical Education at Princeton University and one of the charter members of the organization, presented incoming ASHA President Ruth Boynton with a photograph depicting the "Forefathers of Student Health." According to the conference proceedings, Raycroft's presentation represented the association's "coming of age" on its twentieth anniversary. The photograph illustrated the members of the Society of College Directors of Physical Education, the organization from which the ASHA evolved. True to the name "forefathers," all of the individuals in the photograph were male and the majority were M.D.s. Singled out for special mention were Edward Hitchcock Jr., director of the first college physical education and hygiene department in the world, which was established at Amherst College in 1861; and George Goldie, director of the first college athletic department in the United States, which was founded at Princeton University in 1869.[1] In his accompanying address, Raycroft praised the wisdom of Amherst President William Augustus Sterns and trustee Nathan Allen for leading the way in protecting and promoting the health of students. Raycroft also honored Hitchcock at length, noting that he was a "man of science," a "student of Agassiz," who was "equipped by the best medical education of the time."[2]

So far, Raycroft's celebration of a "great man of science" should not be surprising to those who are familiar with the marginal place women occupied in science and medicine before civil rights and affirmative action laws passed in the 1960s and 1970s. When I first came across the "forefathers" photograph and article, I found it rather odd that this presentation was made to the organization's first female president. Boynton's reaction to Raycroft's words is unknown because she did not leave any written record of these meetings. Raycroft found it "disappointing" that other elite male colleges failed to follow Amherst's example. In contrast, Raycroft observed that business leaders such as Matthew Vassar, whose name was given to the first women's college opened in 1865, and Henry Fowle Durant, who founded Wellesley in 1875, "gave full consideration in both the earliest plans and later developments, to provisions for the promotion and protection of student health."[3] Hitchcock acknowledged that female institutions led the way in preserving student well-being, stating in the 1870s that the "health-influence" of women's colleges was "better than in gentlemen's schools of kindred grade."[4] In light of these progressive views on health at the women's colleges, said Raycroft, pioneer resident physicians and physical directors at the women's colleges, such as Alida C. Avery and Delia F. Woods at Vassar and Emile Jones and Ida F. Parker at Wellesley deserved the same recognition as Hitchcock. Raycroft, however, chose to focus on male leaders, declaring in a footnote that the history of health and physical education at female institutions "would make a very valuable and interesting paper on this little known phase of Student Health and Physical Education."[5]

Since the early 1970s, women's historians have produced a wealth of scholarship on women physicians and how medical views about women's abilities shaped and hindered women's access to higher education and professional experiences in medicine. These historians have shown that even the forward-thinking individuals who created the early women's colleges could not extract themselves from the gender issues of the era. Despite the efforts of women physicians such as Mary Putnam Jacobi, described in Carla Bittel's essay (Chapter 1 in this volume), even supporters of women's higher education were influenced by the work of Harvard medical professor Edward H. Clarke and other physicians who warned of the dangers of advanced study on female physiology.[6] Therefore, founders of women's colleges, and later coeducational institutions, were among the first to institute widespread health services. Consequently, the position of college physician was a common career path for female physicians in the nineteenth century.[7]

There has been comparatively little work on women in college health during the twentieth century, although Toby Appel's work on women physiology professors and Jana Nidiffer's work on prominent deans of women offers an excellent starting point. According to Appel, "for most of the nineteenth century, 'physiology' was nearly synonymous with personal hygiene and social reform." Therefore, the first generation of women physiology teachers were college physicians, who were hired not only to care for students but to teach mandatory courses in physiology. By the 1920s, physiology was transformed into an experimental science, "characterized by highly technical experimental research on the physical and chemical basis of vital functions in animals," and "the older association of physiology and hygiene became outmoded." Appel describes how the academic physiologists with doctoral training "took over from the remaining college physicians who were teaching physiology."[8] Similarly, Jana Nidiffer shows how the office of dean of women evolved out of the duties of women physicians appointed to oversee the health and morality of female students at coeducational institutions. The first such dean of women was Eliza Mosher, who served at the University of Michigan from 1896 to 1902. At the University of California, Berkeley, the first female administrator was Mary Ritter, who served as both physician and unofficial overseer of campus to women undergraduates. According to Nidiffer, eventually these hybrid positions were superseded by a newer generation of college-educated women, who "forged a new professional identity for themselves as the first senior women administrators on coeducational university campuses."[9]

What happened to the female college physicians in the twentieth century? This chapter will explore some answers to this question by focusing on the careers of two women physicians who became leaders in the field of college health. Ruth Evelyn Boynton, who in addition to being the first female president of the ASHA (later renamed the American College Health Association, or ACHA), was also the first woman to head a university health service that served both men and women. In 1967, the association created the Ruth Boynton Award, which honors those who have provided distinguished service to the organization. Dorothy Boulding Ferebee, an African American woman physician at Howard University, like Boynton, directed a coeducational health service. I will argue that the field of college health illustrates both the possibilities and problems for women physicians in the twentieth century. As in the nineteenth century, the position of physician was one of the few academic careers open to women with advanced professional training, even as other positions in academia and college administration disappeared. To retain these positions, however, women physicians had to discard traditional notions of feminine "sympathy" toward the sick in favor of a masculine notion of professionalism that emphasized scientific rigor and objectivity.[10] As Ellen S. More has observed elsewhere, I will show that Boynton and Ferebee were part of a "vanguard of women physicians" who "managed to reconcile an interest in maternal-child health or public health, the traditional interests of women doctors, with recognition as researchers and academic physicians."[11]

Ruth Evelyn Boynton (1896–1977)

Reconstructing Ruth Boynton's life and career is tricky because she left behind few personal papers. Some clues to her life come from a profile written for the *Journal-Lancet* by former University of Minnesota colleague William P. Shepard shortly after Boynton's retirement in 1961.[12] Boynton's articles on the history of college health also illuminate some of her thinking about major turning points in the development of the field.[13] Initially, I used Boynton's articles to understand key events in the history of college health. The archival record for college health is spotty, and it was helpful to draw on the institutional memory of the field's early pioneers. I soon realized these recollections were a form of commemorative practice that revealed much about the state of the field and the professional outlook of Boynton and her colleagues in college health at the time they were written.[14] For Boynton's generation, "coming of age" meant moving beyond the field's origins in physical education and more firmly grounding college health in the principles of scientific medicine. Although physical educators, coaches, and athletic directors were allowed to participate in the ASHA, physicians in the organization were adamant about distinguishing between their work as medical professionals and the work of physical educators and coaches. According to Boynton, while the organization allowed any individual or institution interested in student health and hygiene to join, its main purpose was to encourage member institutions to increase the availability of medical care to its students.[15]

Boynton's perspective represented a major shift in the focus of college health programs as well as the training of those who directed college health departments. Influenced by the "new public health" that grew out of the bacteriological discoveries of the early twentieth century, college health programs increas-

Ruth Evelyn Boynton. Reproduced with permission from the University of Minnesota Archives.

ingly focused on individual health care and hygiene education, rather than on the physical education programs that had dominated previous efforts to protect the health of student bodies.[16] This emphasis on the "new public health" meant that directors of college health programs should be medical doctors trained in the most up-to-date methods of preventive medicine and personal hygiene.

Boynton's background and training ensured that she would be one of the leaders in the emerging field of college health. Boynton was born in La Crosse, Wisconsin, in 1896, the daughter of Ervin and Nellie Alice (Parker) Boynton. When Shepard asked what she thought persuaded her to go into medicine, Boynton said that her hometown had a husband and wife medical practice who attended to her family's health care needs. The wife in the practice, Mary P. Houck, mainly treated women and children, and cared for the final illnesses of two of Boynton's brothers and Boynton's mother, who died while Boynton was in graduate school. According to her interview with Shepard, although Houck never specifically mentioned medicine as a profession for a woman, Boynton said that Houck's "impressive example" was one of the factors that helped her decide in high school to pursue medicine as a career.[17]

Boynton also benefited from expanding opportunities for female education in the early twentieth century, as well as new notions of what it meant to be an educated woman. When Boynton entered the University of Wisconsin in 1914, the institution was, like many coeducational land-grant universities, trying to adjust to the explosion in female enrollments that occurred during the Progressive Era. Between 1890 and 1920, the number of female undergraduates in the United States grew from 56,000 to 282,900. The greatest growth was at coeducational institutions, especially public land-grant universities where tuition was free or inexpensive. By 1920, 81.3 percent of women were in coeducational institutions. At Wisconsin, President Charles Van Hise was committed to coeducation, but like his counterparts at other land-grant institutions, was concerned that the liberal arts were becoming "feminized" as male undergraduates abandoned the humanities in favor of the sciences and engineering. When Van Hise suggested single-sex classrooms, he came under vicious attack from both female alumnae and their male supporters. Eager to make up for this slip, Van Hise sought to find a new kind of dean of women, who would stand in sharp contrast to the "wise and pious matrons" who oversaw the dormitory system for women students. Determined to hire someone who was both a scholar and a seasoned administrator, Van Hise chose Lois Matthews, an associate professor of history at Wellesley and a protégée of Frederick Jackson Turner. According to Turner, Matthews was an "exceptional woman in the solidity and masculine quality of her judgment and ability to envisage a question; while she has the essentially womanly qualities that one demands of a woman." Matthews was appointed dean of women and associate professor of history in 1911 and soon demonstrated her determination to make sure female students were welcomed into the intellectual life of the university. Matthews was especially concerned that too many women students judged themselves according to a male standard of "social availability," that is, physical attractiveness, rather than for their professional or intellectual accomplishments. Matthews also believed in encouraging women students to pursue careers besides the profession of teaching, which was becoming so dominated by women that wages were barely above subsistence levels. To accomplish this goal, Matthews held vocational conferences for female students and brought in women professionals in journalism, medicine, and library work, as well as famous social workers from nearby Chicago. When the United States entered the First World War in April 1917, opportunities for women at

Wisconsin expanded even further, as female undergraduates took over positions in campus organizations men had vacated. Matthews formed a Women's War Work Council to encourage women to pursue new professional opportunities, including medicine and other service professions, which Matthews defined as "Christianity applied to everyday living."[18]

Although Boynton did not comment on her undergraduate experiences, it is likely that studying in an educational environment that encouraged women's intellectual development and commitment to social welfare greatly influenced Boynton's choice of career. After completing her bachelor's degree at the University of Wisconsin in 1918, Boynton began her medical studies at the University of Minnesota, another midwestern institution that had recently gone through a fitful adjustment to the growing presence of women on campus. Like Wisconsin, female students had profited from the presence of a strong women's dean dedicated to "opening minds and opening doors" for college-educated women.[19] The Minnesota medical school was somewhat hospitable toward female students and began admitting a few women shortly after it opened in 1883. In fact, half the graduating class in 1887 was female. Nevertheless, Minnesota, like other coeducational medical colleges, did not always treat women students seriously. Many members of the all-male faculty insinuated "that admitting women to the increasingly competitive medical school was a wasted effort because they would eventually marry and give up medicine." To build a hospitable community for them, the female students in 1901 started a chapter of Alpha Epsilon Iota, a national medical fraternity that had started at the University of Michigan ten years earlier. The purpose of the organization was "to promote good fellowship, to maintain a high order of scholarship and professional achievement" and "to foster a spirit of moral and social helpfulness." Boynton was a member of AEI while she was a medical student and was a strong supporter of the organization while she was a member of the health service staff; she helped the organization incorporate in 1944.[20]

When Boynton entered the University of Minnesota in September 1918, the institution had just created a student health service for both male and female students. Although Minnesota, like other coeducational institutions, appointed physicians to perform physical examinations of incoming students and to teach hygiene and physical education, they were reluctant to provide medical care to students because they feared raising the wrath of local physicians' groups, who were already disturbed by the spread of "contract practice" in factories, mining camps, and other industries. In the case of female students, colleges and universities were able to overcome these objections due to lingering notions about the allegedly negative effect of higher education on female health. At the turn of the century, there was no corresponding justification for medical care for male college students. Although medical concerns about the dangers of mental strain and over study on male physical health began to appear in the mid-nineteenth century, these ideas were applied mostly to boys in primary and secondary schools.[21] Since the average age of male college students in the late nineteenth century was eighteen, male students were considered men, not boys, and were therefore believed to be immune to the physical dangers of study faced by younger boys. Occasionally, an epidemic or high absence rate due to illness would prompt an institution to pay increased attention to male student health care, but most colleges and universities in the late nineteenth and early twentieth centuries, particularly those in urban areas, believed male students were mature enough to find medical care on their own.[22]

At Minnesota, the student health service arose after more than a decade of petitioning from students and concerned faculty members. In 1904, in the wake of a typhoid epidemic in southeastern Minnesota, students requested the creation of a "sickness fund" to pay for medical and hospital care for students who became ill. The Regents refused to approve the measure but did appoint a University Health Committee to oversee public health matters on campus. Eventually, these activities included overseeing the safety of the water supply, including the swimming pool, free smallpox and typhoid vaccinations for students, and courses in hygiene. It was not until Marion Le Roy Burton became president of Minnesota in 1917 that serious consideration was given to the creation of a health service. Burton's earlier work as president of Smith College had convinced him of the necessity of providing health care and preventive medicine for female students. He therefore felt that providing a similar service to both male and female students at Minnesota was "indispensable."[23] In March 1918, Burton convinced the Regents to commit $5,000 of the university budget to create an infirmary, with students paying $3 per semester as a health fee.[24] In July, the Regents approved the appointment of John Sundwall, formerly at the University of Kansas, to head the new health service. Years later, President Burton wrote, "There seemed something almost providential" about starting the health service just before the influenza pandemic hit the campus in the fall of 1918.[25]

The Minnesota student health service was part of a larger effort to provide a "fair chance for the boys" by giving them the same level of medical care available to women. Their arguments reinforced gender prescriptions of the day. Like many other public health measures of the twentieth century, the efforts of college physicians were aimed at protecting future "soldiers and mothers."[26] Raycroft, for example, observed that lack of attention to male health before the 1910s was a major cause of the high number of rejections for the Selective Service examinations during the First World War.[27] Supporters of health services for men also drew on concepts about adolescence and early adulthood promoted by the renowned psychologist G. Stanley Hall. According to Hall, "One of the last sentiments to be developed in human nature is the sense of responsibility." Therefore, college students needed the same guidance and protections as did younger adolescents.[28] This notion of the college's duty in loco parentis had already been established for female students. By the late 1910s and early 1920s, college administrators and physicians were making the same case for male students.[29]

Boynton was among the hundreds of students who were stricken by influenza that fall. According to annual reports, during the first year of operation, the health service saw two thousand cases of influenza—eight hundred of whom were civilian students—with twenty deaths.[30] Boynton did not record her memories of the epidemic, but it is likely that her experiences shaped her decision to enter the field of public health. After receiving her M.D. in 1921, Boynton was hired by the new health service director, Harold S. Diehl, to oversee the health of women students at the university (Diehl and Shepard took care of the male students). Boynton described her first few years at the health service as a "baptism of fire," during which she and her colleagues handled successive waves of epidemic influenza, scarlet fever, and smallpox. Severe respiratory infections were also a common problem, as were numerous cases of tuberculosis. To enhance her knowledge and skills in preventive medicine, Boynton pursued advanced graduate training, leading to a master's in public health, awarded in 1927. According to Shepard, Boynton's training led to a more lucrative position at the University of Chicago, where she served as assistant professor of medicine and chief medical advisor for women from 1927 to 1928. The University of Minnesota lured her back and promoted her to associate professor in preventive medicine and public health in 1931 and to a full professor in 1938.[31]

Given her extensive experience and qualifications, Boynton seemed like a natural choice to replace Diehl when he left the health service in 1936 to become dean of the University of Minnesota College of Medical Sciences. Yet, Diehl expressed some reservations about promoting Boynton to the directorship. In a letter to President L. D. Coffman, Diehl wrote: "In view of the predominance of male students in the University and male physicians on our staff, I am not certain whether or not a woman can satisfactorily administer the department; but if any woman can do it Dr. Boynton will succeed."[32]

Backhanded compliments like these were common in an era when women were still struggling to make it in a man's world. Like Mary Putnam Jacobi before her, Boynton earned the respect of her male colleagues by adopting a male model of professionalism based on scientific medicine and research. At the same time, she distanced herself from women physical educators, whom she considered unqualified to deal with the medical concerns of students. Shepard's profile observes that Boynton had more professional conflicts with the director of physical education for women than she did with male physicians in the health service. Shepard also praised her for embodying the characteristics "of the good physician with empathy rather than sympathy, objectiveness, and understanding of human nature's nobility and its foibles."[33] A review of Boynton's extensive publications shows that she was careful to balance her work as a clinician with that of scientific researcher. Boynton conducted a number of respected studies not only on women's health issues but also on epidemic diseases such as tuberculosis that affected all students. In total, she published more than eighty single or coauthored papers.

Boynton and other women health care professionals proved to be powerful allies in the struggle for professional legitimacy in college health, as the field was fighting a turf war against the mainstream medical profession. During the 1920s and 1930s, the American Medical Association condemned college health services as "a most obnoxious form of contract practice," a medical monopoly that cut into the business of physicians in private practice. The AMA argued that college students could easily afford the services of private physicians, and that a college was "under no more obligation to supply medical care" than it was "to supply clothing, food or any other necessities."[34] In reply, leaders in the ASHA continued to call for a "fair chance for the boys" by pointing to psychological theories about adolescent development. For example, Seneca Egbert of the University of Pennsylvania claimed that college men, particularly freshmen, were too immature to make major decisions regarding their health on their own and tended to avoid seeking medical advice because they feared the physician's verdict or the expense of health care. "For these reasons, if for no others," Egbert claimed, "it would be the positive and moral duty of the educational authorities, to whom the parents have entrusted the physical as well as mental care of their immature sons and daughters, to provide as promptly and conveniently as possible competent medical care and advice for sick students."[35]

Providing a "fair chance for the boys" posed problems for women physicians. In many ways, their professional dilemmas resembled those described by Martha Verbrugge's work on female physical educators during the twentieth century. According to Verbrugge, by emphasizing differences in female bodies and character, women physical educators "gained a gym of their own, but sacrificed resources and clout in a system that privileged men."[36] Similarly, to justify their role in college health, women physicians continued to argue that female students had special needs and bodily differences that could only be addressed by a female physician. They further claimed that it was inappropriate for male physicians to examine female students. Emphasizing the special needs of female students became especially important during the Great Depression as many colleges and universities merged male and female health services in an attempt to cut costs.[37]

Beginning in 1929, women physicians within the ASHA discussed the possibility of holding special sessions on the hygiene of menstruation and menstrual abnormalities, as well as on the unique health administration issues of women's colleges.[38] At the same time, women physicians in college health, like other women physicians mentioned in Naomi Rogers's essay (Chapter 9 in this volume), insisted that special sessions on female health should not be a precursor to a rival organization for women physicians or even a separate segment within ASHA. As Marjorie Jeffries Wagoner, associate physician at Bryn Mawr told president Warren Forsythe in 1930, although problems of particular interest to female students "should have a place on the regular program," she also saw "no good reason" for the women to meet separately, except informally and for social reasons.[39] Therefore, Boynton and other women in college health insisted that despite the special concerns of women students female professionals should not be segregated from male members of the ASHA and should have equal status to male physicians at the campuses on which they served.

Boynton supported not only gender equality but helped pave the way for black physicians to gain admission to the organization. Before the 1940s, the situation of black physicians in college health was similar to that described by Vanessa Gamble in her book on the black hospital movement. According to Gamble, the history of black medical institutions "shows how black physicians made a place for themselves within the profession of medicine between 1920 and 1945, a time when few of them had options beyond the separate, but never equal, black medical world."[40] In the case of college health, physicians and other health care professionals at historically black institutions held regional

University of Minnesota, freshmen physicals for women, 1920s. Reproduced with permission from the University of Minnesota Archives.

conferences on African American students' health problems beginning in the late 1930s and then formed a separate organization called the National Student Health Association (NSHA) in 1940.[41] At the same time, Boynton helped "promote a closer relationship" between the ASHA and the NSHA by inviting Paul B. Cornely, director of Howard University Student Health Service and executive director of the NSHA, to attend the ASHA's annual meeting. Boynton also served as the ASHA's emissary to the NSHA and eventually helped merge the two organizations during the early 1940s.[42]

Black women physicians found it more difficult to "make a place for themselves" in the medical profession because of the double barriers of racism and sexism. Historically black colleges and universities held brighter professional possibilities for women physicians and other professionals. According to Karen Anderson, because the African American community could not afford separate institutions for women, most female students, faculty, and administrators were at coeducational colleges and universities. Even at these institutions, says Anderson, black women faced sex discrimination and even harassment by male faculty and students.[43] Those who succeeded, like their white counterparts, emphasized the relevance of their work to the health of both male and female students.

Dorothy Boulding Ferebee (ca. 1898–1980)

Dorothy Boulding Ferebee's life is easier to reconstruct, as she left behind a large collection of personal papers at Howard University and an oral history conducted as part of the Schlesinger Library's Black Women Oral History Project. Ferebee was born in Norfolk, Virginia, probably in October 1898 (the exact date is not known; the state of Virginia did not provide birth certificates for African Americans at that time). Ferebee benefited greatly from having the opportunity to study at educational institutions in Massachusetts, which were more welcoming to nonwhite students. After graduating from Samuel C. Armstrong High School in Norfolk in 1911, Boulding moved to live with her great-aunt Emma in the West End of Boston, where she attended Girls High School and received a second diploma in 1915. In the fall of 1916, she entered Simmons College, also in Boston, where she became a star athlete in track and field, basketball, and field hockey. Her success in athletics no doubt convinced her of the importance of exercise to female physical well-being. At that time, Simmons was mainly a secretarial school, although it did offer science courses taught by faculty from the Massachusetts Institute of Technology, an opportunity that gave her excellent preparation for her aspirations in medicine.[44]

After graduating cum laude from Simmons in 1920, Ferebee entered Tufts University College of Medicine. In 1924, she received her M.D., graduating first in her class of 137 students. Despite her exceptional abilities, Ferebee encountered both racism and sexism at Tufts. As one of five women, Ferebee found that "we women were always the last to get assignments in amphitheaters and clinics. And I? I was the last of the last because not only was I a woman, but a Negro too."[45] Ferebee found internships at white hospitals closed to her, so she did her internship and residency in obstetrics at Freedman's Hospital, the hospital affiliated with Howard University College of Medicine in Washington, D.C. In 1927, she was appointed to the clinical faculty in obstetrics at Howard, and two years later, became medical physician to women undergraduates at Howard University.

Like Boynton, Ferebee's success was aided by the support of male mentors who were able to transcend the sexist attitudes of the day. Foremost among Ferebee's supporters was the health service's founder and director Paul B. Cornely, who persuaded her to work at Howard by telling her, "I think you will find here almost as many needs as those that you talk about for Mississippi, that although you haven't done it yet, we hear you talk a great deal about it."[46]

Ferebee's talk about Mississippi referred to the Mississippi Health Project, a series of health clinics for rural sharecroppers in the Mississippi Delta organized by volunteers from Alpha Kappa Alpha, the oldest black sorority in the nation, and supervised by Ferebee. In her history of black women's health activism during this period, Susan Smith describes the Mississippi Health Project as an attempt by black middle-class women to bridge the gap between themselves and the less fortunate by serving as "health missionaries" to the black poor.[47] Although the economic circumstances of the AKA women separated them from the sharecroppers they aimed to help, this did not mean these women and other Howard students enjoyed the same health status as white undergraduates. A shortage of health care facilities that served blacks meant that even middle-class students suffered higher rates of serious diseases such as tuberculosis and syphilis than white youth. Segregation in the District of Columbia meant that Howard students frequently had difficulty finding suitable housing that met minimal modern standards of health and safety. Moreover, Howard students encountered the same white prejudices about black health and sexuality as encountered by impoverished blacks, since white society seldom recognized class differences between African Americans. These prejudices included the assumption that their alleged loose morals meant that blacks were more likely to have sexually transmitted diseases than whites. As a result, black women's colleges were in some ways even more rigid than white ones when it came to regulating student social behavior. To combat stereotypes, health reformers also made an explicit link between good health and middle-class respectability.[48]

At black colleges such as Howard, administrators made sure students' lives were closely supervised and that they observed proper habits of personal hygiene and morality. Many black colleges built dormitories for students and made physical education mandatory. Administrators were especially concerned with protecting female students, given the long history of black women's sexual victimization by white men.[49]

By the 1920s, however, undergraduates at Howard and other black colleges were rebelling against the paternalistic policies aimed at preserving their health and respectability. Students demanded an end to compulsory chapel attendance, dress codes, mandatory physical education, and strict rules regulating social interaction between male and female students.[50] To protect female students from the excesses of the new "manners and morals" among Jazz Age undergraduates, Howard President J. Stanley Durkee appointed Lucy Diggs Slowe as dean of women. By May 1923, Slowe requested more dormitory space for women students so that she could better oversee their behavior. She believed that women who lived off-campus were less likely to follow university rules regarding student behavior and thereby "bring the university into disrepute." Yet Slowe also criticized excessive parietal regulations as "demeaning to students and detrimental to the development of independence and leadership skills in women." Instead, she favored creating an environment that would help students internalize appropriate standards of conduct and deportment. According to Slowe, "proper housing is one of the most potent influences in the education of college students." Slowe made sure that the dormitories were "designed to be complete homes for women of refinement," where students in conjunction with faculty mentors planned "systematic programs of cultural activities," which "not only develop the individuality and initiative of the students, but they reduce to a minimum opportunities for unsocial behavior which frequently results in disciplinary action."[51]

At the same time, protection of black student bodies extended into providing increased health services and facilities conducive to promoting student health and safety. The creation of a comprehensive health service was part of a larger effort to improve the physical plant of the university. This movement intensified after Mordecai Wyatt Johnson became Howard's first black president in 1926. The previous year, a study of the Howard campus indicated that many of the institution's facilities were outdated and even hazardous. Students had already raised complaints about the condition of the athletic facilities and dormitories. Shortly after assuming the presidency, Johnson began a major fundraising effort, which included appealing to Congress for financial support. In 1928, Congress agreed to give the same appropriations to Howard as it did to state institutions under the Morrill Land Grant Act of 1890. These funds allowed the university to make major renovations in existing buildings, including eliminating fire and safety hazards, as well as construct new classroom buildings, dormitories, and athletic facilities. The larger operating budget also allowed Howard to employ Cornely and Ferebee as campus physicians.[52]

Ferebee recalled that initially she and Cornely encountered difficulties from some members of the Howard faculty. According to Ferebee, "the board of

Dorothy Boulding Ferebee (third from right). Dorothy Ferebee Papers, Moorland-Spingarn Research Center, Howard University.

directors and some of the instructors of Howard University really were not terribly impressed with the needs of the Health Service. They didn't know what it meant."[53] Like their counterparts at white institutions, these faculty and administrators assumed that students could receive all the health care they needed at Freedmen's Hospital. According to Cornely, while using local hospitals might seem to be "the logical solution" to student health issues, in fact "it appears that services given to students suffer" in such a situation. Hospitals such as Freedmen's tended to stress medical treatment and care rather than "prevention, health education and the promotion of good health habits," which Cornely felt were the primary missions of a college health service.[54]

Fortunately, both Ferebee and Cornely had the full support of President Mordecai Johnson, who helped convince the faculty and trustees of the need for a health service that emphasized prevention, health education, and medical care. Because of this support, Ferebee and Cornely were able to ensure that every student who entered Howard had a health examination, which Ferebee believed was "the beginning of really the outstanding work, we think, of the Howard University Health Service." As his duties in the College of Medicine and the School of Public Health expanded, Cornely gradually turned over more of the daily responsibilities of the health service to Ferebee. Finally, Cornely told Ferebee, "Now you know what Howard University Health Service can do, we want you to take over." In 1949, Cornely left the health service completely to serve as professor in the College of Medicine and head of the Department of Preventive Medicine and Public Health, and Ferebee assumed the directorship of the health service.[55]

Ferebee recalled that her appointment as director prompted a "great deal of jealousy" among male physicians and faculty at Howard, who no doubt resented the promotion of a woman to such a position. According to Ferebee, President Johnson, in contrast, was "a very staunch ally" who "was absolutely superb in seeing that both men and women whose qualifications were recognizable received the kind of recognition to which they were entitled."[56] Ferebee's relationship with Johnson stood in striking contrast to the other prominent woman administrator on campus, Dean Slowe. Johnson repeatedly refused Slowe's requests for raises and budget increases, removed her from policymaking committees in the university, and insisted that she live on campus rather than with close friend Mary Burrill, an English teacher at Dunbar High School who had been Slowe's companion for many years. Slowe no doubt further raised the ire of the president, as well as male faculty members and administrators, by calling for equal opportunity and self-determination for women on campus, and for complaining of sexual harassment of female students by male faculty members.[57] Although other administrators suffered under Johnson's attempts to consolidate decision making in the president's office, Slowe's militancy made her even more vulnerable to attack. Eventually, the conflict between Slowe and Johnson came to a head in 1937, when Slowe was terminally ill and the president demanded she return to work or else lose her job.[58]

Johnson's treatment of Slowe caused an uproar among students and alumni, who were already criticizing Johnson for being a tyrant and for continuing to take a paternalistic approach to student life, especially that of female students. Shortly after Slowe's death, the Executive Committee of the Alumni Association issued a pamphlet entitled *The Case against President Johnson*, in which they condemned the president's "hate, ill will, and malice" toward the late dean.[59] These protests led Johnson to appoint a committee on the status of women at Howard in the early 1940s, chaired by history professor Merze Tate. Not surprisingly, the committee found that women were clustered in the lowest ends of the professional ranks and were paid considerably less than men. Johnson's willingness to take some measures to rectify the situation contributed to Ferebee's appointment as director of the health service.[60]

The "chilly climate" for women at Howard probably contributed to Ferebee's willingness to cross professional boundaries to improve the campus life for women at the university. Unlike Boynton, Ferebee worked well with Maryrose Reeves Allen, the chair of the Department of Physical Education for Women, which could be attributable to Ferebee's experience as a college athlete. In her oral history, Ferebee described how Allen's support was instrumental to her success as director of the health service. Ferebee recalled how she and Allen "worked diligently to see if we could, with what little money we had, make it the finest Health Service possible."[61]

Support from AKA also played a critical role in Ferebee's success as health service director. Members of the sorority saw the improvement of student health programs at historically black colleges as a natural extension of their "medical missionary" work in the rural South and were enthusiastic supporters of Ferebee's work to improve the health of young people. Among the most significant collaborative efforts between Ferebee and AKA was work on sickle cell anemia conducted in the Department of Pediatrics in Howard's College of Medicine.[62] Like other health care professionals, pediatricians at Howard saw growing numbers of black children with the disease in their practices and consequently came to see sickle cell anemia as central to their clinical work. According to historian Keith Wailoo, this interest in pediatric sufferers led to a new model of the disease, which "placed the Negro child (rather than Negro blood) at the center of disease identity." For both patients and physicians, a critical symptom of the disease was the sickle cell crisis, or recurrent bouts of extreme joint and abdominal pain that brought patients to pediatric clinics such as the one at Howard. By the early 1960s, this sickle cell crisis became a symbol for the failure of white-dominated medical institutions to meet the needs of the black community and formed the centerpiece of civic groups dedicated to improving African American health and welfare.[63] Research institutes like the one at Howard were hailed as "rays of hope" that "promised to wipe away society's ignorance" about the disease as well as providing "a new window on the impact of chronic illness on the downward economic drift of patients."[64]

As a campus physician, Ferebee frequently saw students in sickle cell crisis with symptoms of chronic fatigue, weakness, and pain so severe that they had to miss classes or leave school entirely. Realizing the effect that this chronic illness could have on her students' economic and social futures, Ferebee was eager to help find ways to combat the disease. Beginning in 1958, Ferebee asked Scott and Ferguson to perform sickling tests on all new students. These tests indicated that approximately 4.6 percent of all entering students had the sickle cell trait. Scott and Ferguson also studied students with diagnosed sickle cell anemia to discern the effect of the disease on adolescent growth and development.[65]

Members of AKA helped support these efforts by giving the research project grants-in-aid through the organization's National Health Grant Program and by pledging "cooperation in an all-out program of public education about the disease." The AKA Health Committee and the research staff at Howard put together a leaflet called "The Sickle Cell Story," which answered basic questions about the disease and where to go for help. The sorority also provided educational programs in secondary schools and colleges and various chapters helped support sickle cell patients by paying for blood transfusion costs or providing transportation to clinics.[66]

Ferebee's career in college student health was a part of what Edward H. Beardsley calls "Southern missionary work" undertaken by black physicians educated in the North during the early twentieth century. Beardsley argues that these "medical missionaries," most of whom were raised or educated in the North and remained connected to national white organizations and foundations, "played a key role—and one no Southern black doctor could have filled—in forcing a recognition of black health problems and initiating a discussion within both races of how best to solve them."[67] In her oral history, Ferebee recounted that her educational experience in a relatively integrated, primarily white society in New England gave her a "dignity and independence" that allowed her to act and react in a different manner from those raised in a segregated community. Other historians have focused on Ferebee's "health missionary work" among the poor. I would argue that her work among college students was just as significant. Ferebee recognized that to uplift the race, college-educated blacks had to be healthy themselves. "I think that the Health Service afforded very good grounding and a good sense of appreciation of the health needs of people," said Ferebee, "and that it was not enough to get an education, but an education in a sound body."[68] Ferebee's connections with various national organizations such as the American College Health Association, American Public Health Association, and Planned Parenthood, made sure that the health concerns of black students reached a national audience. In 1971, the ACHA awarded Ferebee the Edward Hitchcock Award in honor of her outstanding work in advancing the health of college students.

Conclusion

The preceding case studies illustrate that college health continued to provide professional opportunities for women physicians in the twentieth century. As in the nineteenth century, the position of college physician gave women one of the few opportunities available to them to become faculty members and medical researchers. Yet, to become department heads, these women had to show that they could also handle the needs of male students—hence their interest in issues such as infectious and chronic diseases that affected both men and women. It helped that opportunities opened up in the post–World War II period for male physicians, many of whom abandoned college health in favor of more prestigious careers in medicine and public health. Boynton managed to reach the rank of full professor at Minnesota and even served as acting director of the School of Public Health while her male colleagues served in the military during World War II. Yet it is also clear that Boynton encountered the same "glass ceiling" as did other women physicians. Boynton's biographer intimates that her advancement opportunities were few and that it took outside offers for her to be promoted to full faculty rank.[69]

Ferebee's oral history illustrates the added challenges women of color faced. Although Boynton never commented on her career trajectory, Ferebee realized she was an exception and was quite outspoken about the discrimination most women faced in academia. "It's very difficult for a woman to become a full professor," she told her interviewer in 1979. "It's very difficult for a woman to become head of a department—she can become an assistant. I do not know whether she'll ever be that. She is sometimes allowed the grade just below the assistant." Ferebee concluded that the "ERA [Equal Rights Amendment] is as much needed today as it was years ago, because women do not have the recognition or receive the appointments and salaries to which they are entitled."[70] Although Ferebee was proud of her accomplishments at Howard, her oral history seems to suggest that she was disappointed that she did not receive the same level of national recognition and status as Paul Cornely, who after leaving Howard achieved international renown in the public health field.[71]

Yet these two women were more successful than other female administrators on campus, namely, deans of women. The situation of Lucy Diggs Slowe, while extreme, signified the end for this job category. By the mid-1940s, it became obvious that colleges were demoting deans of women and replacing them with a dean of students, who was usually a man. As the position of dean of women disappeared, says Jana Nidiffer, "a void was created" as women students lost someone with whom they could confide problems unique to them.[72] Although Boynton and Ferebee did not discuss this explicitly, one can imagine that they and other women health professionals on campus helped fill this void by offering female students a place to talk about sexuality, menstruation, and other female matters that they could not discuss with a male campus physician. In contrast to deans of women, then, women doctors on campus were able to hold onto their positions because of a continuing medical interest in the special health needs of women.

These needs became more complicated during the 1960s and 1970s, when female students demanded access to birth control and abortion at college health centers. As Sandra Morgen demonstrates in her essay (Chapter 7 in this volume), the relationship between women physicians and this larger women's health movement was a complicated one. Although Boynton never commented on birth-control issues, the health service at Minnesota began offering contraception to female students in the mid-1960s, albeit only to married or engaged women.[73] Some female physicians, like Ferebee, were outspoken proponents of reproductive rights for women. Other female physicians proved to be very conservative about contraception. For example, Gertrude Mitchell at the University of California, Berkeley, supported abstinence.[74] Conversely, some male physicians, such as Philip Sarrel at Yale, Robert Gage at the University of Massachusetts, and Takey Crist at the University of North Carolina, played a critical role in helping student groups lobby for reproductive health services on campus.[75] The diversity of opinion among male and female college physicians on issues of contraception and sexuality provides another example of what Morgen describes as the fallacy of essentialist notions of gender differences in the practice of medicine.

NOTES

1. Illustration, "Pioneers in College Health, Physical Education and Athletics," *Proceedings of the Twenty-Second Annual Meeting of the American Student Health Association*, 1941, *40*: frontispiece.

2. Joseph Raycroft, "History and Development of Student Health Programs in Colleges and Universities," *Proceedings of the Twenty-First Annual Meeting of the American Student Health Association*, 1940, *39*: 37–43.

3. Ibid., 38–39.

4. Edward Hitchcock, quoted in Mary O. Nutting, "Mount Holyoke," in *Education*

of American Girls Considered in a Series of Essays, ed. Anna C. Brackett (New York: G. P. Putnam's Sons, 1874), 324.

5. Raycroft, "History and Development," 39.

6. Cynthia Eagle Russett, *Sexual Science: The Victorian Construction of Womanhood* (Cambridge, MA: Harvard University Press, 1989); Joan Jacobs Brumberg, " 'Something Happens to Girls': Menarche and the Emergence of the Modern American Hygienic Imperative," *Journal of the History of Sexuality*, 1993, *4*: 99–127; Carol Smith-Rosenberg, *Disorderly Conduct: Visions of Gender in Victorian America* (New York: Knopf, 1985); Mabel Collins Donnelly, *The American Victorian Woman: The Myth and the Reality* (Westport, CT: Greenwood Press, 1986); and G. J. Barker-Benfield, *The Horrors of the Half-Known Life: Male Attitudes toward Women and Sexuality in Nineteenth-Century America* (New York: Harper and Row, 1976).

7. Regina Morantz-Sanchez, *Sympathy and Science: Women Physicians in American Medicine* (New York: Oxford University Press, 1985); Ellen S. More, *Restoring the Balance: Women Physicians and the Profession of Medicine* (Cambridge, MA: Harvard University Press, 2001).

8. Toby Appel, "Physiology in American Women's Colleges: The Rise and Decline of a Female Subculture," *ISIS*, 1994, *85*: 26–56.

9. Jana Nidiffer, *Pioneering Deans of Women: More Than Wise and Pious Matrons* (New York: Teachers College Press, 2000), 2, 56–57.

10. Morantz-Sanchez, *Sympathy and Science*.

11. More, *Restoring the Balance*, 181.

12. William P. Shepard, "Ruth Evelyn Boynton, M.D.," *Journal-Lancet*, 1964, *84*: 355–358.

13. Ruth Boynton, "The Development of Student Health Services," *Student Medicine*, 1952, *1*: 4–8; Boynton, "Historical Development of College Health Services," *Student Medicine*, 1962, *10*: 294–305; Boynton, "The First Fifty Years: A History of the American College Health Association," *Journal of the American College Health Association*, 1971, *19*: 269–285.

14. Pnina Abir-Am and Clark A. Elliott, eds., *Commemorative Practices in Science: Historical Perspectives on the Politics of Collective Memory*, *Osiris* 2nd ser., 14 (Chicago: University of Chicago Press, 1999).

15. Boynton to O. F. Hedley, Heart Disease Investigations, U.S. Public Health Service, November 29, 1937, American Student Health Association Records, SC 146, Stanford University Archives, Stanford, California (hereafter referred to as *ASHA Records*), Box 7, Folder 2.

16. Nancy Tomes, *The Gospel of Germs: Men, Women and the Microbe in Modern Life* (Cambridge, MA: Harvard University Press, 1998), 237–245.

17. Shepard, "Ruth Evelyn Boynton," 355–356.

18. Nidiffer, *Pioneering Deans of Women*, 107–123.

19. Ibid., 79.

20. Leslie A. Loveless, "A Fraternity for Women," *Minnesota Medicine*, September 2003, Vol. 86, www.mmaonline.net/publications/MNMed2003/September/Loveless.html, accessed July 26, 2005.

21. John Duffy, "Mental Strain and 'Overpressure' in the Schools: A Nineteenth-Century Viewpoint," *Journal of the History of Medicine and Allied Sciences*, 1968, *23*: 63–79.

22. Boynton, "Historical Development of College Health Services," 295–296.

23. J. Arthur Myers, *Masters of Medicine: A Historical Sketch of the College of Medical Sciences University of Minnesota, 1888–1966* (St. Louis, MO: Warren H. Green, 1968), 592.

24. University of Minnesota Board of Regents Minutes, March 15, 1918, courtesy of Paul Rupprecht, Director, Boynton Health Service.

25. Tim Brady, "The Great Flu Epidemic," *Minnesota Magazine*, January–February 2005, www.alumni.umn.edu/3Jan20057.html, accessed July 26, 2005.

26. Theda Skocpol, *Protecting Soldiers and Mothers: The Political Origins of Social Policy in the United States* (Cambridge, MA: Harvard University Press, 1992).

27. Raycroft, "History and Development," 40.

28. G. Stanley Hall, "Student Customs," *Proceedings of the American Antiquarian Society*, n.s., 1900, *14*: 105.

29. Seneca Egbert, "Medical Service in Colleges and Universities," *Proceedings of the American Student Health Association*, 1927, *8*: 35. See also Warren E. Forsythe, *Health Service in American Colleges and Universities* (Ann Arbor: University of Michigan, 1926), 8.

30. Myers, *Masters of Medicine*, 594.

31. Shepard, "Ruth Evelyn Boynton," 357.

32. Harold S. Diehl to University of Minnesota President L. D. Coffman, July 9, 1935, University of Minnesota, Health Service Papers, 1918–1943, AD5.1, Folder 5.

33. Shepard, "Ruth Evelyn Boynton," 355–358.

34. "Editorial: School and College Contracts," *AMA Bulletin*, 1927, 22: 1.

35. Egbert, "Medical Service in Colleges and Universities," *Proceedings of the American Student Health Association*, 1927, *8*: 35.

36. Martha Verbrugge, "Gender, Race, and Equity in the Gym: Women's Physical Education at Howard University and the University of Nebraska, 1920s–1950s." Paper presented at the Annual Meeting of the American Association for the History of Medicine, Madison, WI, April 29–May 2, 2004. See also Verbrugge, "Gym Periods and Monthly Periods: Concepts of Menstruation in American Physical Education, 1900–1940," in *Body Talk: Rhetoric, Technology, Reproduction*, ed. Mary M. Lay, Laura J. Gurak, Clare Gravon, and Cynthia Myntti (Madison: University of Wisconsin Press, 2000), 67–97; and Verbrugge, "Recreating the Body: Women's Physical Education and the Science of Sex Differences in America, 1900–1940," *Bulletin of the History of Medicine*, 1997, *71*: 273–304.

37. The ASHA records contain numerous letters from women physicians searching for work because they had been laid off in the wake of "restructuring." See, for example, Sarah Parker White to Marion Hague Rea, March 2, 1933, ASHA Records, Stanford University, Box 5, Folder 4.

38. Marjorie Wagoner to Smiley, January 12, 1929, ASHA Records, Stanford University, Box 6, Folder 11.

39. Dr. Marjorie Jeffries Wagoner, Associate Physician, Bryn Mawr to Forsythe, March 11, 1930, ASHA Records, Stanford University, Box 6, Folder 10. See also Marion Hague Rea to D. F. Smiley, January 21, 1929, ASHA Records, Stanford University, Box 6, Folder 11.

40. Vanessa Northington Gamble, *Making a Place for Ourselves: The Black Hospital Movement, 1920–1945* (New York: Oxford University Press, 1995), xii.

41. The name National Student Health Association was modeled after the National Medical Association, which was formed by blacks excluded from the whites-only American Medical Association.

42. *Health Problems in Negro Colleges: Proceedings of the First Regional Conference for Health Workers in Negro Colleges, Atlanta, Georgia, April 7 and 8, 1939* (New York: National Tuberculosis Association, 1939); *Health Problems in Negro Colleges: Proceedings of the Second Regional Conference of Negro College Health Workers, Nashville, Tennessee, April 5 and 6, 1940* (New York: National Tuberculosis Association, 1941); *Health Problems in Negro Colleges: Proceedings of the First and Second Annual Meetings of the National Student Health Association* (New York: National Tuberculosis Association, 1943).

43. Karen Anderson, "Brickbats and Roses: Lucy Diggs Slowe, 1883–1937," in *Lone Voyagers: Academic Women in Coeducational Universities, 1870–1937*, ed. Geraldine Joncich Clifford (New York: Feminist Press, 1989), 284.

44. Merze Tate, Interview with Dorothy Boulding Ferebee, December 28 and 31, 1979, *Black Women Oral History Project* (Cambridge, MA: Schlesinger Library, Radcliffe College, 1984), 441–443.

45. Quoted in Susan L. Smith, *Sick and Tired of Being Sick and Tired: Black Women's Health Activism in America, 1890–1950* (Philadelphia, PA: University of Pennsylvania Press, 1995), 153.

46. Tate, Interview with Ferebee, p. 457.

47. Smith, *Sick and Tired*, 149–151.

48. Ibid., 18.

49. Margaret A. Lowe, *Looking Good: College Women and Body Image, 1875–1930* (Baltimore: Johns Hopkins University Press, 2003), 13–15.

50. Martin Summers, *Manliness and Its Discontents: The Black Middle Class and the Transformation of Masculinity, 1900–1930* (Chapel Hill: University of North Carolina Press, 2004), 272, 244, 276.

51. Anderson, "Brickbats and Roses," 286–287, 306–307. See also Linda M. Perkins, "Lucy Diggs Slowe: Champion of the Self-Determination of African-American Women in Higher Education," *Journal of Negro History*, 1996, *81*: 89–104.

52. Harry G. Robinson and Hazel Ruth Edwards, *The Long Walk: The Placemaking Legacy of Howard University* (Washington, D.C.: Moorland-Spingarn Research Center, Howard University, 1996), www.howard.edu/longwalk/.

53. Tate, Interview with Ferebee, pp. 457–458.

54. Paul B. Cornely, "The Status of Student Health Programs in Negro Colleges in 1938–39," *Journal of Negro Education*, 1941, *10*: 155.

55. Tate, Interview with Ferebee, 457–458.

56. Ibid., 472.

57. Patrica Bell-Scott, "To Keep My Self-Respect: Dean Lucy Diggs Slowe's 1927 Memorandum on the Sexual Harassment of Black Women," *NWSA Journal*, 1997, *9*: 70–77.

58. Anderson, "Brickbats and Roses," 290.

59. Rayford W. Logan, *Howard University: The First Hundred Years, 1867–1967* (New York: New York University Press, 1969), 336.

60. Tate, Interview with Ferebee, p. 472.

61. Ibid., 458.

62. Marjorie A. Parker, *Alpha Kappa Alpha through the Years, 1908–1988* (Chicago: Mobium Press, 1990), 195–198.

63. Keith Wailoo, *Drawing Blood: Technology and Disease Identity in Twentieth-Century America* (Baltimore: Johns Hopkins University Press, 1997), 153–154.

64. Keith Wailoo, *Dying in the City of the Blues: Sickle Cell Anemia and the Politics of Race and Health* (Chapel Hill: University of North Carolina Press, 2001), 148–149.

65. Ferebee, "Some Health Problems of College Students at Howard University," *Journal of the American Medical Women's Association*, 1960, *15*: 1067–1070.

66. Parker, *Alpha Kappa Alpha*, 196–198.

67. Edward H. Beardsley, *A History of Neglect: Health Care for Blacks and Mill Workers in the Twentieth-Century South* (Knoxville: University of Tennessee Press, 1987), 94.

68. Tate, Interview with Ferebee, p. 449.

69. Shepard, "Ruth Evelyn Boynton," p. 357.

70. Tate, Interview with Ferebee, 472–473.

71. "Paul Bertau Cornely," in *Notable Black American Scientists*, ed. Kristine Krapp (Detroit: Gale Research, 1999), 80–83.

72. Nidiffer, *Pioneering Deans of Women*, 152.

73. "The Pill on Campus," *Newsweek*, October 11, 1965, 92–93.

74. "No Birth Control," *Daily Californian*, February 3, 1967, 3.

75. For more on the history of birth-control services for college students, see Heather Munro Prescott, *Student Bodies: The Influence of Student Health Services on American Society and Medicine* (Ann Arbor: University of Michigan Press, 2007), chap. 7.

Opportunities and Obstacles for Women Physicians in the Twenty-First Century

Erica Frank, Elizabeth Fee, Manon Parry, and Ellen S. More

Clearly, women physicians worked hard to win a larger presence in the American medical profession. Beginning in the mid-nineteenth century, they battled hidebound social norms and, at times, even some of their own inclinations, to achieve their goals. They pioneered not merely new roles within medicine but new ways of *being* a physician. By the beginning of the twenty-first century, women from diverse backgrounds had moved into every area of the profession. Still, if some of the most pernicious assumptions about women physicians have been overturned by these more recent additions to their ranks, a close analysis of the situation of women physicians today reveals that some of the themes we explore in this collection still resonate. The Women Physicians' Health Study, conducted by a team at Emory University led by Erica Frank, documents the current climate and points to future challenges and possibilities. This chapter will highlight the study's numerous findings and will reflect on the prospects for women physicians.

The Women Physicians' Health Study, which began in 1993 and 1994, provides the first comprehensive, scientific examination of women physicians in the United States. The participants in the study, between the ages of thirty and seventy, constituted a 10 percent sample of the country's women physicians (n = 4,501 respondents; the sample excluded physicians in training).[1] The results refute some persistent and negative beliefs about women physicians but raise concerns about ongoing challenges regarding equity, job satisfaction, and career advancement. They also pose the interesting possibility that the generally excellent health and health-prevention behaviors of women physicians may act as positive influences on their patients, increasing patient receptivity to health education and lifestyle changes. These findings may suggest new approaches to medical education and preventive medicine.

Diversity and Discrimination

As the essays in this collection proclaim, we have learned to deconstruct the unitary category of "woman" to account for ways in which other aspects of identity such as race, class, and sexuality shape opportunities and experiences for all women. Today, as in the past, some inequality persists. The double barrier of racial and sexual discrimination faced by women of color in the history of American medicine lingers, clearly illustrated by the demographics of the profession. At the time of the Women Physicians' Health Study, African American women were underrepresented in relation to their proportion of the country's female population, at only 4 percent of the total number of women physicians. Latinas and Hispanic women were similarly underrepresented at 5 percent. Asian American women, conversely, were overrepresented in relation to their numbers in the population, making up 13 percent of the total. The numbers of women from underrepresented minorities—many of whom care for the underserved—have not increased at the same rate as those of white women physicians and make up only 2 percent of all physicians, a point reinforced by Sandra Morgen (Chapter 7 in this volume).[2] Yet foreign-born physicians were overrepresented in proportion to the prevalence of foreign-born individuals in the population, constituting 24 percent of the total number of women physicians.[3] This prompts two concerns, most obviously the "drain" of trained health care practitioners and other professionals from other countries, which leads to a lack of qualified personnel elsewhere—women physicians are clearly part of this global trend. Second, the migration of foreign-born physicians raises questions about the status of women physicians in other countries, and its effect on their decision to move. Although we do not focus on medical systems outside of the

United States in this collection, no doubt many of these essays would reveal new insights when considered alongside histories of women physicians elsewhere.[4]

According to the Women Physicians' Health Study, among women physicians in the United States, black and Hispanic women physicians were more likely than white women to work in primary care (62% and 58%, respectively, vs. 49% of whites) and to practice in urban areas (72% for both blacks and Hispanics vs. 54% for whites).[5] Black women expressed the lowest rates of career satisfaction among respondents. The study did not pose questions about the possible reasons for such dissatisfaction, but it is likely that racial discrimination and difficult working conditions both play a role.

Women physicians overall tended to be politically liberal: 37 percent described themselves as "liberal" (28%) or "very liberal" (9%), 37 percent are politically "moderate," and only 27 percent are "conservative" (21%) or "very conservative" (6%).[6] The same group conducting the Women Physicians' Health Study found similarly liberal trends in a study of U.S. medical students surveyed between 1999 and 2003. Thirty-one percent called themselves "liberal" and 9 percent "very liberal," 33 percent were "moderate," and only 26 percent self-characterized as "politically conservative" (21%) or "very conservative" (5%).[7] Those formulating policies where they wish to have physician support should note these numbers. Women physicians' religious attributes were also different from the rest of the U.S. population. They are less likely to be Christian (61% vs. 85%) and more likely to be Jewish (13% vs. 2%), Buddhist (1% vs. 0.3%), Hindu (4% vs. 0.4%), or atheist/agnostic (6% vs. 0.6%).[8]

The Women Physicians' Health Study found a number of differences between lesbian and heterosexual women physicians. Physicians who self-identified as lesbian were more likely than women who identified as heterosexual to describe themselves as "liberal." They were also more likely to be childless and to report less stress at home and more likely to report histories of depression and sexual abuse. Overall, they also weighed more, were more likely than heterosexual respondents to have had a recent mammogram, and reported lower household incomes.[9]

In addition to many positive conclusions generated by the data, the study also uncovered evidence of ongoing intimidation and harassment of women physicians. Nearly half of women physicians who responded reported having experienced gender-based harassment, especially divorced or separated women physicians and those specializing in historically male specialties.[10] Overall, 37 percent reported experiencing sexual harassment, with a higher prevalence among those who were younger, U.S.-born, or divorced or separated. Gender-based and sexual harassment occurred most commonly while in training rather than at later career stages and was associated with depression and suicide attempts. With homophobia still a problem in American society, lesbian women were more likely than women who identified as heterosexual to report lifestyle-related harassment and days of bad mental health in the past month. Among lesbians, 41 percent (vs. 10% of heterosexuals) reported lifestyle-based harassment; likewise, 62 percent of blacks reported ethnic harassment (vs. 31% of Asians, 20% of Hispanics, and 6% of whites).[11] Clearly, there remains a great deal to be done to make the profession more hospitable to the diversity of women physicians.

Despite these notable problems, women physicians overall seem to have reached a period of relative career satisfaction and to be consolidating their positions within the ranks of medicine. The blatant hostility toward women physicians faced by Elizabeth Blackwell and her successors is far less common, as is the occasional intimation of violence from male colleagues experienced by Ann Preston and others. Yet female scientists still face special scrutiny, despite the generations of work since Mary Putnam Jacobi first gained renown as a scientific investigator. In the mid-nineteenth century, biologically based explanations of women's unsuitability for professional careers were put forth to exclude women from medical education and practice. Opponents argued that educating women would make "flabby the fibre" of the body, undermine the "vigor of nations," contribute to the "deterioration of the human species," and render women "unfit for homely duties and aims of common life."[12] Although such claims are uncommon today in most nations, discouraging assumptions persist, such as the notion that women physicians are more likely to suffer depression than their male colleagues.

Significantly, data from the Women Physicians' Health Study refuted many such misconceptions. For example, unlike other studies suggesting pervasive psychopathology among women physicians, the data showed that depression is about as common among women physicians as it is among other women in the United States, and suicide attempt rates may be lower. Prior studies have found markedly higher rates for suicide among American women physicians (with odds ratios as high as 4:1). While these have been based on small samples (and show some publication bias), subsequent meta-analyses have also shown elevated rates.[13] The Women Physicians' Health Study data showed that about 1.5 percent of women physicians have attempted suicide (vs. 4.2% for all U.S.

women), and 19.5 percent self-identify as having ever been depressed (vs. estimates ranging from 7%–25% for all U.S. women).[14] These data therefore counter an especially prevalent and toxic misapprehension about women physicians' mental health status: while women physicians' suicide completion rates may be more like those of men physicians, depression is about as common among women physicians as among other U.S. women, and suicide attempt rates appear to be lower than they are for other U.S. women.

Balancing Personal Life with a Career in Medicine

Opportunities to pursue a career in medicine without giving up cherished elements of one's personal life seem to have expanded significantly over the past few decades, although not all specialties are equally flexible. Dr. "Zak" Zakrzewska and Dr. "Mom" Chung were among many women in the nineteenth and twentieth centuries who sacrificed personal preferences and desires for the sake of their ideals and ambitions. Data gathered in the Women Physicians' Health Study would suggest such sacrifice is less necessary and less common today. As More and others have also noted, becoming well-educated and well-paid through a career in medicine does not preclude marriage and childrearing.[15] With an average age of forty-two, women physicians were more likely to be married (73%) and less likely to have never married (13%), been separated, divorced (9%), or widowed (1%) than other U.S. women.[16] That women in medicine are more likely to marry than women outside the profession often surprises women undergraduates and medical students who are concerned about their ability to balance being a physician and having a family.

The 70 percent of women physicians with children (with an average of 1.6 children per physician) expressed higher career satisfaction than women physicians without children; more children were also associated with greater career satisfaction.[17] Given the sizeable literature devoted to the problem of "role strain," it is striking that women with children expressed more career satisfaction than those without children. To the extent that women physicians manage to "have it all," how do they do so? The balance, where it is struck, may result from using their high incomes (a mean of $109,000 in 1993–1994)[18] to delegate domestic tasks such as cooking and housework.[19] Women physicians in the United States averaged only five hours per week preparing meals, and five hours on other housework, compared with the nine hours per week on meal preparation and eighteen hours on other housework of all U.S. married women employed full time. Women who are not full-time, paid workers spend even more time on household tasks. However, women physicians did not delegate childcare nearly as much as other domestic tasks. Those with children under eighteen spent a median of twenty-four hours per week on childcare. The extent to which domestic tasks are absorbed by women physicians themselves, shared with their partners, or outsourced to others is a changing equation. And the extent to which women physicians use their high incomes to pay a fair wage to others to complete these tasks has not been studied. Also unstudied is what women physicians think about paying others (typically other women) to do what society usually considers to be "women's work," and how much this situation may parallel the hierarchical workplace relationship between women (and men) physicians and (typically women) nurses and other mid-level providers.

Regarding gender hierarchies in the workplace, an interesting avenue for further research would be the interaction between women physicians and other women in health care. As scholars have dug deeper into the history of the complex interrelationships between women in the medical profession, stories of competition and discrimination have emerged. Women physicians have not always embraced alternative or laywomen healers with whom they might have made common cause on behalf of better health care for the public. This is one of the themes of Sandra Morgen's account of conflicts within women's health centers. Nor have women doctors always contributed to racial harmony among their ranks, as Vanessa Gamble and others have shown.[20] These insights, perhaps overlooked in earlier "heroic" accounts, enrich our understanding of the different experiences of women physicians and the ways they negotiated or failed to negotiate the pressures of the culture and their careers. But what can we say about these issues for women in the profession today? Visitors to the exhibition *Changing the Face of Medicine* were sometimes heard repeating the stereotypical assertion that women physicians do not get on well with other women they work with. Such anecdotal evidence suggests, for example, that women physicians may still be perceived as aggressive in their efforts to be taken seriously, including an expressed resentment at being mistaken for nurses. Yet women physicians have also often drawn on the support of the women around them when facing discrimination or abuse, and some began their careers as nurses or auxiliary health professionals. In recent years, women physicians and women in other health care fields may have found themselves sharing expertise as some of physicians' traditional responsibilities have been reassigned to nurses, nurse practitioners, and physician assistants. Further study could show us how these changes play out on hospital wards and in the care of individual patients and help to separate fact from fiction regarding the interrelationships among women in health care.

Women Physicians and the Cultures of Medicine

Using the Women Physicians' Health Study as a starting point as we evaluate the progress and the pitfalls over the past few decades, we find that the story of women's progress in medicine is unfinished across a number of dimensions. In the 1970s, the future looked bright for the women physicians of the "equal opportunity" generation—those who, in rapidly climbing numbers, entered medical school in the wake of legal challenges to the gender-discriminatory practices of many admissions committees.[21] This generation of women physicians and students carried high expectations for themselves. By 1988, pediatrician Perri Klass could write—albeit controversially, "Many women doctors believe that women do medicine differently, that there are advantages to the way they approach their patients. . . If this is in fact true, and not just a convenient prejudice (and one I still blush to acknowledge in print), then the effect of women on the medical profession may be larger and more far-reaching than we have yet imagined."[22]

Have women fulfilled a promise to transform health care through the example of empathic, patient-centered, woman-friendly doctoring, as some members of an earlier generation of women physicians confidently expected? In the face of the many contradictions in the American health care system today, we conclude this volume with questions as well as answers. To what extent are men and women different in their approach to medicine? Is it true, for example, that men place a somewhat greater value on the technical aspects of medicine while women tend to place a higher value on their relationships with patients? If men and women were unconstrained by the structures of medical institutions in the ways they practiced medicine, would their choices differ? And given that medical practice is highly regulated in most institutional workplaces, have the different preferences of male and female physicians been subsumed by the pressure of the institutional and financial constraints of the system?

Women physicians clearly do some things differently. Overall, they undertake primary care and care of the underserved at higher rates than men. In keeping with their own preferences and traditions as well as with gender-linked social expectations and urgent social need, a disproportionate percentage of women physicians gravitate to primary care specialties.[23] Women also spend more time, on average, with patients than do men.[24] As data from the American Medical Association show (see table C.1), 60 percent of women physicians practiced within specialties that require high levels of patient contact, that is, internal medicine, pediatrics, family medicine, obstetrics-gynecology, and psychiatry in 2000. This trend has not changed appreciably since the 1970s. However, given the values within and tacit hierarchies among medical specialties, it may be that these choices of specialty have hindered many women physicians' chances to become leaders and to shape health care policy.[25]

In the data from the Women Physicians' Health Study,[26] women physicians were most likely to choose pediatrics (16% of respondents), general internal medicine (12%), or psychiatry (11%), similar numbers to those obtained nationally from the American Medical Association (see table C.1). Nearly all (95%) were residency trained, and 65 percent were board certified; they worked an average of thirty-seven clinical hours and eleven nonclinical hours per week.[27] Most were in group (26%) or hospital-based (23%) practices.[28] Although 31 percent of women physicians, if given the choice, would "maybe/probably/definitely" not again decide to become a physician, and 38 percent of women physicians would "maybe/probably/definitely" prefer a different specialty, they were typically satisfied with their careers (84% always, almost always, or usually satisfied).[29] While we do not know of comparable numbers for other occupations or for male physicians, these appear (in absolute terms) to be positive findings overall. Least likely to be satisfied with their careers were younger physicians, those with most work stress or a history of being harassed, and those with least work control. Control of one's work was an especially powerful factor. Those with the most perceived control of their work environment were eleven times more likely to be satisfied with their work environment than those with the least control.[30]

A few other general professional characteristics should also be noted. The study refuted the idea that student loan indebtedness predicts specialty choice. There was no relationship between being a primary care physician and indebtedness, even when adjusted for age and race/ethnicity.[31] Most (71%) physicians participated in either pro bono work (a median of two hours per week among the 52% participating), nonmedical volunteering (two hours per week among the 43% participating), or both (27% of the physicians, averaging four hours of pro bono and two hours of nonmedical volunteering).[32] They averaged thirteen

Table C.1 Top Specialty Choices of Women Physicians (numbers and percentages)

Specialty	1993–1994 (WPHS)	1990[a] (AMA)	2000[a] (AMA)
Pediatrics	807 (16%)	15,675 (15%)	30,322 (16%)
Internal medicine (general)	470 (12%)	19,171 (18%)	37,073 (19%)
Internal medicine (subspecialist)	337 (8%)	Unavailable	Unavailable
Psychiatry	570 (11%)	8,170 (8%)	11,648 (6%)
Family medicine	347 (8%)	10,602 (10%)[b]	22,739 (12%)[b]
Obstetrics-gynecology	313 (8%)	7,551 (7%)	14,124 (7%)
Anesthesiology	274 (6%)	4,608 (4%)	7,335 (4%)
Pathology	229 (4%)	3,716 (4%)	5,408 (3%)
Radiology	163 (3%)	Unavailable	Unavailable
General practice	150 (4%)	Unavailable	Unavailable

Note: Findings from the Women Physicians Health Study (WPHS) and the American Medical Association (AMA).

[a] Erica Frank, J. Rock, and D. Sara, "Characteristics of Female Obstetrician-Gynecologists in the United States," *Obstetrics and Gynecology*, 1999, 5, no. 1; American Medical Association, *Physician Characteristics and Distribution in the US*, 2008 Edition. Available at: www.ama-assn.org/ama/pub/category/12916.html, accessed May 21, 2008.

[b] Includes total number for general practice and family practice.

hours per month on continuing medical education, most commonly by reading medical journals (one hour per week).[33]

What are the characteristics of women physicians' practices when we compare different specialties? In comparison with other women physicians, family physicians were more likely to be U.S. born and liberal, to earn less, and to have less work stress.[34] Similarly, pediatricians worked fewer hours and earned less, expressed less work stress and career dissatisfaction, and were more likely to have children than other women physicians.[35] A third primary care specialty, general internists, reported more severe work stress and a greater desire to change their specialty if reliving their lives than did other primary care physicians. Subspecialized internists were more likely to report working excessive hours, having lower incomes, and having more nonclinical work than did other subspecialists.[36] Overall, internists reported less career satisfaction than did other specialists.[37] Obstetrician-gynecologists assumed fewer domestic responsibilities, earned more, were more likely to be in group practice, and were especially likely to counsel their patients about breast cancer and hormone replacement therapy.[38]

Psychiatrists were older, in poorer health, less likely to be married, more likely to be current or ex-smokers, and more likely to be liberal.[39] They had similar career satisfaction and income but worked fewer hours and had higher satisfaction with their specialty than other women physicians.

Among the higher-earning specialties and subspecialties, dermatologists were less likely to have as much training, confidence, or interest in routine prevention-related screening or counseling than did primary care physicians (though they did often counsel about skin cancer prevention).[40] Dermatologists were more likely to be born in the United States, white, and married to a physician than were other U.S. women physicians. They were also more likely to be board certified, in solo or two-person practices, and have higher incomes, and they experienced less work stress (including greater career satisfaction, fewer nights on call, and a lower likelihood of feeling that they were overworked). Fewer emergency physicians were residency trained than were other U.S. women physicians (73% vs. 95%).[41] They worked fewer hours (averaging thirty-eight per week vs. forty-three) but earned considerably more and had similar stress and satisfaction levels. Radiologists worked more hours, and had far higher incomes, but had less career satisfaction and work control and were more likely to feel overworked.[42] This imbalance could be relatively easy to correct by working less, though one radiologist has described their high hourly billings as trapping the specialists "in a gilded cage." Surgeons were relatively few in number (compared with their numbers and proportions among men physicians), younger than other women physicians, and more likely to be born in the United States, white, unmarried, and childless.[43] Surgeons worked considerably more hours and more on-call nights but also earned considerably more and had similar feelings of work control, stress, and satisfaction compared with other women physicians. Among younger women physicians, a higher proportion are entering surgery; even this historically male-dominated specialty is now changing.[44] Although the issue is not explored in this volume, one would expect to see a similar trend in other areas that have been especially resistant to women, such as the military.

Academic physicians were more likely than their colleagues working outside of universities to be white and born in the United States and to identify themselves as "liberal." They were also more likely to be board certified and to work in urban areas, to work fewer clinical hours a week (twenty-eight versus thirty-nine for nonacademic physicians) and more nonclinical hours (nineteen versus four), and to earn less than nonacademicians.[45] Academicians reported less work control and were more likely to report working too much, but they were more satisfied with their specialty choices than were nonacademicians.

Work satisfaction is especially pertinent because it has often been surmised that women are more likely than men to weigh lifestyle factors such as regular hours of work when considering medical specialties.[46] Yet recent research suggests that the priorities of medical students of both sexes have begun to converge since the 1980s. In recent decades, a growing number of women and men are choosing specialties with more predictable hours and less demanding schedules, gravitating toward radiology, anesthesiology, and dermatology, and away from primary care fields and general surgery.[47] Whether because of women's influence on the culture of medicine, both men and women are choosing more flexible specialties.

Women Physicians: The Relationship between Physician Health and Patient Health

Women's health may be the sphere in which women physicians have changed the culture of medicine most profoundly. The importance of focusing on the health of neglected groups such as women and children was well understood by women physicians in the past, as Susan Wells (Chapter 8) and Heather Munro Prescott (Chapter 12) each make clear. There is also a strong relationship between the health of physicians, the health issues that affect their own lives, and their advice to patients.

The Women Physicians' Health Study found that women physicians' health was excellent when compared with that of women in the general population and even when compared with other high socioeconomic status women.[48] Physicians have extensive health-related knowledge and have high socioeconomic status, including all three of its key markers: (1) educational attainment, (2) income level, and (3) occupational prestige. Given that these variables are all powerful predictors of health,[49] physicians' health may therefore be a "gold standard," demonstrating the health outcomes achievable in positive circumstances. But the main reason physicians' health matters is that the frequency of physicians' prevention counseling to patients is positively correlated to their personal health practices.[50] We may be able to improve the health of entire patient populations by improving the health practices of medical students and physicians. Data from a four-year, sixteen–medical school follow-up study showed that this is achievable.[51]

Comparing physicians' health habits reported by the Women Physicians' Health Study with those of a representative national sample of women of high socioeconomic status, and with women in the general population from the U.S. Centers for Disease Control and Prevention's Behavioral Risk Factor Surveillance System, women physicians' health behaviors are excellent.[52] Unlike women in the general population and even other women of high socioeconomic status, their reported behaviors exceeded the year 2000 national goals established by the U.S. Public Health Service.[53] Examined behaviors and screening habits included the following: tobacco use, abstaining from smoking for at least one day in the past year, seat belt use, blood stool testing, sigmoidoscopic intestinal exam, cholesterol measurement, blood pressure measurement, Papanicolaou testing for cervical cancer, breast examination by a clinician in women younger than forty, mammography, average age at onset of smoking (for those who were ever cigarette smokers), and average alcohol use. Women physicians were less likely to smoke cigarettes (3.7% were smokers) than other women of high socioeconomic status (8%) and other women in the general population (25%). Physician ex-smokers had quit at a younger age, and the few who still smoked consumed fewer cigarettes per day. Most women physicians drank alcohol (73% had consumed some alcohol in the past month), drinking an average of twice a week, and 1.4 drinks per episode. But like other women of high socioeconomic status, only 0.1 percent admitted to drinking more than four drinks on one occasion in the past month. Women physicians ate less fat and ate more fruits and vegetables (3.5 servings per day) than women in the general population (though they reported eating fewer fruits and vegetables than other women of high socioeconomic status).[54]

Another set of health-related variables were unusually positive in this group. Between 5 percent and 8 percent of women physicians (vs. 1%–2% of the general population) were vegetarian.[55] Half (50.2%) took a multivitamin/mineral supplement, and 36 percent did so regularly; 38 percent took calcium (26% regularly).[56] Among those over the age of 50, 35 percent took occasional aspirin and 6 percent took aspirin at least daily (a significantly higher rate than the general population).[57] Nearly all (96%) exercised; 49 percent exercised enough[58] to meet the American College of Sports Medicine's recommendations of at least thirty minutes of activity, at least three times per week.[59] The one area in which women physicians' health behaviors did not generally exceed those of other women of high socioeconomic status was in their screening practices. Although women physicians were consistently more likely to receive screening tests than were women in the general population, they had similar results to those of other high-socioeconomic-status women.[60] Physicians with a family history of cancer were more likely to report recent cancer screening.[61] Importantly and interestingly, pregnant physicians' health status was better than age-matched nonpregnant physicians, and their work amount, desire to work less, perceived work control, career satisfaction, and work stress did not differ from age-matched colleagues.[62]

Premenopausal U.S. women physicians were more likely to use contraceptives than were other premenopausal U.S. women and to do so in ways consistent both with delaying fertility and reducing total fertility (presumably because they have good knowledge of and access to contraceptive options and are highly motivated to plan their families to advance their personal and professional goals).[63] And in this 1993–1994 survey, postmenopausal physicians were more likely than other U.S. women to report using hormone replacement therapy (though this has likely changed in light of newer data on negative side effects of hormone replacement).[64] They were also more likely to perform radon testing than were other Americans (18% vs. a range of 3% to 9% for the general population)[65] and were half as likely to report owning a firearm (16.5%) than were other U.S. women.[66] Fewer women physicians than other U.S. women stated that they had been victims of domestic violence (3.7%) or sexual abuse (4.7%).[67] And finally, 39 percent said that the health care they received was excellent, 37 percent thought it very good, 19 percent good, 4 percent fair, and only 1 percent reported it to be of poor quality.[68] Physicians' assessments of their personal care may be especially useful because they know what constitutes ideal care, though the care they receive may not be representative (in fact, it may be better than others receive).

Women who practiced preventive health care themselves were found to significantly outscore other physicians in prescribing preventive measures for patients, a finding of potentially great importance. The study examined fourteen different counseling and screening practices: these clinical practices were significantly and consistently correlated with practicing a related health habit oneself (and with being a primary care practitioner).[69] For example, significant associations were found between physicians' fat consumption and their likelihood of counseling patients about lowering cholesterol through dietary changes. Physicians' personal practices regarding breast self-exams were also strongly correlated with their performance of clinical breast examinations, and their personal sunscreen use was positively related to their providing skin cancer counseling. Significant associations were found, in addition, between personal and clinical practices relating to exercise, alcohol, tobacco, flu vaccine, and hormone replacement. Patient counseling/screening was sometimes also positively correlated with current attempts to improve the physician's own health-related habits, and with physicians' race/ethnicity, geographic region, practice site, and amount of continuing medical education.

The study also found that physicians' parents' educational attainment (especially their mothers') modestly affected some health habits (higher fruit and vegetable intake, having a regular physician, and a lesser likelihood of owning a gun) but did not affect their exercise habits, fat consumption, compliance with prevention recommendations, current health status, or histories of smoking, hypertension, or dyslipidemia (overproduction or deficiency of lipids in the bloodstream). This shows that many negative effects of parental socioeconomic status can be significantly alleviated after only one generation of high education, income, and occupational attainment. This finding suggests the continuing health benefits of helping those from lower socioeconomic backgrounds succeed educationally and professionally.[70]

Women Physicians and the Medical Profession

At the time of publication, health care in the United States is in disarray. The treatment options U.S. physicians can offer their patients are constrained by the prices of drugs, by inequitable access to care and facilities, and by the limits of health insurance coverage. By 2005, 46.6 million Americans were living without health insurance, but even those with coverage were facing rising costs. Fifty percent of the people who filed for bankruptcy in the United States did so, in part, because of medical expenses.[71] Despite the large expenditures on health care in the United States, evidence suggests that white, middle-aged Americans are less healthy than their British counterparts.[72] Poorer Americans are even worse off, as the study of health disparities has amply demonstrated. Unable to receive regular screenings, preventive interventions, or early treatment, many will find that their conditions worsen, causing long-term illness or disability. Lacking any protection against income loss, these patients will fall further into poverty. The problems of U.S. medicine, including access to care, are issues of attitudes, economics, and politics that go well beyond gender dynamics within the profession. Thus, as we reflect on the present state of medical care in the United States, we may consider the extent to which the movement of women into the mainstream of medicine has changed patients' experiences. What sort of prophets were those women physicians to whom Perri Klass turned for insight back in 1988 and who claimed that the entrance of women into medical careers would humanize medical practice? Is it possible that she was asking the wrong questions? Is it possible that women do make excellent physicians but that the mere presence of more women in medicine is not enough to produce a better health care system in the face of multiple constraints?

At the top of the profession, in academia and policymaking circles, women still remain a minority. To transform the culture of medicine takes more than good doctoring; it requires strategic interventions by those in positions of influence. It is true that women physicians are a steadily increasing proportion of the profession, likely to reach 30 percent near the end of this decade, and also that young women are showing continued interest in medicine as a career. Since 2003, they have represented close to 50 percent of applicants to medical schools.[73] Yet after graduation, women physicians still confront what *Annals of Internal Medicine* called "Unequal Pay for Equal Work."[74] A large 2004 study of medical academia reported in *Annals* found that, "among the 1814 faculty respondents . . . female faculty were less likely to be full professors than were men with similar professional roles and achievement." Accounting for a wide range of professional activities, hours worked, length of time in rank, and department type, the study's authors found slower rates of promotion and lower rates of compensation for women throughout their sample. Data collected by the Association of American Medical Colleges back them up. In 2005 women represented about one-third of medical school faculty, and only 17 percent of full professors and 11 percent of department chairs.[75] To quote from the website of the Society for Executive Leadership in Academic Medicine (SELAM), "Despite their greater numbers matriculating at schools of medicine and dentistry, women are still significantly underrepresented within the topmost administrative ranks of our nation's academic health centers. At the same time, there is a widely acknowledged need to diversify leadership in academic medicine and improve cultural and gender sensitivity in medical training and healthcare delivery."[76] While women physicians have become within the past two decades deans of medical schools and the heads of the Public Health Service, the National Institutes of Health, and the Centers for Disease Control and Prevention, most top positions are still held by men.

Since women are gaining the numerical strength to influence the medical profession—they account for 27 percent of all physicians today—what is preventing them from gaining a proportional share of the leadership? The situation has undeniably improved since the days when the medical school admissions process posed the main barrier to a career as a physician for a woman.[77] After graduating from medical school, women physicians now have a range of interesting career opportunities. Yet, in still higher proportions than men, their careers take pathways leading away from influence over the structures and policies of American health care. Historically, many women physicians have served as advocates for vulnerable groups, a choice that may run counter to career advancement. Other potential hindrances include insufficient mentoring, inflexible promotion schedules that conflict with childbearing and child-rearing timetables, and the difficulties women face in taking a better job when their families would rather not move.[78]

Some might say that women are responsible for their own attenuated career progress—after all, no one is forced to marry or to have children. And, if they do make these choices, it is their responsibility to negotiate equitable arrangements within their own households. But such a view is too narrow. Even those women who do not attenuate their career progress in response to family needs frequently find that they are on the "slower track" compared with men who start out in similar positions.[79] The obstacles still confronting women physicians are not primarily caused by unwillingness to pursue their careers to the fullest. Rather, women physicians confront the lingering presence of gender-bound, unrealistically high expectations for working women with families, to carry both sets of responsibilities as if each were a full time job. Ironically, they may also still face the subtle discrimination resulting from *lower* expectations of those same women in their capacity as professionals. A thoroughgoing reform of expectations about parenting—not just mothering—is in order and affordable childcare facilities should be provided so that both parents can contribute to the labor market while sharing responsibility for parenting.

Without active recruitment into positions of leadership and other societal changes, women will not move up the hierarchy in proportion to their numbers. Discouraged by the inhospitable culture and inflexible schedule of some of the more prestigious specialties, many will be less likely to advance to the highest levels of leadership. Yet, if you cannot reach a position with decision-making responsibilities, you cannot make system-wide policy decisions that will benefit other women physicians, the medical profession, and American health care.

Women throughout the professions share many of the concerns discussed in these essays, such as the need for female mentors and leaders for education, networking, employment, and further training. Most face barriers to the highest paid and most prestigious areas of specialization. Certainly, women in the wider world of health care occupy a broad range of careers, although here we focus specifically on women physicians. The interactions among women across these areas of medicine, and across other areas of health and welfare, may offer important lessons for building a coalition of reformers to take on the failures of medical care in our society, and the social and economic forces that make certain groups more vulnerable to disease and that undermine their quality of life.

Perhaps we can also learn from history. The essays in this collection raise issues not only of professional identity, boundary crossing, deferred or denied personal fulfillment, and the gendering of career pressures and expectations but also offer examples of female solidarity and health care activism. Medicine has changed dramatically from the days of Elizabeth Blackwell, and today's challenges are different. There is much to be done, and only when women have achieved parity or near parity in their professional lives will they achieve parity in formulating the response to the next struggle. From positions of leadership, fulfilling the hopes and aspirations of the generations of women who preceded them, they will be well placed to make health care and health education equitable, accessible, affordable, and appropriate to the needs of the entire population.

NOTES

1. The Women Physicians' Health Study is a study of U.S. women M.D.s initiated by Erica Frank, to better understand characteristics of U.S. women physicians, and the clinical and other implications of those characteristics. The population database is the American Medical Association's Physician Masterfile, a record of all U.S. M.D.s. The study was primarily funded by the American Heart Association and the U.S. Centers for Disease Control and Prevention and secondarily by the National Institutes of Health, Wyeth-Ayerst, and Merck Pharmaceuticals. A bibliography linking to publications based on the study is available at http://med.emory.edu/WPHS/.

Some of the most fruitful clinical research of the past decade has come from studying convenience samples of men physicians and women nurses. See Charlene F. Belanger, Charles H. Hennekens, Bernard Rosner, and Frank E. Speizer, "The Nurses' Health Study," *American Journal of Nursing*, 1978, *78*: 1039–1040; Charlene F. Belanger, Frank E. Speizer, Charles H. Hennekens, Bernard Rosner, Walter Willett, and Christopher Bain, "The Nurses' Health Study: Current Findings," *American Journal of Nursing*, 1980, *80*, 1333; JoAnn E. Manson, Julie E. Buring, S. Satterfield, and Charles H. Hennekens, "Baseline Characteristics of Participants in the Physicians' Health Study: A Randomized Trial of Aspirin and Beta-Carotene in U.S. Physicians," *American Journal of Preventive Medicine*, 1991, 7: 150–154; Charles H. Hennekens and Kimberley Eberlein for the

Physicians' Health Study Research Group, "A Randomized Trial of Aspirin and Beta-Carotene among U.S. Physicians," *Preventive Medicine*, 1985, *14*, 165–168.

One study examined several thousand men and women physicians' substance use habits. See Patrick H. Hughes, Nancy Brandenberg, Dewitt C. Baldwin, Carla L. Storr, Kristine M. Williams, and David V. Sheehan, "Prevalence of Substance Use among U.S. Physicians," *Journal of the American Medical Association*, 1992, *267*: 2333–2339. The few prior physicians' health studies have been small, have involved especially low numbers of women (typically considerably fewer than 100) and have not attempted to describe this interesting and important population in detail. The Women Physicians' Health Study is the first extensive survey of a large, random sample. Erica Frank, "The Women Physicians' Health Study: Background, Objectives, and Methods," *Journal of the American Medical Women's Association*, 1995, *50* (2): 64–66.

2. *Physician Characteristics and Distribution in the US, 2003* (Chicago: American Medical Association, 2003), table 14. The count of underrepresented minorities excludes Asian women. In addition, it is likely a significantly undercounted figure since the racial/ethnic identification of almost ninety thousand women physicians was counted as "unknown."

3. Erica Frank, Richard Rothenberg, W. Virgil Brown, and Hilda Maibach, "Basic Demographic and Professional Characteristics of U.S. Women Physicians," *Western Journal of Medicine*, 1997, *166*: 179–184.

4. See Sue Kilminster, Julia Downes et al., "Women in Medicine—Is There a Problem? A Literature Review of the Changing Gender Composition, Structures, and Occupational Cultures in Medicine," *Medical Education*, 2007, *41*: 39–49, for an overview of global trends.

5. Giselle Corbie-Smith, Erica Frank, and Herbert W. Nickens, "The Intersection of Race, Gender, and Primary Care: Results from the Women Physicians' Health Study," *Journal of the National Medical Association*, 2000, 92: 472–480.

6. Percentages may exceed 100 percent total where fractions have been rounded up. Erica Frank, "Political Self-Characterization of U.S. Women Physicians," *Social Science and Medicine*, 1999, *48*: 1475–1481.

7. Erica Frank, Jennifer S. Carrera, and Shafik Dharamsi, "Political Self-Characterization of U.S. Medical Students," *Journal of General Internal Medicine*, 2007, 22(4): 514–517.

8. Erica Frank, Mary L. Dell, and Rebecca Chopp, "Religious Characteristics of U.S. Women Physicians," *Social Science and Medicine*, 1999, *49:* 1717–1722.

9. Donna J. Brogan, Katherine O'Hanlan, Lisa Elon, and Erica Frank, "Health and Professional Characteristics of Lesbian and Heterosexual Women Physicians," *Journal of the American Medical Women's Association*, 2003, *58*: 10–19.

10. Erica Frank, Donna J. Brogan, and M. Schiffman, "Prevalence and Correlates of Harassment among U.S. Women Physicians," *Archives of Internal Medicine*, 1998, *158*: 352–358.

11. Giselle Corbie-Smith, Erica Frank, Herbert W. Nickens, and L. Elon, "Prevalences and Correlates of Ethnic Harassment in the U.S. Women Physicians' Health Study," *Academic Medicine*, 1999, 74(6): 695–701; D. J. Brogan, Erica Frank, L. Elon, S. P. Sivanesan, and Katherine O' Hanlan, "Harassment of Lesbians as Medical Students and Physicians," *Journal of the American Medical Women's Association*, 1999, *282*: 1290–1292.

12. Jennifer Phegley, *Educating the Proper Woman Reader: Victorian Family Literary Magazines and the Cultural Health of the Nation* (Columbus: Ohio State University Press, 2004).

13. Erica Frank and Arden D. Dingle, "Self-Reported Depression and Suicide At-

tempts among U.S. Women Physicians," *American Journal of Psychiatry*, 1999, *156* (12): 1887–1894; Eva S. Schernhammer and Graham A. Colditz, "Suicide Rates among Physicians: A Quantitative and Gender Assessment (Meta-Analysis)," *American Journal of Psychiatry*, 2004, *161*: 2295–2302; Erica Frank, Holly Biola, and Carol A. Burnett, "Mortality Rate and Causes among U.S. Physicians," *American Journal of Preventive Medicine*, 2000, *19*: 155–159; Sarah Lindeman, Esa Laara, Helina Hakko, and Jouko Lonnqvist, "A Systematic Review of Gender-Specific Suicide Mortality in Medical Doctors," *British Journal of Psychiatry*, 1996, *168*: 274–279.

14. Frank and Dingle, "Self-Reported Depression and Suicide Attempts," 1887–1894.

15. Ellen S. More, *Restoring the Balance: Women Physicians and the Profession of Medicine, 1850–1995* (Cambridge, MA: Harvard University Press, 1999).

16. Erica Frank et al., "Basic Demographic and Professional Characteristics of U.S. Women Physicians."

17. Erica Frank, Julie E. McMurray, Mark Linzer, and Lisa Elon, "Career Satisfaction of U.S. Women Physicians," *Archives of Internal Medicine*, 1999, *159*: 1417–1426.

18. Frank et al., "Basic Demographic and Professional Characteristics of U.S. Women Physicians."

19. Erica Frank, Lynn Harvey, and Lisa Elon, "Family Responsibilities and Domestic Activities of U.S. Women Physicians," *Archives of Family Medicine*, 2000, *9*: 134–140.

20. Vanessa Northington Gamble, "When Racism Trumps Sisterhood: A History of Racial Discrimination and White Women Physicians," delivered at the symposium, "Women Physicians, Women's Politics, Women's Health: Emerging Narratives," National Library of Medicine, Bethesda, Maryland, March 10–11, 2005; idem, "Subcutaneous Scars," *Health Affairs*, 2000, *19* (1): 164–169; idem, "The Provident Hospital Project, An Experiment in Race Relations," *Bulletin of the History of Medicine*, 1991, *65* (4): 457–475. Also see Darlene Clark Hine, Elsa Barkley Brown, and Rosalyn Terborg-Penn, eds., *Black Women in America: An Historical Encyclopedia*, 2 vols. (Bloomington: Indiana University Press, 1994); Marian Gray Secundy, "To Have a Heritage Unique in the Ages: Voices of African American Female Healers," in *The Empathic Practitioner: Empathy, Gender, and Medicine*, ed. Ellen Singer More and Maureen A. Milligan (New Brunswick, NJ: Rutgers University Press, 1994), 222–236.

21. Diane Magrane et al., eds., *Women in U.S. Academic Medicine Statistics and Medical School Benchmarking 2003–2004* (Washington, D.C.: Association of American Medical Colleges, 2004), table 1, p. 11.

22. Perri Klass, "Are Women Better Doctors?" in *Baby Doctor: A Pediatrician's Training* (New York: Ballantine Books, 1992), 262–283. The original article appeared as the cover story in the *New York Times Magazine*, April 19, 1988.

23. Emily M. Lambert and Eric S. Holmboe, "The Relationship between Specialty Choice and Gender of U.S. Medical Students, 1990–2003," *Academic Medicine*, 2005, *80* (9): 797–802.

24. Debra L. Roter and Judith A. Hall. "Why Physician Gender Matters in Shaping the Physician-Patient Relationship," *Journal of Women's Health*, 1998, 7 (9): 1093–1097.

25. More, *Restoring the Balance*, 221–222; Debra L. Roter, Judith A. Hall, and Yutaka Aoki, "Physician Gender Effects in Medical Communication: A Meta-Analytic Review," *Journal of the American Medical Association*, 2002, *288*: 756–764; Christine Laine and Barbara J. Turner, "Unequal Pay for Equal Work: The Gender Gap in Academic Medicine," *Annals of Internal Medicine*, 2004, *141* (3): 238–241.

26. Frank et al., "Basic Demographic and Professional Characteristics of U.S. Women Physicians."
27. Ibid.
28. Ibid.
29. Frank et al., "Career Satisfaction of U.S. Women Physicians," 1417–1426.
30. Ibid.
31. Erica Frank and Shamiram Feinglass, "Student Loan Debt Does Not Predict Choosing a Primary Care Specialty for U.S. Women Physicians," *Journal of General Internal Medicine*, 1999, *14* (6): 347–350.
32. Erica Frank, Jason Breyan, and Lisa Elon, "Pro Bono Work and Nonmedical Volunteerism among U.S. Women Physicians," *Journal of Women's Health*, 2003, *12* (6): 589–598.
33. Erica Frank, Grant Baldwin, and Alan M. Langlieb, "Continuing Medical Education Habits of U.S. Women Physicians," *Journal of the American Medical Women's Association*, 2000, *55* (1): 27–28.
34. Erica Frank and Lawrence J. Lutz, "Characteristics of Women U.S. Family Physicians," *Archives of Family Medicine*, 1999 (8): 313–318.
35. Erica Frank and Lillian Meacham, "Characteristics of Women Pediatricians," *Clinical Pediatrics*, 2001, *40*: 17–25.
36. Erica Frank, Tricia Kunovich-Frieze, and Giselle Corbie-Smith, "Characteristics of Women Internists," *Medscape General Medicine*, 2002, *4*: 1.
37. Ibid.
38. Erica Frank, John Rock, and Danielle Sara, "Characteristics of Female Obstetrician-Gynecologists in the United States," *Obstetrics and Gynecology*, 1999, *5* (1): 659–665.
39. Erica Frank, Lisa Boswell, Leah J. Dickstein, and Daniel P. Chapman, "Characteristics of Female Psychiatrists," *American Journal of Psychiatry*, 2001 *158* (2): 205–212.
40. Erica Frank and Sareeta R. Singh, "Personal and Practice-Related Characteristics of a Subsample of U.S. Women Dermatologists: Data from the Women Physicians' Health Study," *International Journal of Dermatology*, 2001, *40*: 393–400.
41. Erica Frank, Vicken Totten, and Louise Andrew, "Characteristics of Women Emergency Physicians," *Internet Journal of Emergency and Intensive Care Medicine*, 2000, *4*: 2.
42. Erica Frank and Kay Vydareny, "Characteristics of Women Radiologists in the United States," *American Journal of Roentgenology*, 1999, *173*: 531–536.
43. Erica Frank, Michelle Brownstein, Kimberly Ephgrave, and Leigh Neumayer, "Characteristics of Women Surgeons in the United States," *American Journal of Surgery*, 1998, *176*: 244–250.
44. More, *Restoring the Balance*; Frank et al., "Characteristics of Women Emergency Physicians."
45. Erica Frank and Patricia Hudgins, "Academic vs. Nonacademic Women Physicians: Data from the Women Physicians' Health Study," *Academic Medicine*, 1999, *74*: 553–556.
46. Emily M. Lambert and Eric S. Hilmboe, "The Relationship between Specialty Choice and Gender of U.S. Medical Students, 1990–2003," *Academic Medicine*, 2005, *80* (9): 797–802.
47. Ibid.
48. Erica Frank, Donna Brogan, Ali H. Mokdad, Eduardo Simoes, Henry Kahn, and

Raymond S. Greenberg, "Health-Related Behaviors of Women Physicians vs. Other Women in the United States," *Archives of Internal Medicine*, 1998, *158*: 342–348.

49. Marilyn W. Winkleby, Darius Jatulis, Erica Frank, and Stephen P. Fortmann, "Socioeconomic Status and Health: Comparing Contributions of Education, Income and Occupation to Cardiovascular Disease Risk Factors," *American Journal of Public Health and the Nation's Health*, 1992, *82*: 816–820.

50. Erica Frank, Richard Rothenberg, Charles Lewis, and Brooke Fielding Beldoff, "Correlates of Physicians' Prevention-Related Practices: Findings from the Women Physicians' Health Study," *Archives of Family Medicine*, 2000, *9*: 359–367.

51. Erica Frank, "Physician Health and Patient Care," *Journal of the American Medical Association*, 2004, *291*: 637.

52. Frank et al., "Health-Related Behaviors of Women Physicians vs. Other Women in the United States."

53. National Center for Health Statistics, *Healthy People 2000 Final Review*, Hyattsville, MD: Public Health Service; 2001; Frank et al., "Correlates of Physicians' Prevention-Related Practices."

54. Frank et al., "Health-Related Behaviors of Women Physicians vs. Other Women in the United States."

55. Randall F. White, Jenna Seymour, and Erica Frank, "Vegetarianism among U.S. Women Physicians," *Journal of the American Dietetic Association*, 1999, *99* (5): 595–598.

56. Erica Frank, Adrienne Bendich, and Maxine Denniston, "Use of Vitamin-Mineral Supplements by Female Physicians in the United States," *American Journal of Clinical Nutrition*, 2000, 72: 969–975.

57. Erica Frank, Lawrence Sperling, and Kevin Wu, "Aspirin Use among Women Physicians in the United States," *American Journal of Cardiology*, 2000, *86*: 465–466.

58. Erica Frank, Kavitha B. Schelbert, and Lisa Elon, "Exercise Counseling and Personal Exercise Habits of U.S. Women Physicians," *Journal of the American Medical Women's Association*, 2003, *58* (3): 178–184.

59. American College of Sports Medicine, *Guidelines for Exercise Testing and Prescription* (Philadelphia: Lea and Febiger, 1986); Frank et al., "Health-Related Behaviors of Women Physicians vs. Other Women in the United States"; Erica Frank, Barbara K. Rimer, Donna J. Brogan, and Lisa Elon, "U.S. Women Physicians' Personal and Clinical Breast Cancer Screening Practices," *Journal of Women's Health & Gender-Based Medicine*, 2000, *9* (7): 791–801; Mona Sarayia, Erica Frank, Lisa Elon, Grant Baldwin, and Barbara McAlpine, "Personal and Clinical Skin Cancer Prevention Practices of U.S. Women Physicians," *Archives of Dermatology*, 2000, *136*: 633–642.

60. Frank et al., "Health-Related Behaviors of Women Physicians vs. Other Women in the United States."

61. Mona Sarayia, Stephen Coughlin, Wylie Burke, Lisa Elon, and Erica Frank, "The Role of Family History in Personal Prevention Practices among U.S. Women Physicians," *Community Genetics*, 2001 (4): 102–108.

62. Erica Frank and Kristen Cone, "Characteristics of Pregnant vs. Non-Pregnant Women Physicians: Findings from the Women Physicians' Health Study," *International Journal of Gynaecology and Obstetrics*, 2000, *69*: 37–46.

63. Erica Frank, "Contraceptive Use by Female Physicians in the United States," *Obstetrics and Gynecology*, 1999, *94* (5): 666–671.

64. Sally E. McNagny, Nanette K. Wenger, and Erica Frank, "Personal Postmeno-

pausal Hormone Replacement Therapy by Women Physicians in the United States," *Annals of Internal Medicine*, 1997, *127*: 1093–1096.

65. Grant Baldwin, Erica Frank, and Brooke D. Fielding, "U.S. Women Physicians' Residential Radon Testing Practices," *American Journal of Preventive Medicine*, 1998, *15*: 1.

66. Erica Frank and Arthur Kellermann, "Firearm Ownership among Female Physicians in the United States," *Southern Medical Journal*, 1999, *92* (11): 1083–1088.

67. Joyce P. Doyle, Erica Frank, Linda E. Saltzman, Pam M. McMahon, and Brooke D. Fielding, "Domestic Violence and Sexual Abuse in Women Physicians: Associated Medical, Psychiatric, and Professional Difficulties," *Journal of Women's Health and Gender-Based Medicine*, 1999, *8* (7): 955–965.

68. Erica Frank and Carolyn Clancy, "U.S. Women Physicians' Assessment of the Quality of Healthcare They Receive," *Journal of Women's Health*, 1999, *8* (1): 1–8.

69. Frank et al., "Correlates of Physicians' Prevention-Related Practices: Findings from the Women Physicians' Health Study"; Erica Frank, Kavitha B. Schelbert, and Lisa Elon, "Exercise Counseling and Personal Exercise Habits of U.S. Women Physicians," *Journal of the American Medical Women's Association*, 2003, *58* (3): 178–184; Frank et al., "U.S. Women Physicians' Personal and Clinical Breast Cancer Screening Practices"; Sarayia et al., "Personal and Clinical Skin Cancer Prevention Practices of U.S. Women Physicians"; Sarayia et al., "The Role of Family History in Personal Prevention Practices among U.S. Women Physicians"; Erica Frank, Elsa H. Wright, M. K. Serdula, Lisa Elon, and Grant Baldwin, "Personal and Professional Nutrition-Related Practices of U.S. Female Physicians," *American Journal of Clinical Nutrition*, 2002, *75*: 326–332; Alyssa Easton, Corinne Husten, Anne Malarcher, et al., "Smoking Cessation Counseling by Primary Care Women Physicians: Women Physicans' Health Study," *Women's Health*, 2001, *32* (4): 77–90; Alyssa Easton, Corinne Husten, Lisa Elon, Linda Pederson, and Erica Frank, "Non-Primary Care Physicians and Smoking Cessation Counseling: Women Physicians' Health Study," *Women's Health*, 2001, *34* (4): 15–29; Anne Malarcher, Alyssa Easton, Corinne Husten, and Erica Frank, "Smoking Cessation Counseling: Training and Practice among Pediatricians," *Clinical Pediatrics*, 2002, *41* (5): 341–349.

70. Erica Frank, Lisa Elon, and Carol Hogue, "Trangenerational Persistence of Education as a Health Risk: Findings from the Women Physicians' Health Study," *Journal of Women's Health*, 2003, *12* (5): 505–512.

71. David Himmelstein, Elizabeth Warren, Deborah Thorne, and Steffie Woolhander, "Illness and Injury as Contributors to Bankruptcy," *Health Affairs Web Exclusive W5–63*, February 2, 2005.

72. James Banks, Michael Marmot, Zoe Oldfield, and James P. Smith, "Disease and Disadvantage in the United States and in England," *Journal of the American Medical Association*, 2006, *295* (16): 2037–2045.

73. American Medical Association, *Physician Characteristics*, 1–2; "Table 3—Women in U.S. Medical Schools over a 10-Year Period," www.ama-assn.org/ama/pub/category/12914.html, accessed February 14, 2006.

74. Laine and Turner, "Unequal Pay for Equal Work."

75. C. Laine and B. J. Turner, "Unequal Pay for Equal Work"; Arlene S. Ash, Phyllis L. Carr, Richard Goldstein, and Robert H. Friedman, "Compensation and Advancement of Women in Academic Medicine: Is There Equity?" *Annals of Internal Medicine*, 2004, *141* (3): 205–214. For 2007 data, see Diane Magrane, Valarie Clark, et al., "Women in

U.S. Academic Medicine Statistics, 2006–2007," accessed at www.aamc.org/members/wim/statistics/stats07/start.htm.

76. Executive Leadership in Academic Medicine (ELAM) website, www.drexelmed.edu/elam, accessed on March 3, 2006.

77. Regina Morantz-Sanchez, *Sympathy and Science: Women Physicians in American Medicine* (New York: Oxford University Press, 1985).

78. More, *Restoring the Balance*, 248–258; "AAMC Project Committee on Increasing Women's Leadership in Academic Medicine," *Academic Medicine*, 1996, *71* (7): 801–810.

79. American Medical Association, *Physician Characteristics*; Laine and Turner, "Unequal Compensation for Equal Work"; Ash et al., "Compensation and Advancement of Women in Academic Medicine."

Contributors

Carla Bittel is an assistant professor of history at Loyola Marymount University. She is the author of "Science, Suffrage, and Experimentation: Mary Putnam Jacobi and the Controversy over Vivisection in Late Nineteenth-Century America," *Bulletin of the History of Medicine*, Winter 2005. She is now completing a book on the science and politics of Mary Putnam Jacobi, to be published by the University of North Carolina Press in 2009.

Elizabeth Fee is chief of the History of Medicine Division of the National Library of Medicine, National Institutes of Health, and part-time professor of history at the Institute for the History of Medicine at the Johns Hopkins Medical Institutions. Her publications include two edited collections, *Women's Health, Politics, and Power: Essays on Sex/Gender, Medicine, and Public Health* with Nancy Krieger (Baywood, 1994) and *Women and Health: The Politics of Sex in Medicine* (Baywood, 1983).

Eve Fine is a researcher at the University of Wisconsin–Madison. She is the author of "Pathways to Practice: Women Physicians in Chicago, 1850–1902" (Ph.D. diss., University of Wisconsin–Madison, 2007).

Erica Frank is a professor and Canada Research Chair at the University of British Columbia. She is the author of "Physician Health and Patient Care" in the *Journal of the American Medical Association* (2004).

Virginia A. Metaxas is a professor of history and women's studies at Southern Connecticut State University. Her most recent publication is "'Licentiousness Has Slain Its Hundreds of Thousands': The Missionary Discourse of Sex, Death, and Disease in Nineteenth-Century Hawai'i," in *Gender and Globalization in Asia and the Pacific: Method, Practice, Theory*, ed. Kathy E. Ferguson and Monique Mironesco (University of Hawai'i Press, 2008).

Regina Morantz-Sanchez is a professor in the Department of History at the University of Michigan. She is the author of *Sympathy and Science: Women Physicians in American Medicine* (University of North Carolina Press, 2000).

Ellen S. More is a professor of psychiatry and head of the Office of Medical History and Archives of the Lamar Soutter Library at the University of Massachusetts Medical School, Worcester. She is the author of *Restoring the Balance: Women Physicians and the Profession of Medicine, 1850–1995* (1999; Harvard University Press, 2001), which in 2003 won the Rossiter Prize from the History of Science Society.

Sandra Morgen is associate dean of the Graduate School and professor of Anthropology at the University of Oregon. She is the author of *Into Our Own Hands: The Women's Health Movement in the U.S., 1969–1990* (Rutgers University Press, 2002).

Heather Munro Prescott is a professor of history and coordinator of Women, Gender, and Sexuality Studies at Central Connecticut State University. She is the author of *Student Bodies: The Impact of Student Health on American Society and Medicine* (University of Michigan Press, 2007).

Robert A. Nye is the Thomas Hart and Mary Jones Horning Professor of the Humanities and a professor of history, emeritus, at Oregon State University. His most recent publication is "Western Masculinities in War and Peace," *American Historical Review*, 2007, April: 417–438.

Manon Parry was co-curator, with Ellen S. More, of *Changing the Face of Medicine: Celebrating America's Women Physicians*, an exhibition at the National Library of Medicine from 2004 to 2006. She also created a traveling version visiting sixty-one libraries across the United States, and has written on women in medicine for *Notable American Women* and the *American Journal of Public Health*.

Naomi Rogers is an associate professor in the Section of the History of Medicine and Women's, Gender, and Sexuality Studies Program at Yale University. She is the author of *The Polio Wars: Elizabeth Kenny and the Golden Age of American Medicine* (Oxford University Press, forthcoming).

Arleen Marcia Tuchman is a professor of history and director of the Center for Medicine, Health, and Society at Vanderbilt University. She is the author of *Science Has No Sex: The Life of Marie Zakrzewska, M.D.*

Judy Tzu-Chun Wu is an associate professor in the Department of History at Ohio State University. Her book on Margaret Chung is entitled *Doctor Mom Chung of the Fair-Haired Bastards: The Life of a Wartime Celebrity* (University of California Press, 2005).

Susan Wells is a professor of English at Temple University. She is the author of "Reading the Written Body" in *Signs: A Journal of Women in Culture and Society* (2008).

Index